Heat Transfer, Combustion and Flow Dynamics in Propulsion Systems

Heat Transfer, Combustion and Flow Dynamics in Propulsion Systems

Editors

Jian Liu
Qingfei Fu
Yiheng Tong
Chaoyang Liu

MDPI

Basel • Beijing • Wuhan • Barcelona • Belgrade • Novi Sad • Cluj • Manchester

Editors
Jian Liu
Research Institute of
Aerospace Technology
Central South University
Changsha
China

Qingfei Fu
School of Astronautics
Beihang University
Beijing
China

Yiheng Tong
Department of Aerospace
Science and Technology
Space Engineering University
Beijing
China

Chaoyang Liu
Science and Technology on
Scramjet Laboratory
National University of
Defense Technology
Changsha
China

Editorial Office
MDPI
St. Alban-Anlage 66
4052 Basel, Switzerland

This is a reprint of articles from the Special Issue published online in the open access journal *Aerospace* (ISSN 2226-4310) (available at: www.mdpi.com/journal/aerospace/special_issues/D55KC8KFXQ).

For citation purposes, cite each article independently as indicated on the article page online and as indicated below:

Lastname, A.A.; Lastname, B.B. Article Title. *Journal Name* **Year**, *Volume Number*, Page Range.

ISBN 978-3-0365-9027-1 (Hbk)
ISBN 978-3-0365-9026-4 (PDF)
doi.org/10.3390/books978-3-0365-9026-4

Contents

aerospace MDPI

Article

Effects of Gas Thermophysical Properties on the Full-Range Endwall Film Cooling of a Turbine Vane

Jian Liu, **Mengyao Xu** and **Wenxiong Xi** *

Research Institute of Aerospace Technology, Central South University, Changsha 410012, China; jian.liu@csu.edu.cn (J.L.); 15130715317@163.com (M.X.)
* Correspondence: 13739076081@163.com

Abstract: To protect turbine endwall from heat damage of hot exhaust gas, film cooling is the most significant method. The complex vortex structures on the endwall, such as the development of horseshoe vortices and transverse flow, affects cooling coverage on the endwall. In this study, the effects of gas thermophysical properties on full-range endwall film cooling of a turbine vane are investigated. Three kinds of gas thermophysical properties models are considered, i.e., the constant property gas model, ideal gas model, and real gas model, with six full-range endwall film cooling holes patterns based on different distribution principles. From the results, when gas thermophysical properties are considered, the coolant coverage in the pressure side (PS)-vane junction region is improved in Pattern B, Pattern D, Pattern E, and Pattern F, which are respectively designed based on the passage middle gap, limiting streamlines, heat transfer coefficients (HTCs), and four-holes pattern. Endwall η distribution is mainly determined by relative ratio of ejecting velocity and density of the hot gas and the coolant. For the cooling holes on the endwall with an injection angle of 30°, the density ratio is more dominant in determining the coolant coverage. At the injection angle of 45°, i.e., the slot region, the ejecting velocity is more dominant in determining the coolant coverage. When the ejecting velocity Is large enough from the slot, the coolant coverage on the downstream endwall region is also improved.

Keywords: endwall film cooling; thermophysical property; turbine vane; coolant coverage

Citation: Liu, J.; Xu, M.; Xi, W. Effects of Gas Thermophysical Properties on the Full-Range Endwall Film Cooling of a Turbine Vane. *Aerospace* **2023**, *10*, 592. https://doi.org/10.3390/aerospace10070592

Academic Editor: Konstantinos Kontis

Received: 28 April 2023
Revised: 19 June 2023
Accepted: 27 June 2023
Published: 28 June 2023

1. Introduction

Film cooling is the most effective method to protect turbine blades from heat load damage by ejecting coolant flows from several discrete cooling holes. Film cooling can use compressed air within gas turbine systems, and reduce temperature on the solid surfaces directly and effectively, although it is associated with reduced aerodynamic efficiency. The benefits of film cooling are that it uses the compressed air and decreases the blade surface temperature directly [1].

The endwall film cooling of the first-stage vane is designed to safeguard the endwall from the thermal load originating from the combustor's exhaust gas. This cooling approach is challenging due to the complex flow characteristics, such as horseshoe vortices and passage vortices, which influence coolant coverage on the endwall [2], particularly concerning the first-stage vane. Prior research has largely concentrated on two difficult areas: the leading edge (LE) region and the vane-pressure side (PS) junction region, both of which are characterized by a strong transverse pressure gradient. Some film cooling results gathered from flat plates are not directly applicable [3,4]. Over recent decades, extensive research has been conducted on endwall film cooling in a linear or annular cascade [5–18]. Simon and colleagues [5,6] carried out film cooling measurements on a contoured endwall with a nozzle guide vane. The suction side (SS) coolant migration vanished and the mixing effect became more apparent at higher coolant flow rates. Meanwhile, Knost and Thole [8] gauged the film cooling effectiveness (η) on the endwall for the first-stage vane. Their observations highlighted two areas, namely the LE region and the PS-endwall junction region, where it was particularly challenging for coolant flows to achieve coverage. Ghosh and Goldstein [9] explored the impact of the inlet skew on heat and mass transfer within a linear cascade. Their findings indicated that the inlet skew had notable effects on the PS due to its interaction with the passage vortex. Shiau et al. [10,11] showcased comprehensive endwall effectiveness contours, employing combinations of layback fan-shaped and cylindrical holes arranged on the vane using a pressure-sensitive paint (PSP) technique in annular cascades. Their research corroborated that PSP is a promising method for measuring high-quality cooling effectiveness (η). In a separate study, Yang et al. [12] analyzed the conjugated heat transfer on the endwall, taking into account the effects of upstream purge flow. Their findings highlighted that the purge flow significantly enhanced the overall cooling effectiveness on the endwall, especially its uniformity. Chen et al. [13,14] studied full-coverage film cooling on a non-axisymmetric contoured endwall. They concluded that the contoured endwall outperformed the flat endwall. Bu et al. [16] established a turbine endwall with a combined regime of jet impingement and film cooling, and investigated four types of geometric parameters. They performed an Analysis of Variance (ANOVA) to scrutinize the primary effect of each parameter and the correlation between them.

Endwall film cooling is intricate due to the intense flow impingement and complex vortex structures. Despite extensive research on film cooling, the optimal configuration of cooling holes on the endwall remains unclear. As inferred from previous studies [19,20], the primary design principles for endwall film cooling are centered around pressure distributions. Liu et al. [21] explored the full-range endwall film cooling of a turbine vane, incorporating different design principles. The principles for the arrangement of cooling holes were based on pressure coefficients, streamlines, and heat transfer coefficient (HTC) distributions. Their findings suggested that designs premised on streamline distribution and HTC distributions result in greater coolant coverage, and the design based on HTC distributions could curtail high-temperature zones. In another study, Pu et al. [22] conducted an experimental investigation of overall cooling in laminated configurations at a turbine vane endwall. They discovered that a dense arrangement of large-diameter impingement holes was effective in enhancing the cooling efficiency of the pressure side corner.

However, most of the prior studies were conducted in low-temperature experimental setups or under simplified working conditions, often overlooking the effect of the thermophysical properties of gas. Recognizing the influence of the thermophysical properties of fluid, some foundational research has been undertaken. Zhang et al. [23] conducted a numerical investigation into liquid-film cooling under high pressure. Considering the gas in a two-phase liquid solution, the property of real gas, and varying thermophysical properties, they established a mathematical model for high-pressure film cooling. In another study, Nematollahi et al. [24] explored the evaluation of the cubic model of real gas in supersonic flow within a wedge-shaped cascade. They developed and compared four additional cubic model equations of real gas, based on the NIST Refprop model. However, they found that the Aungier-Redlich-Kwong equations of the state model provided superior accuracy. Yuan et al. [25] devised a new model for real gas to quantify and predict gas leakage in high-pressure gas pipelines. This model offers two significant enhancements: it employs the mass fraction to determine the proportion of leaked gas flow to in-pipe gas flow, and it incorporates real-gas thermodynamics into the fundamental governing equations.

In recent years, various studies focusing on the flow structures of high temperature and high-pressure gas within engine applications have considered the impact of realistic working conditions or real gas models [26–29]. Salvadori et al. [26] delved into the performance of high-pressure turbine endwall film cooling under realistic inlet conditions, such as non-uniformities of total temperature and velocity profile. They found that the inlet swirl restricts coolant coverage, as it alters the development of the horseshoe vortex (HV) and generates a stronger passage vortex (PV), which negatively affects platform cooling. Aiming to bolster cooling performance in the leading edge region, Wen et al. [27] examined the effects of leading edge film cooling arrangements via adiabatic and aero-thermal coupled simulation. Their research confirmed that the trenched configuration could be applied to the first-stage turbine vane to elevate inlet temperature and enhance engine efficiency. Bai et al. [29] probed into the effects of axisymmetric convergent contouring and blowing ratio on endwall film cooling and vane pressure side surface phantom cooling performance under real gas turbine operating conditions. Their findings suggested that the optimal endwall contouring shapes serve as an effective technical approach to reduce coolant depletion. Additionally, several fundamental research endeavors have been carried out to contemplate the effect of gas physical properties [23–25,30,31].

For cooling a turbine blade, lots of experiments have been performed in a low temperature test rig ignoring the variations of gas thermophysical properties. Considering the gas thermophysical properties, the blowing ratio and density ratio are changed in the real operation condition. For endwall film cooling, the interactions between different cooling holes are strong, so it is necessary to investigate the effect of gas thermophysical properties. Therefore, the effects of gas thermophysical properties on full-range endwall film cooling of a turbine vane are investigated in the present study. Six kinds of full-range endwall film cooling designs are considered with the validated turbulence model. Three kinds of gas thermophysical properties models are considered, i.e., the constant property gas model, real gas model, and ideal gas model. Flow structures and temperature fields of endwall film cooling are displayed and analyzed by three-dimensional numerical calculations.

2. Description of Computational Domain

For the simulation of endwall film cooling, a simplified guide vane is utilized, drawn from Knost's thesis [20] and magnified nine times from the original size. The computational domain is depicted in Figure 1, with relevant design and operational parameters outlined in Table 1. This domain comprises a main flow passage and a section for the coolant supply plenum. The main flow passage is constructed between two guide vanes, with an upstream extended part and a downstream extended part added to accommodate flow developments. The mainstream adopts a mass flow inlet with a rate of 1.6 kg/s and a Reynolds number of 128,600 (with dynamic viscosity selected at a temperature of 1700 K). The reference length corresponds to the vane's chord, which spans 594 mm. Results obtained under the same flow conditions as in Knost's work [20] serve as the benchmark for comparisons. Film cooling holes, angled at 30°, are distributed on the endwall. Coolant is supplied from a plenum, mirroring the real working conditions for endwall film cooling. In addition to the film cooling holes, a slot is located upstream of the holes, offering supplementary protection for the endwall. A similar slot is referenced [5,6], resulting from the assembly of two endwall components. Positioned 0.31C upstream of the vane are the slot aids in the coolant supply from the slot and the film cooling holes. Accordingly, two parameters, S and F, are defined to represent the coolant mass flow rate from the slot and the film cooling holes in relation to the mainstream flow rate.

Table 1. Related Geometric parameters and flow conditions.

Geometric Parameters		Flow Conditions	
Scale	9	Inlet Reynolds number	128,600
Span of the vane	S_p = 552.42 mm	Inlet mass flow rate	1.6 kg/s
Chord of the vane	C = 594 mm	S	1–2%
Pitch between adjacent vanes	P = 457.38 mm	F	1–2%
Coolant injection angle	α = 30°	Inlet mainstream turbulence intensity	8%

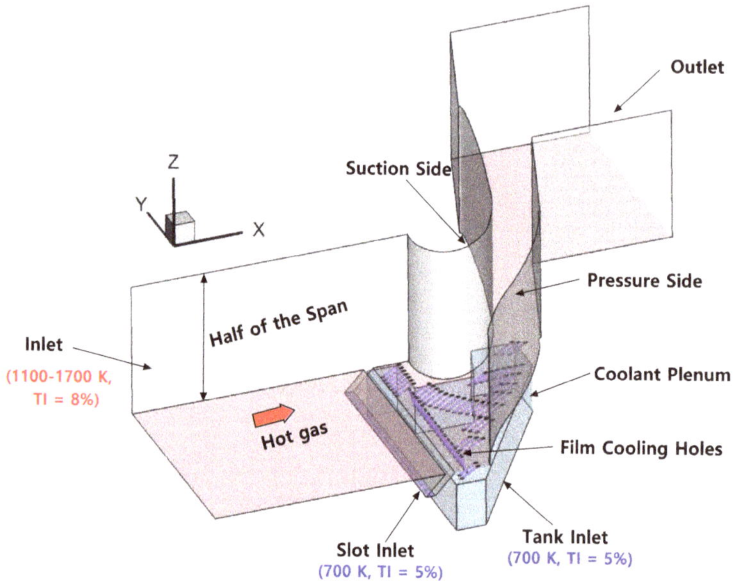

Figure 1. Schematic of the computational domain.

$$S = \frac{Slot\ mass\ flow\ rate}{Mainstream\ mass\ flow\ rate} \quad (1)$$

$$F = \frac{Film\ holes\ mass\ flow\ rate}{Mainstream\ mass\ flow\ rate} \quad (2)$$

In this paper, the S and F are set as 1–2%, respectively. The effects of mass flow rate ratio are considered.

The arrangement of endwall film cooling holes is based on varied principles, resulting in six different configurations, as shown in Figure 2. Pattern A and Pattern B have their roots in previous arrangements derived from references [20,32]. Pattern B, in particular, considers the effects of the slot that forms when two endwall components are assembled. Pattern C, Pattern D, and Pattern E are designed on the basis of pressure distributions, streamline distributions, and heat transfer coefficient distributions, respectively, with the aim of enhancing coolant coverage. These distributions are calculated without factoring in the impact of coolant flows from endwall film cooling holes. Pattern F, meanwhile, is also predicated on pressure coefficients and introduces a four-hole pattern. Leveraging the fractal theory, this four-hole pattern provides advantages in enhancing local coolant coverage. For Patterns A-E, film cooling holes are organized with a pitch ratio (P_c/D) of 3. For Pattern F, the distance separating two opposing film cooling holes is four times their diameter ($4D$). All film cooling holes are cylindrical with an injection angle of 30°, and are arranged on the endwall with a length ratio of 10. Regardless of the design, the mass flow rate remains constant, with the total number of cooling holes ranging between 40 and 50.

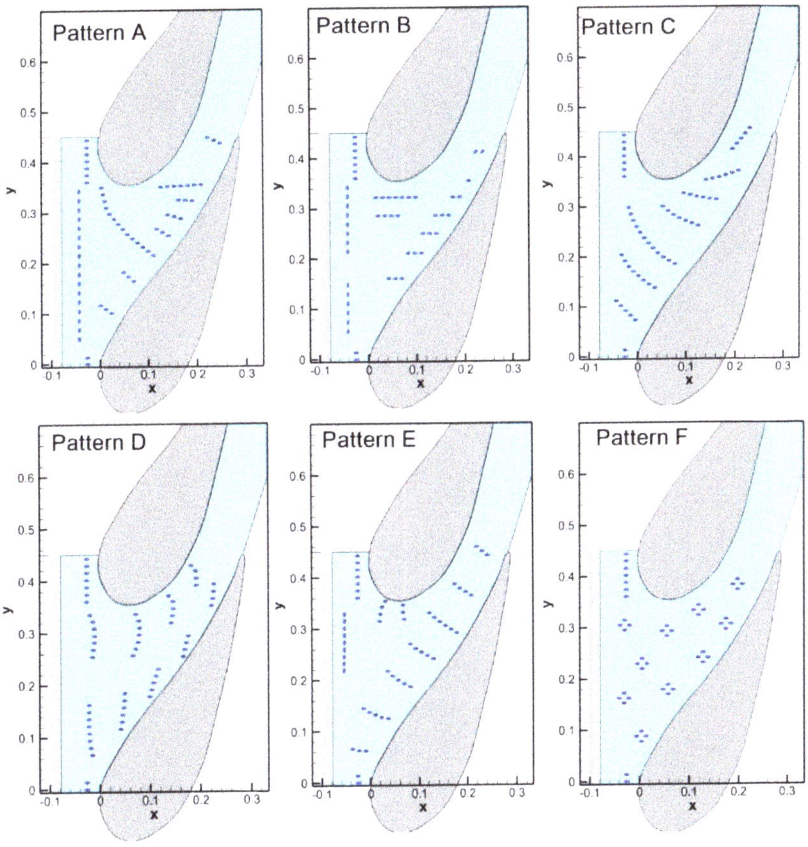

Figure 2. Different patterns of film cooling holes arranged on the turbine endwall.

3. Computational Method and Procedure

3.1. Mesh Details

This paper employed ANSYS Fluent 19.1 to generate the meshes, due to the complexity of the computational domain. The domain is divided into several sub-domains for the purpose of mesh generation, as detailed in Figure 3. These sub-domains are interconnected through interfaces which facilitate the transfer of result data. These interfaces ensure accurate data transfer between two distinct mesh surfaces through interpolation. Structured meshes are utilized within each sub-domain in an effort to decrease mesh quantity and enhance computational accuracy. Due to the employment of the k-ω SST turbulence model, meshes located near the walls are densely packed, with wall y plus approximately equal to one. For each sub-domain, unique mesh strategies such as O-block and Y-block are implemented to boost mesh quality. Furthermore, the region encompassing the interfaces also features exceptionally dense meshes, ensuring accurate and seamless data transfer.

Figure 3. Overview of the generated structured meshes.

A mesh independence study is performed by Pattern D at film cooling mass flow ratio = 2% and slot mass flow ratio = 1%. Four kinds of mesh systems are designed with a grid number of 5.36 million, 7.63 million, 10.29 million, and 13.69 million, respectively. The mesh systems are built based on node increase factors of 0.8, 0.9, 1.0, and 1.1. The pressure drop between the inlet and the outlet, and averaged film cooling effectiveness are used for the comparisons shown in Table 2. The pressure drops and film cooling effectiveness are obtained by an area-averaged method. The definition of η is below.

$$\eta = \frac{T_w - T_g}{T_c - T_g} \tag{3}$$

where T_{w} is the wall temperature, T_{g} is the hot gas temperature, and T_{c} is the coolant temperature.

From the table, the pressure drops predicted by different mesh systems have pretty good agreement, and the difference is within 1%. For averaged film cooling effectiveness, the difference between two mesh systems is within 2%. Considering the requirements of wall treatment function, calculation accuracy and computational efforts, the mesh system with a grid number of 10.29 M is selected.

Table 2. Comparisons of averaged pressure drop and cooling effectiveness (Pattern D).

	Scale Factor	Total Meshes	ΔP (Pa)	η_{avg}
Mesh 1	0.8	5,344,239	361.4	0.327
Mesh 2	0.9	7,552,357	361.5	0.326
Mesh 3	1.0	10,281,718	361.5	0.323
Mesh 4	1.1	13,746,964	361.5	0.322

3.2. Turbulence Model

The selection of turbulence model is important when solving the complex flow problem. In this work, the *k*-ω SST model is selected to deal with wall bounded flows on the turbine endwall. The *k*-ω SST model combines the advantage of the *k*-ω and *k*-ε models, which exhibit good accuracy in the prediction of wall shear flows.

The comparison of the predicted results with the experimental data is shown in Figure 4. Besides the film cooling effectives, the basic vane parameter, i.e., pressure coefficients and endwall heat transfer have been validated in previous work [21], and the related experimental data is obtained from [33]. Figure 4a compares η contours predicted by the *k*-ω SST model with the experimental data captured by the IR camera. The film cooling holes on the endwall are arranged according to experimental Model 1 and Model 2 in [33]. From the figure, the η distribution contour has good agreement with the experimental data. The large errors are found in the vane-endwall connection regions where the IR camera has poor accuracy. A large quantity of coolant ejects from the upstream slot and pushes the coolant flows from endwall cooling holes to the suction side. For this pattern of film cooling holes, the coolant upstream the LE has difficulty being ejected out, and more coolant is more easily ejected from the downstream cooling holes where the ambient pressure is small. Figure 4b presents the comparison of lateral averaged η arranged in Model 1 and Model 2. Each data point in the figure is obtained by averaging a cluster of result data in the spanwise direction. Overall, the trend of averaged η along the streamwise direction predicted by the numerical calculation matches the experimental data well, and the error is within 10%. A larger error can be found in the upstream region, where the coolants are mainly ejected from the slot. In the simulation, the inlet boundary condition of the slot is the set uniform velocity, but it is difficult to achieve this in the experiment. The different ambient pressure causes the different ejection pattern, and a relatively large error in the peak value. Based on the results, the numerical results predicted by the *k*-ω SST model have provided enough accuracy and can be used in the lateral calculation.

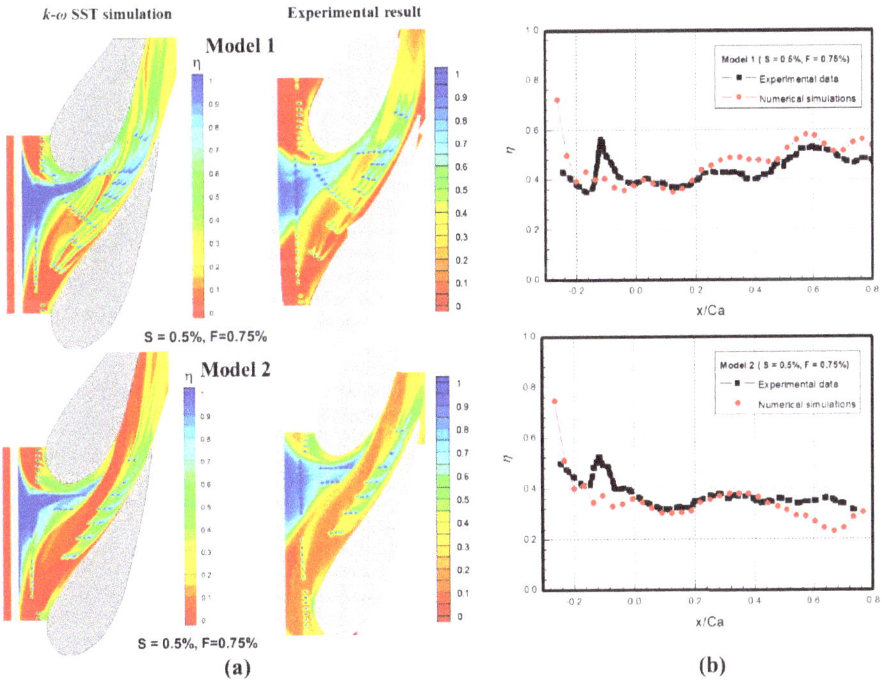

Figure 4. Validation of the turbulence model by comparing with the experimental data of Pattern A. (**a**) comparisons of the film cooling effectiveness contour. (**b**) comparison of lateral averaged film cooling effectiveness.

3.3. Gas Thermophysical Properties

The density of an ideal gas law in an incompressible flow is calculated by

$$\rho = \frac{P_{op}}{\frac{R}{M_w}T} \quad (4)$$

where, R = the universal gas constant; M_w = the molecular weight of the gas; P_{op} = the operating pressure.

Ideal gas law is the simplest mathematical thermodynamic equation to connect temperature, pressure, and other items. However, it becomes increasingly inaccurate at higher pressures and lower temperatures. The Redlich-Kwong state equation [34] is improved to solve this problem, and the original form is

$$P = \frac{RT}{V-b} - \frac{\alpha_0}{V(V+b)T_r^{0.5}} \quad (5)$$

where P = absolute pressure (Pa); R = universal gas constant. V = specific molar volume (m^3/kmol); T = temperature (K); T_r = reduced temperature T/T_c; where T_c is the critical temperature; α_0 and b are constants related to the fluid critical pressure and temperature.

For the viscosity, Sutherland's law is used, and its form is below with three coefficients.

$$\mu = \mu_0\left(\frac{T}{T_0}\right)^{3/2}\frac{T_0+S}{T+S} \quad (6)$$

where μ = the viscosity in kg/m·s; T = the static temperature in K; μ_0 = reference value in kg/m·s; T_0 = reference temperature in K; S = an effective temperature in K (Sutherland constant). For air at moderate temperatures and pressures, $\mu_0 = 1.716 \times 10^{-5}$ kg/m·s, T = 273.11 K and S = 110.56 K.

Thermal conductivity is obtained by kinetic theory.

$$k = \frac{15}{4}\frac{R}{M_w}\mu\left[\frac{4}{15}\frac{c_p M_w}{R} + \frac{1}{3}\right] \quad (7)$$

where R is the universal gas constant, M is the molecular weight, μ is the material's specified or computed viscosity, C_p is the material's specified or computed specific heat capacity.

Specific heat capacity is obtained by kinetic theory.

$$c_{p,j} = \frac{1}{2}\frac{R}{M_{w,j}}(f_i+2) \quad (8)$$

where f_i is the number of degrees of freedom for the gas species.

For the constant gas properties, the thermophysical properties are obtained from Table 3 below. For the calculation case with constant gas properties, the gas properties are obtained by the temperature of the mainstream.

Table 3. Gas thermophysical properties at 1 Mpa from database of NIST.

Temperature (K)	Density (kg/m^3)	Conductivity (10^{-2} W/m·K)	Capacity (kJ/kg·K)	Dynamic Viscosity (10^{-5} Pa·s)
700	5.023	5.066	1.0781	3.33
1100	3.199	7.327	1.1631	4.44
1400	2.515	8.918	1.2154	5.17
1700	2.071	10.551	1.2677	5.85

3.4. Boundary Conditions and Solver

The boundary setting is the same with the experiments in Ref. [20]. The mainstream inlet is set as (1100–1700 K) with a turbulence intensity of 8%. The mass flow rate of the inlet is set as 1.6 kg/s, and corresponding Re is 128,600 when the inlet air is set at 1700 K (the dynamic viscosity is 5.85×10^{-5} Pa·s). The inlet region is built with an upstream extended channel to develop the inlet mainstream. In the calculation, three kinds of gas thermophysical properties models are applied in the solver, i.e., the constant property gas model, real gas model, and ideal gas model. For the constant property gas, the gas thermophysical model is determined by the mainstream. The corresponding thermophysical model is provided in Section 3.3. Half of the vane height is used in the computation domain to reduce the mesh quantity, and the symmetric boundary is set for the top wall. The periodic boundary condition is set for two sides, except the vane surfaces. The outlet is set as pressure outlet with an ambient pressure of 1 mPa. The coolant of the endwall film cooling holes is supplied by the plenum connected to the endwall, with a mass flow rate ratio of 1–2%. The temperature is set as 700 K, with a turbulence intensity of 5%. Mass flow inlet is chosen for the upstream slot, and flow mass rate ratio is set as 1–2% of the mainstream. The turbulence intensity setting in this work is based on an engine-like condition which is different from the experiment in Ref. [20].

The SIMPLEC method is employed to couple the pressure and velocity fields. The second upwind scheme is utilized for the discretization of pressure, momentum, turbulent kinetic energy, specific dissipation rate, and energy equations. The convergence is determined based on the residuals and average temperature on the endwall. Absolute criteria for continuity, x-velocity, y-velocity, z-velocity, k and ω items are set at 10^{-5}, and for the energy equation, it is set at 10^{-8}. To guarantee convergence, the error in the average temperature on the endwall between two iterations should be less than 10^{-4}.

4. Results and Discussions

Figure 5 compares the endwall η contours for different patterns at S = 1% and F = 2%. The real gas model is chosen to deal with gas thermophysical properties. From the figure, coolant flow from endwall cooling holes interacts with coolant flows from the upstream slot, and a large coolant coverage region is found in the flow passage between two vanes. The arrangement of cooling holes on the endwall also affects the flow path of the passage vortex. In Pattern B and Pattern E, the shedding coolant flows develop along the mainstream and are easier to impinge on the SS of the vane. For different designs, the coolant coverage in the "tough" region is different, Pattern B, Pattern D, Pattern E, and Pattern F have better coolant coverage in the pressure-vane junction regions. The coolants have difficulty being ejected out from the LE regions of Pattern A, Pattern C, and Pattern D.

Considering gas thermophysical properties, comparisons of endwall η contours for different patterns are shown in Figure 6. Three kinds of gas thermophysical models are applied, respectively, the real gas model, ideal gas model, and constant property gas model. Pattern A and Pattern D, which are, respectively, based on pressure coefficient distributions and limiting streamline distributions, are used for comparisons. From the figure, the coolant injection is obviously strengthened in the case of constant property gas. It is indicated that the case of constant property gas has smaller coolant coverage in the wake region compared with the other cases. For the cases of real gas and ideal gas, there is larger coolant coverage in the wake region of the cooling holes. The cases of real gas and ideal gas have larger gas density at high temperature. To ensure the same mass flow rate, a smaller coolant injection velocity is obtained for the cases of real gas and ideal gas. However, the difference between the case with the real gas model and the case with the ideal gas model is very small because the temperature and pressure is far from the critical point.

Figure 7 compares the lateral averaged η of Pattern A and Pattern D with different gas thermophysical properties. From the figure, larger averaged η is found in the slot region for the case with constant property gas. This trend is totally different from the coolant coverage of the cooling holes for the case with constant property gas. The injection angle of 45° is for the shedding flow of the slot, while the injection angle of 30° is for the shedding flow of the cooling holes. For the coolant with the injection angle of 30°, the gas density is more dominant in determining coolant coverage, and the regions of the cooling holes for the cases with real gas and ideal gas have larger coolant coverage. However, for the injection angle of 45°, i.e., the slot region, the ejecting velocity is more dominant in determining coolant coverage, and the case with constant property gas has larger coolant coverage. In the region of $0 < x/C < 0.3$, the cases with the real gas model and ideal gas model have large coolant coverage. With the mixed flow developing on the endwall, the difference in coolant coverage for different cases becomes smaller, and they overlap together.

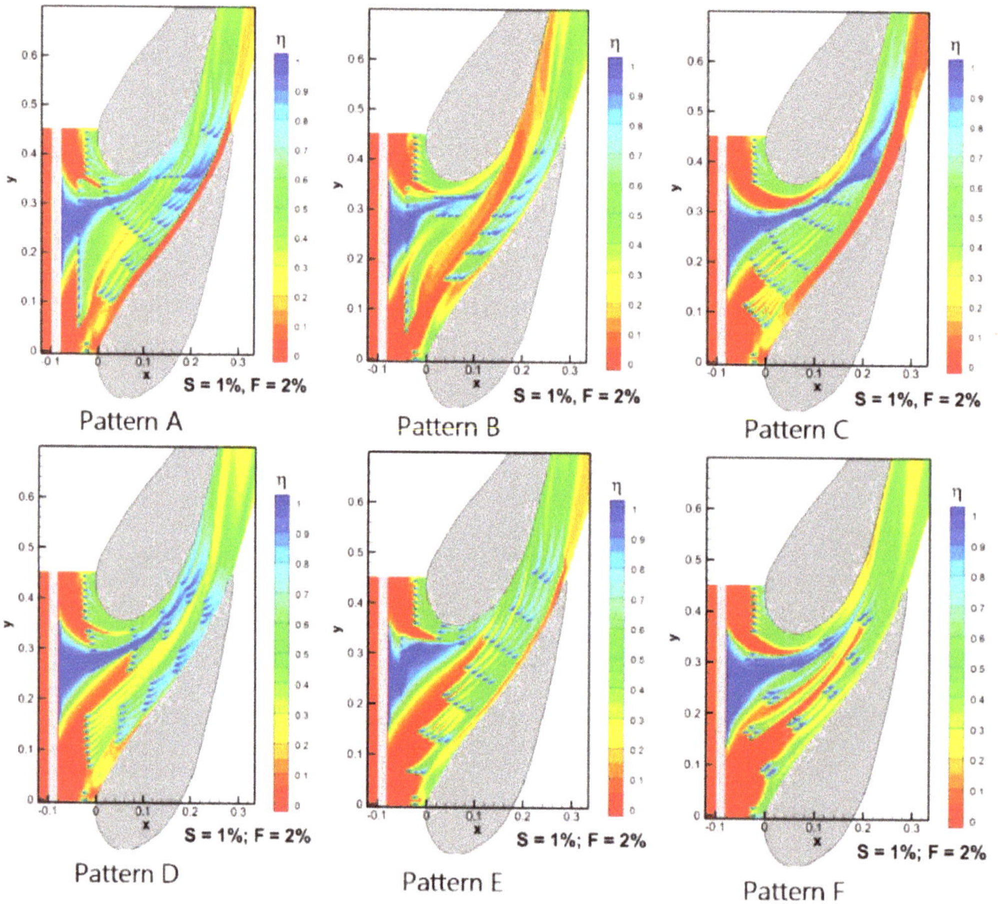

Figure 5. Endwall η contours for different patterns at S = 1% and F = 2% using the real gas model.

Figure 8 compares the endwall η contours for Pattern D at different cooling hole mass flow ratios with different gas thermophysical properties, i.e., the real gas model and constant property gas model. The mainstream temperature is set at 1700 K. The *S* is set at 1%, and *F* ranges from 1–2% in the considered cases. Overall, the coolant coverage is increased with the increased *S*, and a relatively full coolant coverage is found at *F* = 2%. However, the distributions of the coolant flow are totally different when different gas thermophysical models are applied, though the same coolant supply is set for all the cases. When *F* is set at 1% in the case of real gas, the coolant cannot eject from the cooling holes at the region around $x/C = 0$. For the case of constant property gas, the coolant coverage is obviously improved. At the temperature of 1700 K, the model of real gas has a larger density than the case of the constant property gas. The gas with smaller density has larger ejecting velocity, and is easily ejected from the upstream region. When the coolant supply is increased from $F = 1.5\%$ to 2%, the coolant is not increased obviously. It is expected that more coolant flows mix with the mainstream, and there is little coolant coverage on the endwall when the coolant with smaller density is applied. When a single cooling hole is obverted, an obvious cooling wake for the cooling holes in the case of the real gas model can be found. The increased F has a more obvious effect in the case with the real gas model. It is indicated that the gas with larger density can protect the wall better when the coolant ejecting velocity is increased. The coolant coverage from the upstream slot is almost the same for all the cases, which indicates that the coolant flows from the downstream cooling holes have no obvious effects on the coolant ejecting from the slot.

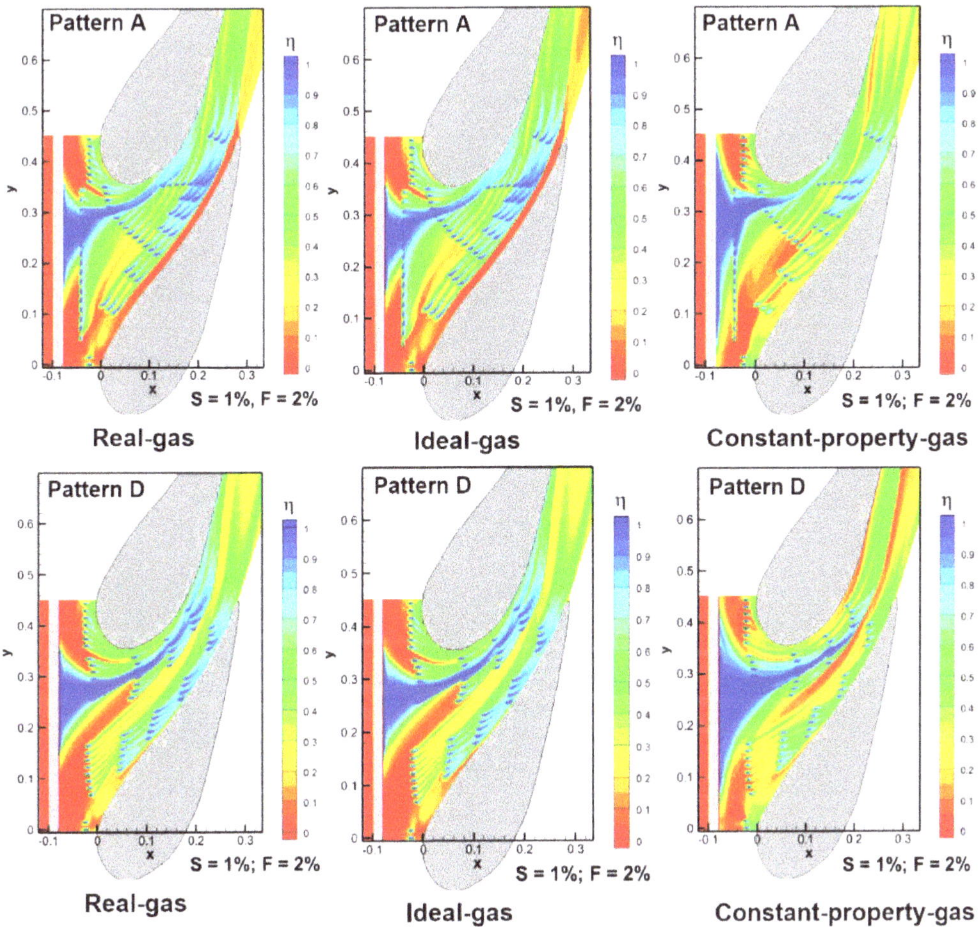

Figure 6. Comparisons of endwall η contours for Pattern A and Pattern D with different gas thermophysical properties.

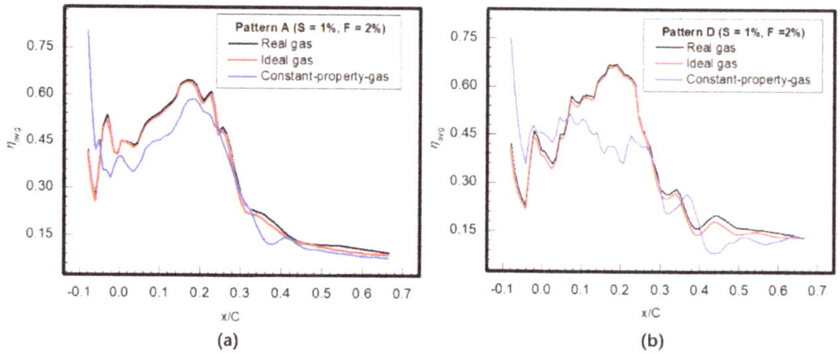

Figure 7. Comparisons of lateral averaged η for Pattern A and Pattern D at different gas thermophysical properties. (**a**) Pattern A; (**b**) Pattern D.

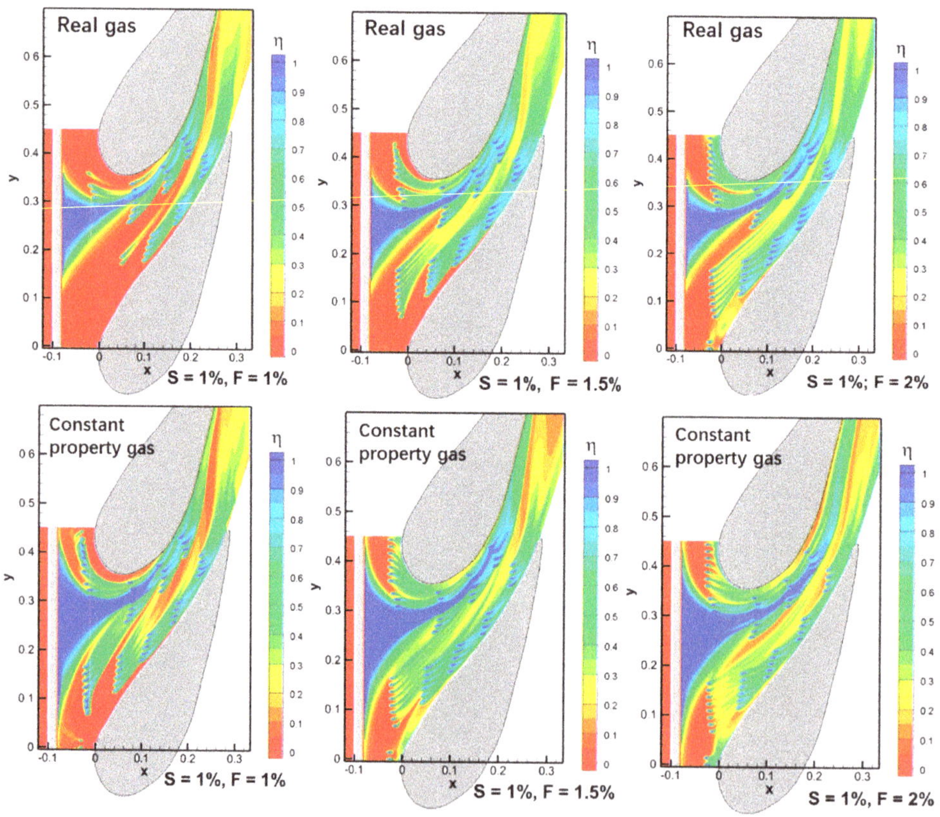

Figure 8. Comparisons of endwall η contours for Pattern D with different gas thermophysical properties at different cooling hole mass flow ratios.

Comparisons of lateral averaged η for Pattern D with different gas thermophysical properties at different cooling hole mass flow ratios are presented in Figure 9. From the figure, the distribution displays a trend of an increase at first, and then a decrease. The larger averaged η is found in the region of $0.1 < x/C < 0.3$, when large quantities of cooling holes are arranged. For the case with the real gas model, averaged η is increased with the increased F. However, the increased trend is weakened when the F is increased from 1.5%

to 2%. In addition, the distribution of averaged η is overlapped when the S is increased for the case with the constant property gas model. This phenomenon is caused by the different gas densities for different gas models at 1700 K. It is expected that the gas with larger density has the advantage of enlarging coolant coverage with the increased F.

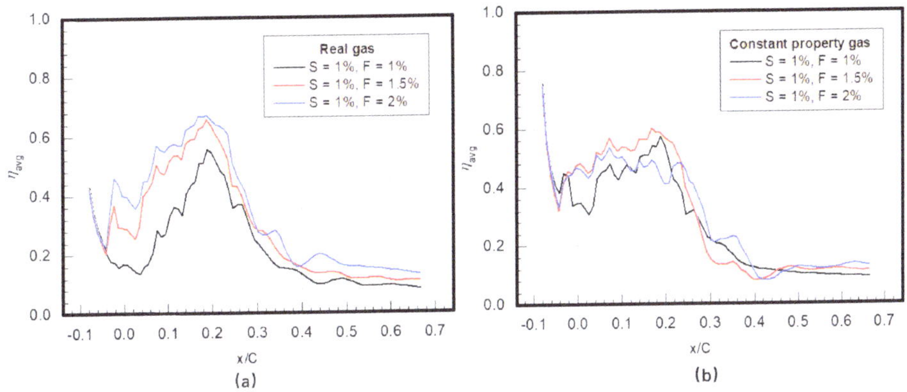

Figure 9. Comparisons of lateral averaged η distributions for Pattern D with different gas thermophysical properties at different cooling hole mass flow ratios. (**a**) For real gas; (**b**) For constant property gas.

Figure 10 compares endwall η contours for Pattern D with different gas thermophysical properties at different slot mass flow ratios. The S ranges from 1–2%, and the F is set as 2%. The mainstream temperature is also set at 1700 K. From the figure, it is clear that the ejected flows from the upstream slot have significant coolant coverage in the middle passage between two vanes. With the increased S, the coolant coverage is enlarged from the middle passage to the LE endwall and develops along the horseshoe vortex in the passage. The enlarged coolant coverage is more evident in the cases with the constant property gas model. However, the increase of coolant coverage is weakened in the case with the real gas model when S is increased from 1.5% to 2%. Similar to the phenomenon in Figure 8, the waking region is more obvious in the cases with the real gas model. For the case with the constant property gas model at the largest S, i.e., S = 2%, a relatively large and strong wake region is also found. It is indicated that when the ejected flows from the slot are very strong at a high velocity, it can change the pressure field of the downstream endwall. Because of the low-pressure field downstream of the shedding flows, it is beneficial for coolant coverage on the endwall.

Comparisons of lateral averaged η for Pattern D with different gas thermophysical properties at different slot mass flow ratios are displayed in Figure 11. Comparing Figures 9 and 11, it is found that the coolant flows from the slot mainly take effect in the region of $-0.1 < x/C < 0.1$, and slightly increased lateral averaged η is found in the region of $0.1 < x/C < 0.3$. In the region of $0.3 < x/C < 0.6$, the distribution trend is different in the cases with the real gas model, which proves the effect of the ejected flows from the slot on the downstream flow field. As expected, the increase of lateral averaged η is not obvious when the S is increased from 1.5% to 2%. However, the η in the cases of constant property gas is gradually improved when the S is increased from 1.0% to 2%. It is indicated that the increased velocity is more effective to improve the coolant coverage for the slot with an injection angle of 45°.

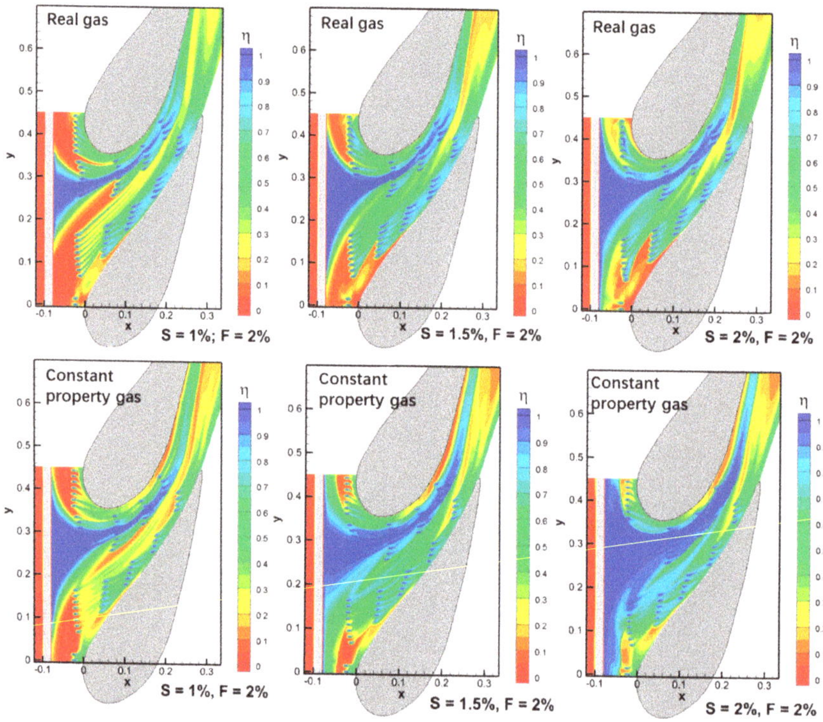

Figure 10. Comparisons of endwall η contours for Pattern D with different gas thermophysical properties at different slot mass flow ratios.

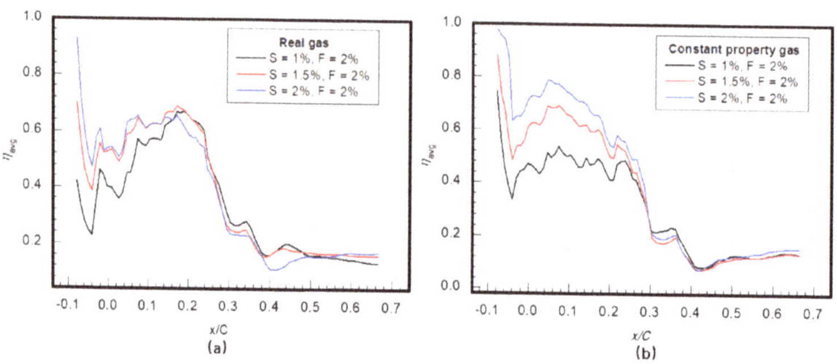

Figure 11. Comparisons of lateral averaged η for Pattern D with different gas thermophysical properties at different slot mass flow ratios. (**a**) For real gas; (**b**) For constant property gas.

Figure 12 compares endwall η contours for Pattern D with different gas thermophysical properties at different mainstream temperatures. The mainstream temperature ranges from 1700 K to 1100 K, with the same turbulence intensity. For the constant property gas model, the thermophysical property is obtained at the mainstream temperature, which can be found in Table 3. For the cases with the real gas model, the η distribution is changed when the mainstream temperature is changed. From the figure, it is found that the η from the upstream slot is enlarged with the decreased mainstream temperature from 1700 K to 1100 K. When the mainstream temperature is decreased, the density difference between the hot gas and the coolant flow becomes small and the mainstream velocity decreases. When the mainstream velocity is decreased, the coolant ejecting from the slot becomes easy and has benefits for coolant flows covering the endwall. It is expected that when the mainstream temperature is decreased to 700 K, the η contours become similar to the cases with the constant property gas. For the cases with constant property gas, the η contours almost stay the same when the mainstream temperature is changed. In addition, the wake region is more obvious in the cases with the real gas model, and it is gradually weakened when the mainstream temperature decreases from 1700 K to 1100 K. This also provides additional evidence that the determination of the coolant coverage of the slot and the cooling holes are different, i.e., the coolant coverage of the slot is more sensitive to the ejecting velocity of the slot and the coolant coverage of the cooling holes is more sensitive to the gas density.

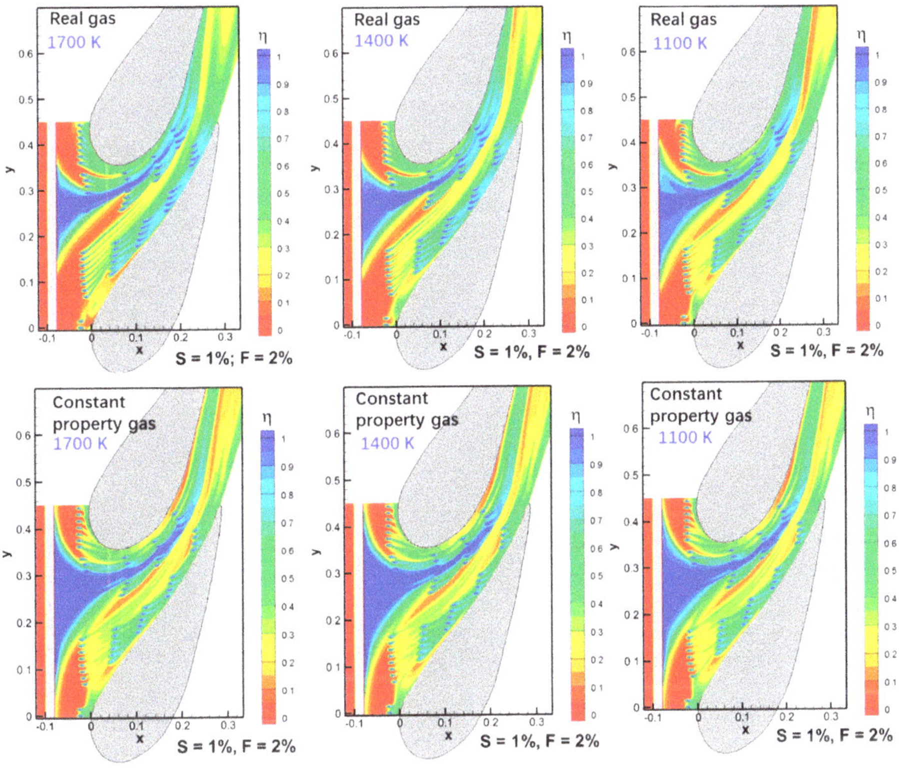

Figure 12. Comparisons of endwall η contours for Pattern D with different gas thermophysical properties at different mainstream temperatures.

Comparisons of lateral averaged η for Pattern D with different gas thermophysical properties at different mainstream temperatures are presented in Figure 13. As shown in the figure, the lateral averaged endwall η is increased with the decreased mainstream temperature in the region of $-0.1 < x/C < 0.1$, where the coolant flows from the slot mainly take effect. While in the region of $0.1 < x/C < 0.3$, the distribution of lateral averaged η is totally different when the mainstream temperature is decreased. For the cases with constant property gas, the distribution of lateral averaged η is almost the same. Therefore, it is concluded that the endwall η distribution is mainly determined by the relative ratio of ejecting velocity and density magnitude of the hot gas and the coolant. The effect of ejecting velocity and density magnitude on the coolant coverage differs at different ejecting angles, such as the slot and the cooling holes.

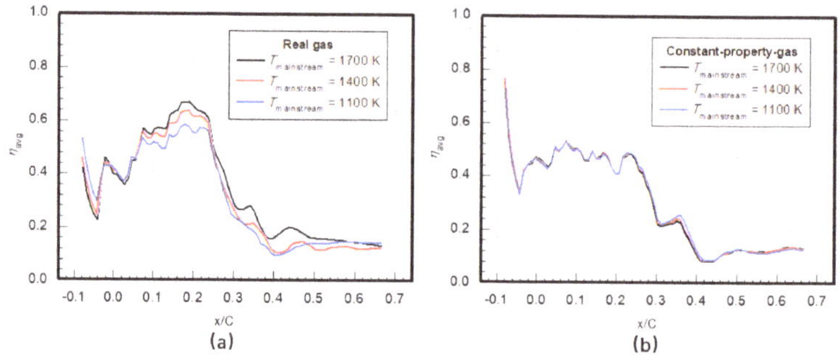

Figure 13. Comparisons of lateral averaged η for Pattern D with different gas thermophysical properties at different mainstream temperatures. (**a**) For real gas; (**b**) For constant property gas.

Three-dimensional streamlines and velocity distributions originating from slot and cooling holes for different patterns are shown in Figure 14. The mainstream is set at 1700 K, and the real gas model is applied for all the cases. The coolant injection from the slot interacts with the mainstream, and the mixing flows are drawn to the SS by the transverse pressure gradient. The coolant flows from the slot are hard to attach on the endwall when it develops along the flow passage. For the coolant from the endwall film cooling holes, it is accelerated along the flow passage, and pushed to the SS by the transverse pressure gradient. However, for different cases, the coolant coverage on the vane-PS junction region is different. From the figure, Patten B, Pattern D, and Pattern F have better coolant coverage in the "tough" region.

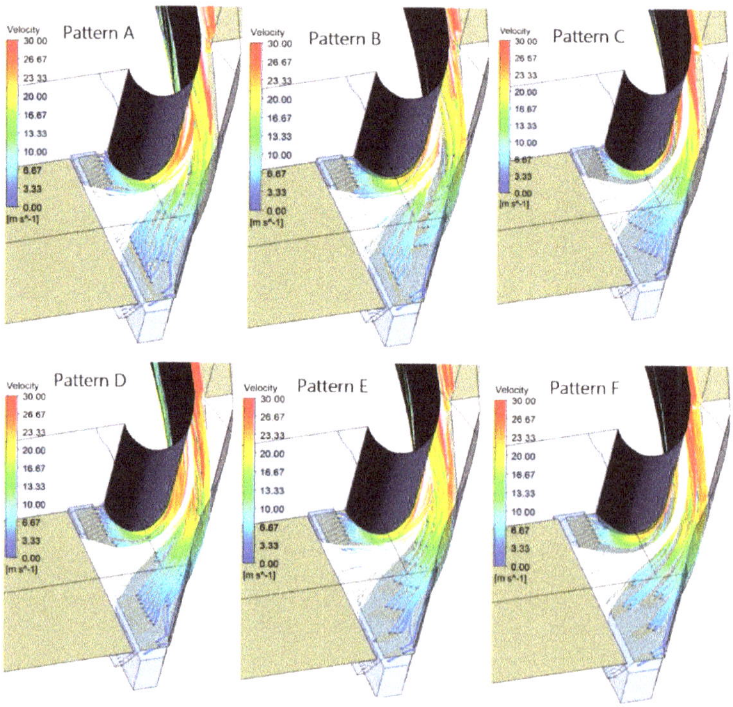

Figure 14. 3D streamlines and velocity distributions originating from slot and cooling holes for different patterns using real gas model.

Three-dimensional streamlines and velocity distributions originating from the slot and the cooling holes with different gas thermophysical properties are shown in Figure 15. It is clear that the coolant ejecting velocity is larger for the case with constant property gas in the figure. In addition, the coolant ejection from the slot is stronger than other cases, and the ejection region is enlarged. The larger ejection velocity has the benefit for coolant coverage in the region of the vane-PS junction. The difference between the cases of real gas and ideal gas is very small because the state of temperature and pressure is far from the critical point.

Figure 16 presents the streamlines originating from endwall film cooling holes interacting with the mainstream. The coolant flow developing along the vane passage is accelerated by the mainstream. It is also observed that a corner vortex is found in the region of the vane-PS junction region, which prevents the coolant from attaching on the endwall. From the PS to the SS, the transverse flow is accelerated and the velocity peak value is found on the suction side. The coolant ejection from the cooling holes in the LE region develops along the PS and is drawn upward by the mainstream. Therefore, the poor coolant coverage in the corner region exists.

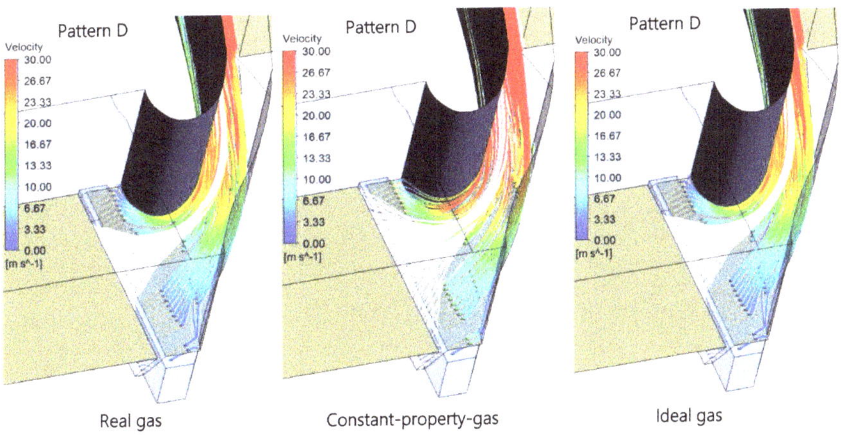

Figure 15. 3D streamlines and velocity distributions originating from the slot and the cooling holes with different gas thermophysical properties.

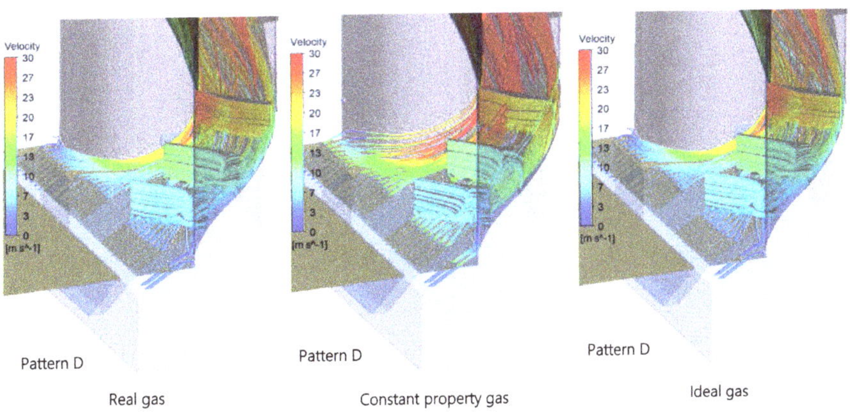

Figure 16. Interaction of 3D streamlines with mainstream flow for Pattern D with different gas thermophysical properties.

5. Conclusions

In the present study, the effects of gas thermophysical properties on full-range endwall film cooling are investigated at engine-like boundary conditions. Six kinds of full-range endwall film cooling designs are considered with the validated turbulence model, i.e., the k-ω SST model. Three kinds of gas thermophysical property models are considered, i.e., the real gas model, ideal gas model, and constant property gas model. Cooling efficiency and flow structures of full-range endwall film cooling are presented and discussed.

The coolant coverage in the PS-vane junction region is improved in Pattern B, Pattern D, Pattern E, and Pattern F with the application of the real gas model, which are designed based on the passage middle gap, limiting streamline, heat transfer coefficient, and the four-holes pattern. The coolants have difficulty being ejected out from the LE regions of Pattern A, Pattern C, and Pattern D because more cooling holes are arranged in the upstream regions.

The endwall η distribution is mainly determined by the relative ratio of ejecting velocity and density of the hot gas and the coolant. For the cases with the constant property gas model, i.e., the relative ratio of ejecting velocity and density is not changed, and the endwall η distribution almost has no changes, though different mainstream temperatures are applied. For the cases with the real gas model and the ideal gas model, they have larger gas density for the coolant flows from the cooling holes. For the case with the constant property gas, it has a larger ejecting velocity for the coolant flows from the slot. The effects of ejecting velocity and density magnitude on the coolant coverage differ at different ejecting angles. With an injection angle of 30°, the gas density is more dominant in determining the coolant coverage in the region of the cooling holes. At the injection angle of 45°, i.e., the slot region, the ejecting velocity is more dominant in determining the coolant coverage. Related conclusions provide good references for conducting endwall film cooling experiments in a room-temperature test rig, which ignores the effects of gas thermophysical properties.

Author Contributions: Conceptualization, J.L.; writing—original draft preparation, J.L. and M.X.; writing—review and editing, W.X.; All authors have read and agreed to the published version of the manuscript.

Funding: This work was supported by the Key Laboratory Continuously Supporting Project (grant No. WDZC6142703202216), the Natural Science Foundation of Hunan Province (Grant No. 2023JJ40733), and the National Natural Science Foundation of China (Grant No. 12272133).

Data Availability Statement: Not applicable.

Conflicts of Interest: The authors declare no conflict of interest.

Nomenclature

Latin characters

C	chord of the vane (m)
D	diameter of cooling holes (m)
F	cooling holes mass flow rate ratio
l	length of cooling holes (m)
P	pressure (pa)
P_c	pitch between two cooling holes (m)
P_v	pitch between adjacent vanes (m)
Re	Reynolds number
S	slot mass flow rate ratio
S_p	span of the vane (m)
T_g	mainstream temperature (K)
T_c	coolant temperature (K)
u	mainstream velocity (m/s)

x	streamwise direction (m)
y	spanwise direction (m)
z	normal direction (m)
Greek symbols	
α	injection angle (°)
λ	thermal conductivity (W/m·K)
η	film cooling effectiveness
μ	fluid dynamic viscosity (Pa·s)
ρ	fluid density (kg/m^3)
Subscripts	
m	average/overall
w	wall
c	coolant
Abbreviations	
HV	horseshoe vortex
LE	leading edge
PS	pressure side
SS	suction side
TI	turbulence intensity

References

1. Han, J.C.; Ekkad, S. Recent development in turbine blade film cooling. *Int. J. Rotating Mach.* **2001**, *7*, 21–40. [CrossRef]
2. Langston, L. Crossflows in a turbine cascade passage. *J. Eng. Gas Turbines Power* **1980**, *102*, 866–874. [CrossRef]
3. Friedrichs, S.; Hodson, H.; Dawes, W. Distribution of film-cooling effectiveness on a turbine endwall measured using the ammonia and diazo technique. ASME 1995 International Gas Turbine and Aeroengine Congress and Exposition. *J. Turbomach.* **1996**, *18*, V004T009A001.
4. Shi, W.; Chen, P.T.; Li, X.Y.; Ren, J.; Jiang, H.D. Uncertainty Quantification of the Effects of Small Manufacturing Deviations on Film Cooling: A Fan-Shaped Hole. *Aerospace* **2019**, *6*, 46. [CrossRef]
5. Oke, R.A.; Simon, T.W.; Burd, S.W.; Vahlberg, R. Measurements in a turbine cascade over a contoured endwall: Discrete hole injection of bleed flow. In *Turbo Expo 2000: Power for Land, Sea, and Air*; ASME: New York, NY, USA, 2000; p. V003T001A022.
6. Oke, R.; Simon, T.; Shih, T.; Zhu, B.; Lin, Y.L.; Chyu, M. Measurements over a film-cooled, contoured endwall with various coolant injection rates. In *Turbo Expo 2001: Power for Land, Sea, and Air*; ASME: New York, NY, USA, 2001; p. V003T001A025.
7. Zhang, L.J.; Jaiswal, R.S. Turbine Nozzle Endwall Film Cooling Study Using Pressure Sensitive Paint. In *Turbo Expo 2001: Power for Land, Sea, and Air*; ASME: New York, NY, USA, 2001; p. V003T001A030.
8. Knost, D.; Thole, K. Adiabatic effectiveness measurements of endwall film-cooling for a first stage vane. In *Turbo Expo 2004: Power for Land, Sea, and Air*; ASME: New York, NY, USA, 2004; pp. 353–362.
9. Ghosh, K.; Goldstein, R.J. Effect of inlet skew on heat/mass transfer from a simulated turbine blade. In Proceedings of the ASME 2011 Turbo Expo: Turbine Technical Conference and Exposition, Vancouver, BC, Canada, 6–10 June 2011; pp. 1677–1687.
10. Shiau, C.C.; Chen, A.F.; Han, J.C.; Azad, S.; Lee, C.P. Full-scale turbine vane endwall film-cooling effectiveness distribution using pressure-sensitive paint technique. *ASME J. Turbomach.* **2016**, *138*, 051002. [CrossRef]
11. Shiau, C.C.; Sahin, I.; Wang, N.; Han, J.C.; Xu, H.; Fox, M. Turbine Vane Endwall film cooling comparison from five film-hole design patterns and three upstream leakage injection angles. In *Turbo Expo 2018: Turbomachinery Technical Conference and Exposition*; ASME: New York, NY, USA, 2018; p. V05CT19A005.
12. Yang, X.; Liu, Z.; Zhao, Q.; Liu, Z.; Feng, Z.; Guo, F.; Ding, L.; Simon, T.W. Conjugate heat transfer measurements and predictions for the vane endwall of a high-pressure turbine with upstream purge flow. *Int. J. Heat Mass Transf.* **2019**, *140*, 634–647. [CrossRef]
13. Chen, P.; Wang, L.; Li, X.; Ren, J.; Jiang, H. Effect of axial turbine non-axisymmetric endwall contouring on film cooling at different locations. *Int. J. Heat Mass Transf.* **2020**, *147*, 118995. [CrossRef]
14. Chen, P.; Xue, Q.; Li, X.; Ren, J.; Jiang, H. Full coverage film cooling on a non-axisymmetric contoured endwall and the baseline endwall. *Int. J. Therm. Sci.* **2020**, *158*, 106562. [CrossRef]
15. Liu, J.; Du, W.; Hussain, S.; Xie, G.; Sundén, B. Endwall film cooling holes design upstream of the leading edge of a turbine vane. *Numer. Heat Transf. Part A Appl.* **2021**, *79*, 222–245. [CrossRef]
16. Bu, H.; Guo, Z.; Song, L.; Li, J. Effects of cooling configurations on the aerothermal performance of a turbine endwall with jet impingement and film cooling. *ASME J. Turbomach.* **2021**, *143*, 061013. [CrossRef]
17. Yang, X.; Zhao, Q.; Liu, Z.; Liu, Z.; Feng, Z. Film cooling patterns over an aircraft engine turbine endwall with slot leakage and discrete hole injection. *Int. J. Heat Mass Transf.* **2021**, *165*, 120565. [CrossRef]
18. Yan, D.J.; Jiang, H.M.; Li, J.L.; Xie, W.B.; Wang, Z.G.; Lu, S.P.; Zhang, Q. Influence of gas-to-wall temperature ratio on the leakage flow and cooling performance of a turbine squealer tip. *Aerospace* **2022**, *9*, 627. [CrossRef]
19. Knost, D.G.; Thole, K.A. Computational predictions of endwall film-cooling for a first stage vane. In *Turbo Expo 2003*; ASME: New York, NY, USA, 2003; pp. 163–173.

20. Knost, D.G. Predictions and Measurements of Film-Cooling on the Endwall of a First Stage Vane. Ph.D. Thesis, Virginia Tech, Blacksburg, VA, USA, 2003.
21. Liu, J.; Du, W.; Zhang, G.; Hussain, S.; Wang, L.; Xie, G.; Sundén, B. Design of Full-Scale Endwall Film Cooling of a Turbine Vane. *ASME J. Heat Transf.* **2020**, *142*, 022201. [CrossRef]
22. Pu, J.; Zhang, T.; Wang, J. Experimental study of cooling air effect on overall cooling of laminated configuration at a turbine vane end-wall. *Case Stud. Therm. Eng.* **2022**, *32*, 101890. [CrossRef]
23. Zhang, H.W.; He, Y.L.; Tao, W.Q. Numerical study on liquid-film cooling at high pressure. *Numer. Heat Tranfer Part A Appl.* **2010**, *58*, 163–186. [CrossRef]
24. Nematollahi, O.; Nili-Ahmadabadi, M.; Kim, K.C. The influence of cubic real-gas equations of state in the supersonic regime of dense gases. *J. Mech. Sci. Technol.* **2020**, *34*, 1581–1589. [CrossRef]
25. Yuan, F.; Zeng, Y.; Khoo, B.C. A new real-gas model to characterize and predict gas leakage for high-pressure gas pipeline. *J. Loss Prev. Process Ind.* **2022**, *74*, 104650. [CrossRef]
26. Salvadori, S.; Ottanelli, L.; Jonsson, M.; Ott, P.; Martelli, F. Investigation of High-Pressure Turbine Endwall Film-Cooling Performance Under Realistic Inlet Conditions. *J. Propuls. Power* **2012**, *28*, 799–810. [CrossRef]
27. Wen, F.B.; Hou, R.; Luo, Y.X.; Wang, S.T. Study on leading-edge trenched holes arrangement under real turbine flow conditions. *Eng. Appl. Comput. Fluid Mech.* **2021**, *15*, 781–797. [CrossRef]
28. Deng, Q.H.; Wang, H.H.; He, W.; Feng, Z.P. Cooling Characteristic of a Wall Jet for Suppressing Crossflow Effect under Conjugate Heat Transfer Condition. *Aerospace* **2022**, *9*, 29. [CrossRef]
29. Bai, B.; Li, Z.G.; Li, J.; Mao, S.; Ng, W.F. The Effects of axisymmetric convergent contouring and blowing ratio on endwall film cooling and vane pressure side surface phantom cooling performance. *ASME J. Eng. Gas Turbines Power* **2022**, *144*, 021020. [CrossRef]
30. Florio, L.A. Effect of gas equation of state on CFD predictions for ignition characteristics of hydrogen escaping from a tank. *Int. J. Hydrog. Energy* **2014**, *39*, 18451–18471. [CrossRef]
31. Song, W.H.; Yao, J.; Ma, J.S.; Li, Y.; Han, W.C. A pore structure based real gas transport model to determine gas permeability in nanoporous shale. *Int. J. Heat Mass Transf.* **2018**, *126*, 151–160. [CrossRef]
32. Nicklas, M. Film-cooled turbine endwall in a transonic flow field: Part ii—Heat transfer and film-cooling effectiveness. *ASME J. Turbomach.* **2001**, *123*, 720–729. [CrossRef]
33. Kang, M.B.; Kohli, A.; Thole, K. Heat transfer and flowfield measurements in the leading edge region of a stator vane endwall. *ASME J. Turbomach.* **1999**, *121*, 558–568. [CrossRef]
34. Poling, B.E.; Prausnitz, J.M.; O'connell, J.P. *Properties of Gases and Liquids*; McGraw-Hill Education: New York, NY, USA, 2001.

aerospace

MDPI

Article

Pressure Characteristics and Vortex Observation in Chiral-Symmetric Space Orthogonal Bifurcation

Zixuan Fang [1], Dingwei Zhang [1], Xiaokang Liu [1], Jingxuan Li [1,2], Lijun Yang [1,2] and Qingfei Fu [1,2,*]

[1] School of Astronautic, Beihang University, Beijing 102206, China; fangzx@buaa.edu.cn (Z.F.)
[2] Aircraft and Propulsion Laboratory, Ningbo Institute of Technology, Beihang University, Ningbo 315800, China
* Correspondence: fuqingfei@buaa.edu.cn

Abstract: In aerospace engine delivery systems, "one-in-two-out" bifurcation structures are commonly used for flow distribution to downstream pipelines. There are two common "one-in-two-out" bifurcation structures in aircraft engines: the planar orthogonal bifurcation and the spatial orthogonal bifurcation. By adjusting the flow supply upstream and the cross-sectional diameter downstream, the flow distribution in the two branches can be adjusted, i.e., the "splitting ratio" changes. In this paper, a dismantling and flexible experimental system is constructed to measure the pressure signals in each channel and use non-linear dynamic analysis methods to extract pressure characteristics. The particle image velocimetry (PIV) technique combined with the fine rope tracing technique is creatively used to observe the vortex structure in the cross section of the downstream branch. The study found that for spatial orthogonal bifurcation, the pressure signal characteristics in each channel are basically the same at larger splitting ratios, regardless of the chirality. As the splitting ratio decreases, the difference in pressure signal characteristics between the two branches gradually becomes evident and becomes related to the chirality. Moreover, unlike the planar orthogonal bifurcation structure, a complete large vortex structure has not been found in the downstream branch of the spatial orthogonal bifurcation structure, regardless of changes in the splitting ratio, and it is unrelated to the chirality.

Keywords: spatial orthogonal bifurcation; chirality; pressure measurement; nonlinear dynamic analysis; vortex structure

Citation: Fang, Z.; Zhang, D.; Liu, X.; Li, J.; Yang, L.; Fu, Q. Pressure Characteristics and Vortex Observation in Chiral-Symmetric Space Orthogonal Bifurcation. *Aerospace* **2023**, *10*, 568. https://doi.org/10.3390/aerospace10060568

Academic Editor: Hossein Zare-Behtash

Received: 29 April 2023
Revised: 12 June 2023
Accepted: 17 June 2023
Published: 19 June 2023

1. Introduction

The "one inlet and two outlets" bifurcation structure is widely used in fluid transportation systems. In the fields of construction, industry, and materials preparation, bifurcation structures are widely used for the distribution of water, steam, and other fluid media [1,2]. In the field of microfluidics, bifurcation structures can also be used for heat exchange assistance [3–5]. In the fields of medicine and biology, similar structures exist in the cardiovascular system [6]. In liquid rocket engines, bifurcation structures are common in propellant supply systems and play a role in flow regulation [7]. According to the relative position of the three channels, bifurcation structures can be divided into two categories: planar orthogonal bifurcation (POB, i.e., the three channels are located in the same plane, also known as "T-junction") and spatial orthogonal bifurcation (SOB, i.e., the three channels are arranged according to spatial orthogonal coordinates). The main research focus of the flow behavior in bifurcation structures is the study of pressure and vortex structures inside the pipes.

Wang et al. [8,9] built an experimental platform that closely resembles a real circular cross-section pipe system, bmeasured the pressure signals inside a T-shaped bifurcation to obtain pressure signals of flow conditions such as bubbly flow and spiral flow and conducted nonlinear dynamical analysis. Their experimental system features two branches with identical cross-sectional areas, resulting in equal flow rates within each branch during steady-state flow. The essential dimensionless parameter under investigation is the extraction rate, defined as the ratio of outlet flow rate W_3 to inlet flow rate W_1. They discovered a strong association between the disturbance behavior of pressure drop across the main and branch channels and the aforementioned parameter W_3/W_1. In practical flow processes, the cross-sectional areas of the two branches are often intentionally designed differently to meet downstream flow requirements, leading to the development of asymmetric T-junctions. In such configurations, the flow rates in the two branches are unequal, necessitating the introduction of a critical parameter: the splitting ratio, which denotes the ratio of the flow rate in branch 1 to that in branch 2. Recently, Fang et al. [10] conducted experiments to study the pressure oscillation phenomena in asymmetric T-junctions and uncovered the influence of the splitting ratio on these oscillations.

In complex and narrow spaces, such as the delivery systems of liquid rocket engines, bifurcations must be carefully arranged with a specific spatial angle to ensure proper placement. This has given rise to the development of spatial bifurcations. This study focuses on the SOB, which is the bifurcation of two branches in space that form a 90-degree angle. Our investigation has revealed a significant dearth of literature on the pressure characteristics of this type of bifurcation structure, despite its importance in the aerospace industry. Chirality is an interesting topic when it comes to space orthogonality. Chirality is observed as a functional difference between structures that can be mirrored and superimposed in space. This phenomenon is widely present in fields such as quantum chemistry, medicine, and nuclear physics [11–15]. Despite the importance of chirality in various fields, there is a lack of research exploring the impact of macroscopic chiral symmetry on fluid flow inside pipelines, especially when dealing with asymmetric branching pipelines with chiral symmetry and differing diameters. Hence, as shown in Figure 1, the main objective of this study is to examine the flow behavior in chiral-symmetric SOB.

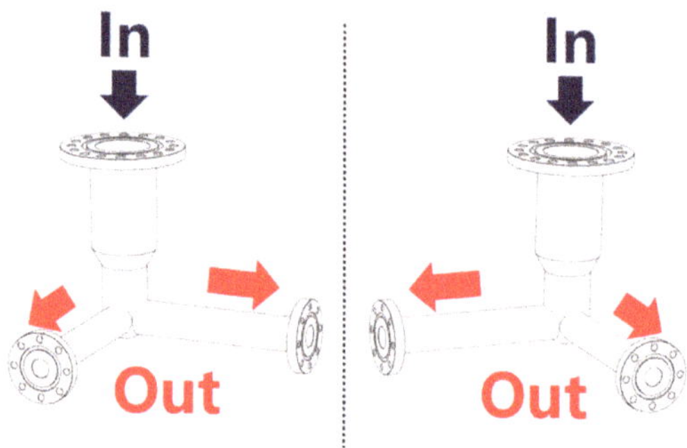

Figure 1. Schematic diagram of the space orthogonal bifurcation with handedness symmetry.

Apart from pressure characteristics, another significant flow feature in bifurcation structures is the vortex structure present in the cross-section. In recent years, there has been a large amount of experimental and numerical simulation research on internal flow in T-junctions in microfluidics. Vigolo [16] was the first to discover, through experiments, the existence of a stagnant flow region at the junction of the bifurcation in a T-junction, where low-density solid particles that pass through this region will "suspend" and remain relatively still. Vigolo believed that there is a region at the bifurcation junction that causes solid particles to stagnate within a specified range of Reynolds numbers, and the mechanism of stagnation is that two vortexes are generated at the bifurcation junction, and the reverse pressure gradient and reflux in the vortex core capture and restrict low-density solid particles near the junction. In order to clearly delineate the shape of the recirculation region, numerical simulation methods are needed to simulate the flow process. Chen [17] was the first to use the direct numeral simulation (DNS) method to study the flow in a T-junction, quantitatively describing the development of vortices inside the pipe. Ault further studied the motion trajectory of electrolyzed water bubbles at the bifurcation point of a POB at different Reynolds numbers and angles through experiments and combined the results of numerical simulation to demonstrate the existence of four recirculation regions at the bifurcation point, each recirculation region is bounded by two stagnation points (SP), one upstream and one downstream [18]. Based on the above research, Chen [19] used the same simulation method to study the flow inside a POB at different angles. Chen found that at the downstream position of the POB, outside of the recirculation region of the vortex, the streamline will form a spiral structure, called "spiral flow". He did not conduct in-depth research and explanation on this phenomenon, but this phenomenon is roughly consistent with Xiong' s computational results [20]. The experimental research on vortex structures in bifurcation structures mentioned above was mainly conducted in small-scale plane channels, and there is a significant difference between the size of the pipeline in actual transport systems and the experimental setup. The vortex behavior inside spatial bifurcation structures has not been considered. Recently, researchers have employed the PLIF (Planar Laser-Induced Fluorescence) method to visualize and study the flow field within POB at high Reynolds numbers [21]. Our research also aims to advance the research on SOB, investigating the internal flow patterns and providing insights for future studies in the aerospace field.

2. Materials and Methods

The basic experimental system of this article is illustrated in Figure 2. We built a modular experimental platform to study the pressure characteristics and cross-sectional vortex structures of SOB. In this figure, 1 is the joint module made of aluminum alloy, responsible for installing sensors and fixing devices; 2 is the diversion module made of aluminum alloy, responsible for diverting upstream fluid into two downstream branches, with internal dimensions identical to the original structure; 3 is the observation module, a glass tube made of high-transparency quartz glass; 4 is a pressure sensor fixed on the joint module; 5 is a slide rail that can easily adjust the relative position of each structural component; and 6 is a sliding block with a locking positioning device. The modules are clamped together by screws. This plan cannot observe the diversion area, but overall, it greatly reduces the shear and bending moments borne by the quartz tube and is flexible to disassemble and assemble without affecting the observation of the flow state in the downstream branch, greatly improving the efficiency of the experiment.

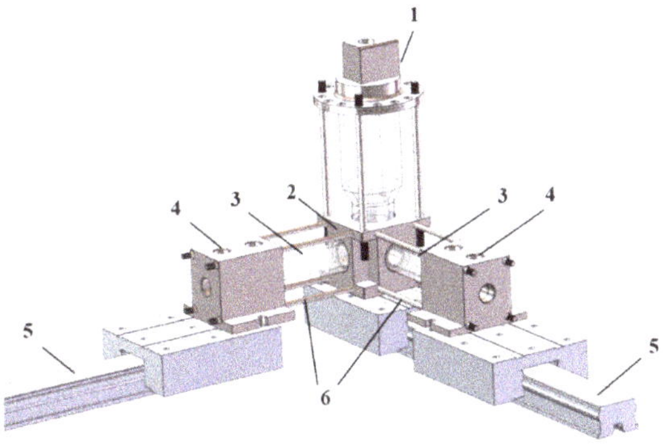

Figure 2. Schematic diagram of the experimental device for spatial orthogonal bifurcating pipeline.

The two downstream branches are selected with inner diameters of 18 mm and 22 mm to ensure the flow can reach a high Reynolds number, and defined as branch 1 and branch 2, respectively. The different inner diameters result in different flow rates in the two branches. Currently, there is no research that describes the effect of such flow rate differences on the formation and development of vortices in the SOB, nor are there any publicly available literature that shows the impact on the formation and development of vortices in the POB. The splitting ratio is defined as the ratio of the mass flow rates in the two branches:

$$\eta = \frac{\dot{m}_1}{\dot{m}_2} \tag{1}$$

In the equation, $\dot{m}_1$ is the mass flow rate of branch 1, $\dot{m}_2$ is the mass flow rate of branch 2, both of which can be directly read from the flow meter in the experimental system.

The focus of this study is the effect of flow rate ratio variation onflow behavior; and therefore, there is no need to consider the influence of surface tension. Furthermore, it is required that the working fluid has good transparency for visualization purposes and be cost-effective. Considering these factors, water was chosen as the experimental fluid, with a density of 1000 kg/m^3. The maximum valve opening of the downstream valves is 20 mm, and the feedback loop of the PLC inside the valve can control the valve opening in real-time, ensuring that the flow rates of the two branches are controlled within the given range. In the experiment, the flow rate of branch 1 is controlled at around 0.4 kg/s, and the flow rate of branch 2 varies in a relatively large range. At the same time, the pressure signals on the main branch and the two branches are collected through pressure sensors, with a sampling frequency of 5000 Hz and a sampling duration of 20 s.

In order to study the effect of chirality on the flow process, chiral symmetric components were also designed: in component 1, branch 1 and branch 2 are clockwise at a 90-degree angle, while in component 2, branch 1 and branch 2 are counterclockwise at a 90-degree angle.

3. Results and Discussion

This study focuses on investigating the influence of splitting ratio on the flow patterns within the SOB. The specific mass flow rates in each branch need to consider the maximum supply capacity of the supply system (plunger-type water pump) and the actual flow scaling in real aerospace propulsion systems. Therefore, in this study, the flow rate in branch 1 is controlled to be around 0.4 kg/s, while the flow rate in branch 2 varies between 0.2 and 0.9 kg/s. This ensures that the dimensionless splitting ratio is distributed between 0.4 and 2. By the way, the Reynolds number in each branch is defined as:

$$Re = \frac{\rho U d}{\mu} \tag{2}$$

where ρ is the density of water, U is the flow velocity in the branch, which can be obtained by measuring the mass flow rate, d is the diameter of the flow branch, and μ is the viscosity of water. After calculation, it can be found that the Reynolds numbers in both branches are of the order of 10^4, indicating turbulent flow. The Reynolds number in the smaller branch remains around 30,000, while in the larger branch, it varies between 10,000 and 50,000.

To begin with, the pressure signals in each channel under different splitting ratios were transformed into the frequency domain using FFT to perform a Fourier transform on the time-domain signals. The power spectral density (PSD) was then used to represent the vibration amplitude at various frequencies. The PSDs in each channel of both workpieces 1 and 2 under different splitting ratios are shown in Figures 3 and 4, respectively. The analysis results only retained the low-frequency range of 0–200 Hz.

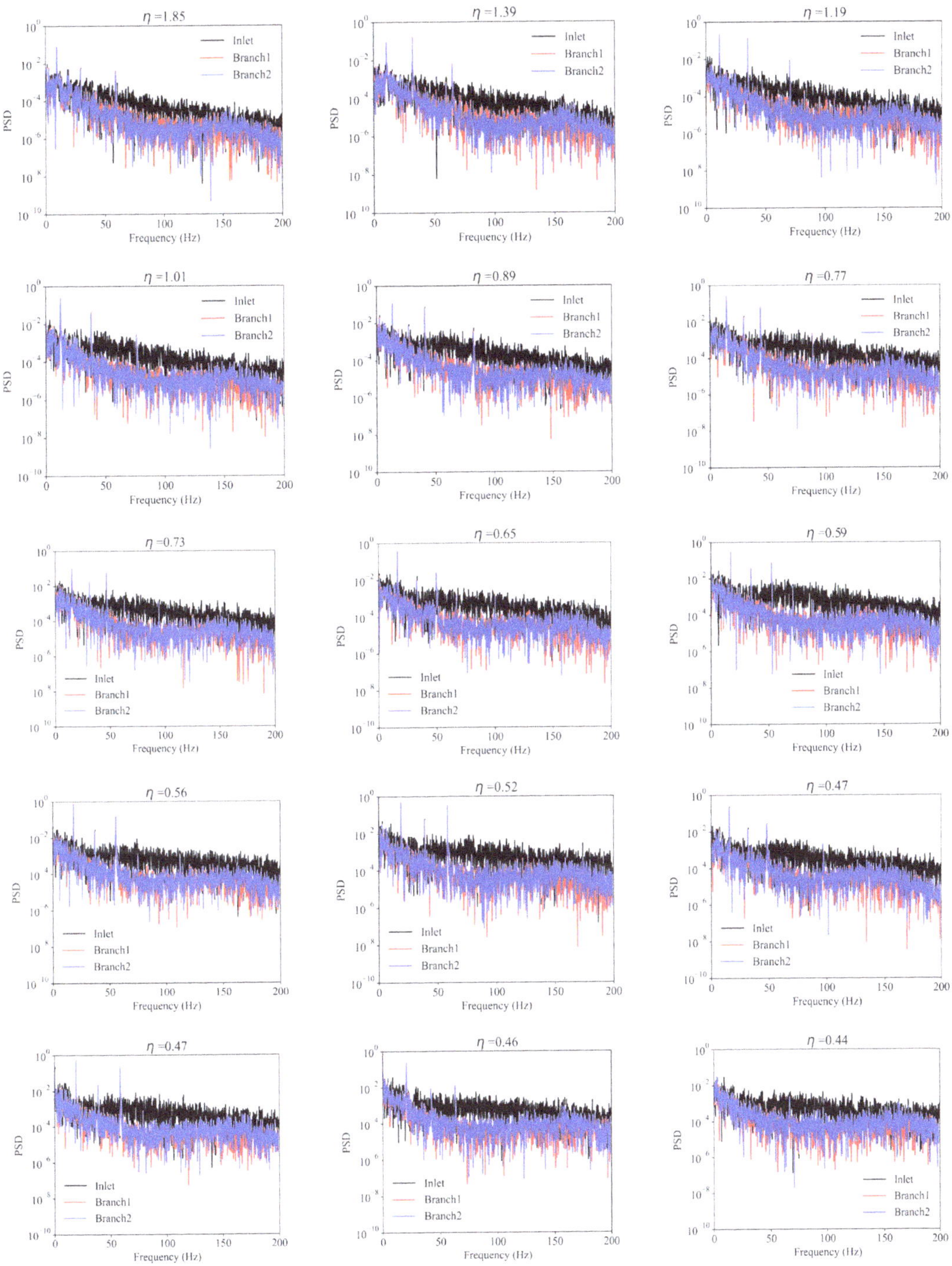

Figure 3. *Cont.*

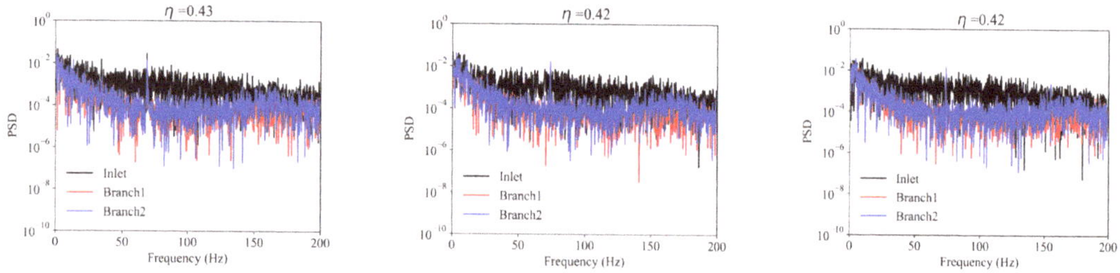

Figure 3. PSD at different splitting ratio of Workpiece 1.

The PSD of the pressure signals in different channels under different splitting ratios for Workpiece 1 is shown in Figure 3. The black, red, and blue lines in the figure represent the PSD of the pressure signals in the inlet, branch 1, and branch 2, respectively. It can be seen that the spectral signals in all three channels are generally in the form of wideband signals, with only peaks appearing near certain frequency values, and the positions of the peaks are basically the same. When the splitting ratio is above 0.5, three peaks can be observed in the frequency spectrum within 0–100 Hz, with two more prominent peaks appearing near 10 Hz and 50 Hz, and the positions of the peaks will shift to a certain extent with the change of the splitting ratio. When the splitting ratio drops below 0.5, there are basically no peaks in the main channel signal, and the peaks in both branches mainly appear near 75 Hz but are still not higher than the corresponding main channel amplitude. At the same time, the pressure amplitude in the main channel is generally higher than that in the branches. This indicates that the frequency domain characteristics of the pressure signal are closely related to the splitting ratio. A larger splitting ratio means that the flow rate in branch 1 is higher than that in branch 2, which means that more fluid enters the smaller-diameter pipe under steady-state conditions, causing the three channels of the bifurcation structure to simultaneously exhibit three "characteristic modes". As the splitting ratio decreases, the fluid entering branch 2 increases and gradually becomes dominant, which will smooth out the peaks in the main channel.

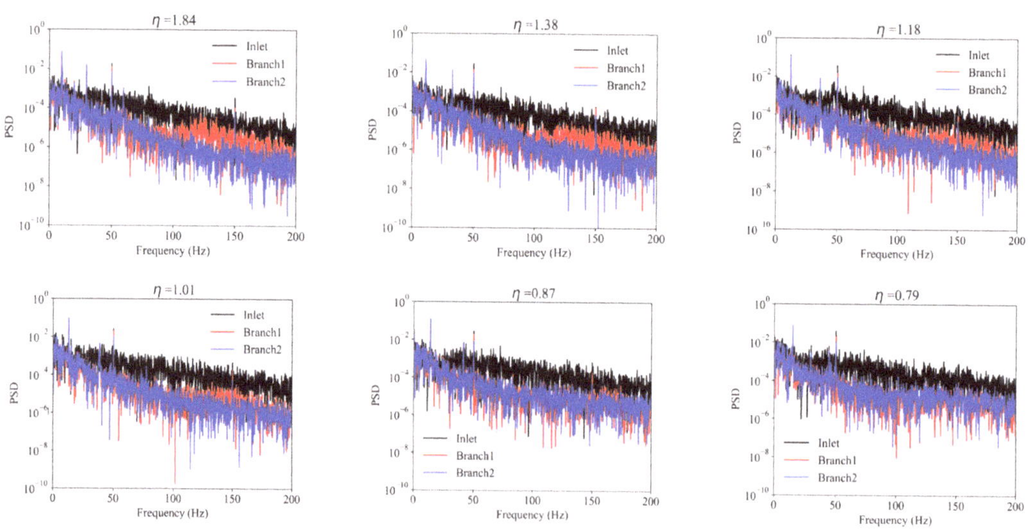

Figure 4. *Cont.*

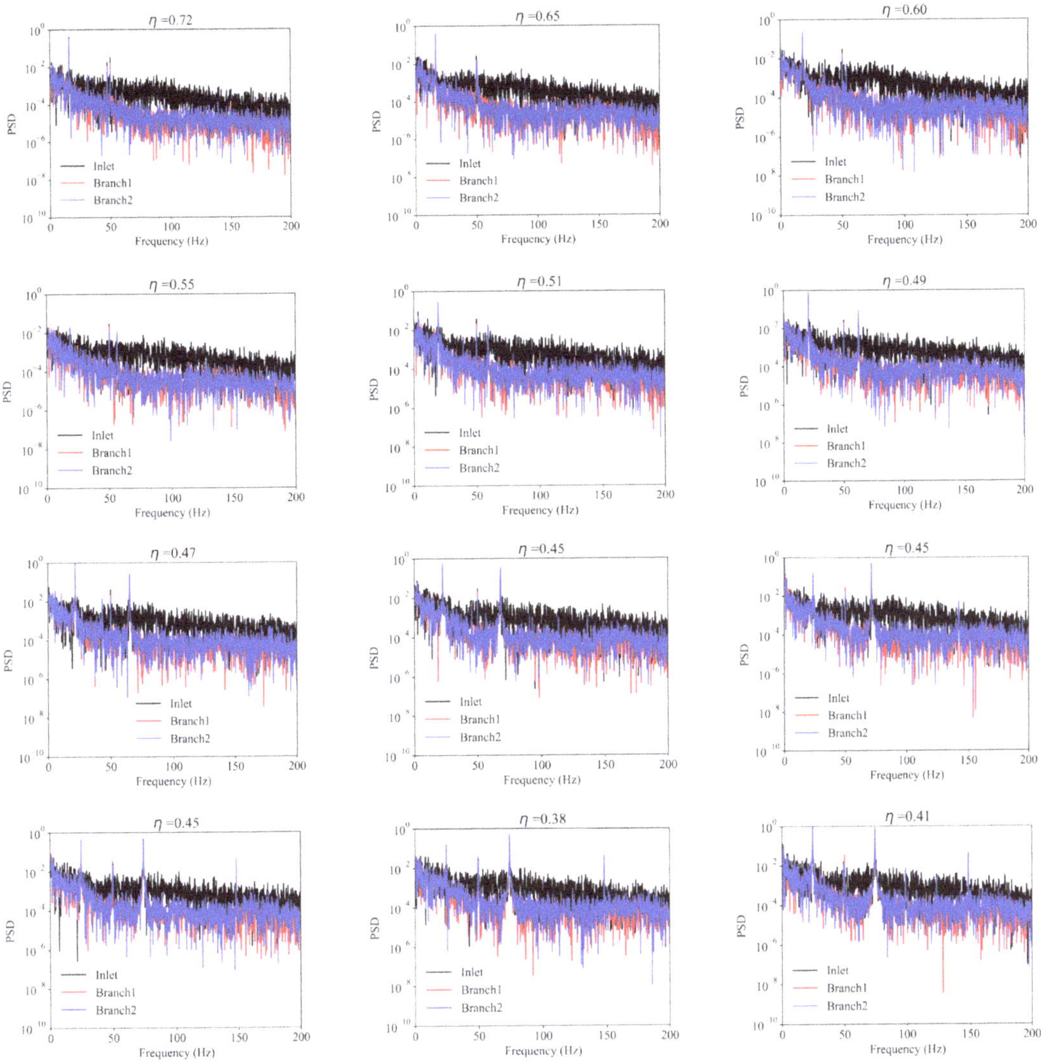

Figure 4. PSD at different splitting ratio of Workpiece 2.

As shown in Figure 4, it can be observed that Workpiece 2 also has peaks under different splitting ratios, and when the splitting ratio is above 0.5, the peaks in each channel also show a trend of being flattened. However, unlike Workpiece 1, when the splitting ratio drops below 0.5 and the flow rate in branch 2 becomes much larger than that in branch 1, the frequency spectrum signals in each channel of Workpiece 2 do not become broadband, but instead form a new peak near 70 Hz, which becomes more prominent with decreasing splitting ratio. The original peaks in each channel also do not degrade into broadband but maintain their characteristics. Obviously, the chiral structure significantly affects the pressure distribution of the flow in the spatial orthogonal bifurcation structure, and counterclockwise rotation generates more "characteristic modes" than clockwise rotation.

Figure 4 also shows that the synchronous variation of the frequency spectra in branches 1 and 2 in Workpiece 2 is weaker than that in Workpiece 1, especially when the splitting ratio is above 1. It can be seen that the pressure amplitude in branch 1 is higher than that in branch 2, which may be related to the larger flow rate in branch 1, but this degree of separation was not observed in Workpiece 1.

The above analysis was performed on the pressure signals from a statistical perspective. To further analyze the differences in the characteristics of the pressure signals, a nonlinear dynamical method is needed to study the behavior of the attractors in phase space in each channel and under different working conditions in Workpiece 1 and Workpiece 2, as shown in Figures 5 and 6. The autocorrelation function is a common statistical measure used to quantify the similarity between a signal and a delayed version of itself. It is typically calculated by multiplying the corresponding values of the signal and its delayed version, and then summing up these products over a certain range of delays. The autocorrelation function provides a more intuitive representation of the periodic characteristics of the data and serves as the basis for subsequent phase space reconstruction.

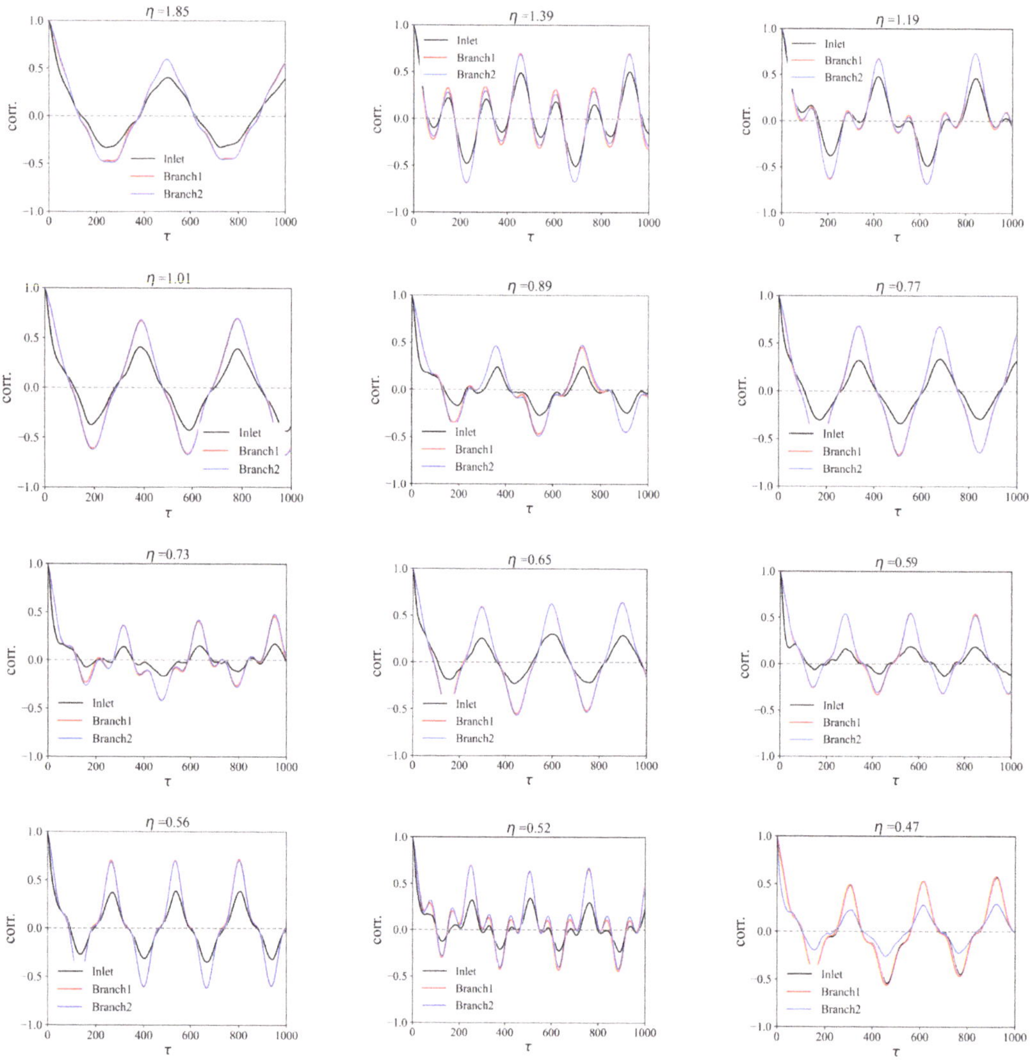

Figure 5. *Cont.*

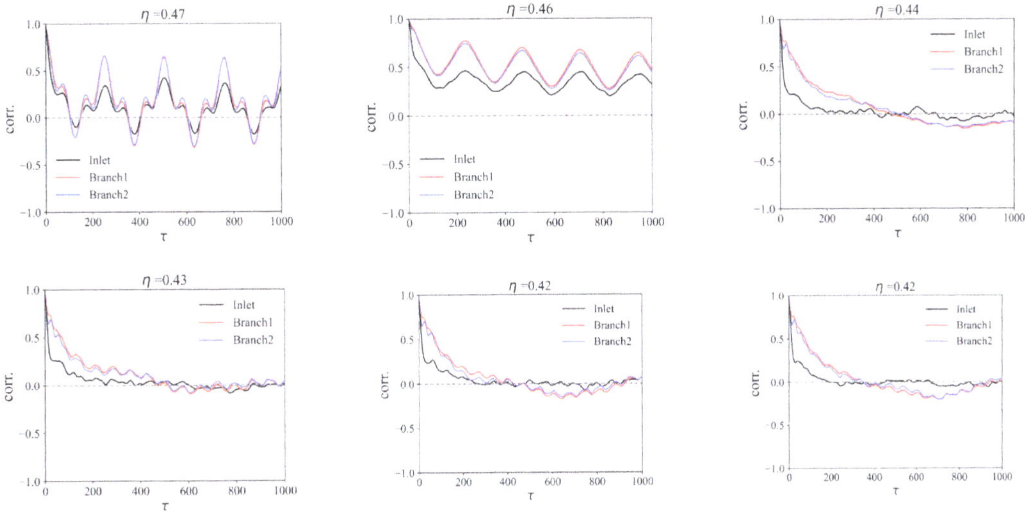

Figure 5. Autocorrelation function at different splitting ratio of Workpiece 1.

As shown in Figure 5, the results of the PSD analysis are consistent with the autocorrelation functions. The main channel and the two branches exhibit similar trends in both periodicity and amplitude, and the pressure signals in branch 1 and branch 2 show basically the same changes. When the splitting ratio is greater than 0.47, the pressure signals in different channels all exhibit good autocorrelation, with periodic crossings of the zero-scale line. When the splitting ratio decreases below 0.47, significant changes are observed in the autocorrelation of the three channels, gradually changing from periodic crossings of the zero-scale line to oscillating and decaying near the zero-scale line. The reason for this phenomenon may be that when the splitting ratio decreases to below 0.47, the attractor shape inside the dynamical system undergoes a transformation, but the specific reason still needs to be further analyzed using phase space trajectories.

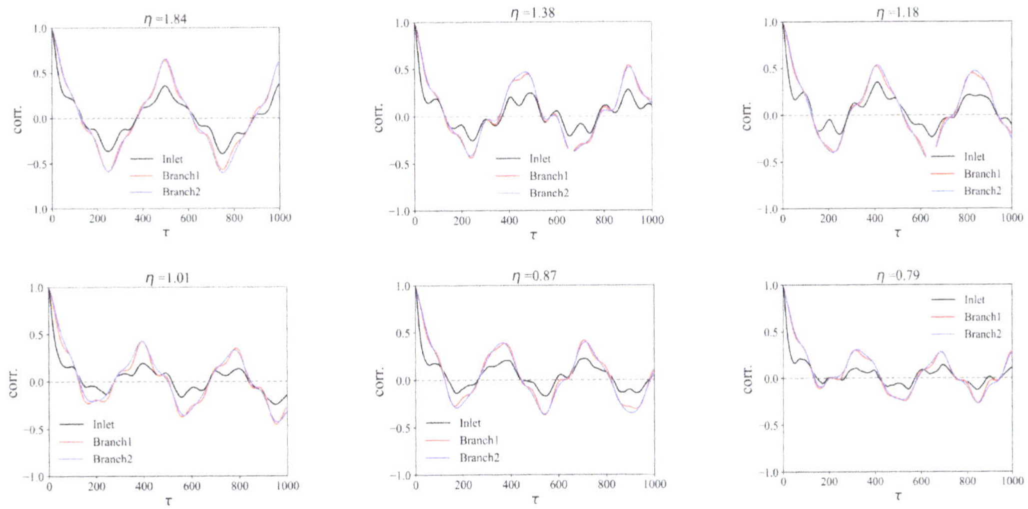

Figure 6. *Cont.*

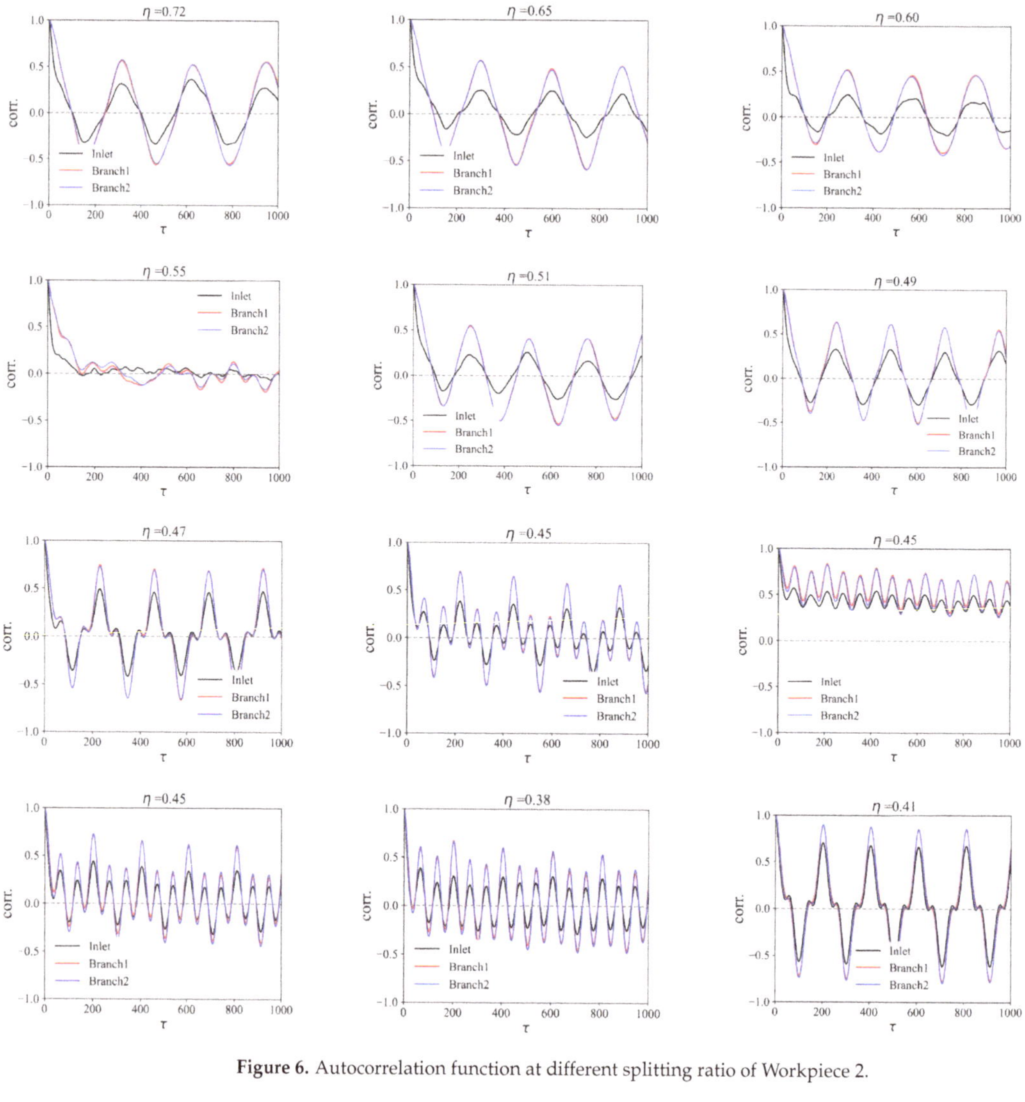

Figure 6. Autocorrelation function at different splitting ratio of Workpiece 2.

From Figure 6, it can be observed that in workpiece 2, as the split ratio decreases, the autocorrelation of the pressure signals in each channel also changes. Particularly when the split ratio is equal to 0.55, it exhibits similar oscillatory decay behavior as in workpiece 1 at small split ratios. However, unlike workpiece 1, the autocorrelation signals in workpiece 2 continue to exhibit strong periodicity below a split ratio of 0.5 instead of oscillatory decay. Similar to the PSD signals, the autocorrelation functions of the two branches in workpiece 2 also do not exhibit complete synchronization when the split ratio is large, and their difference is much more significant than in workpiece 1. However, with the increase in flow rate in branch 2, their synchronization is corrected.

After obtaining the autocorrelation functions, in order to achieve the optimal visualization effect of the phase space trajectory, the time delay is usually selected as the value corresponding to the first crossing of the zero-scale line in the autocorrelation function. In this paper, the time delay is uniformly selected as 200. The reconstructed phase space trajectories of workpiece 1 and workpiece 2 are shown in Figures 7 and 8, respectively.

In Figure 7, it can be observed that the attractor shapes of the pressure signals in the three channels are similar in phase space, with the attractor volume of the main channel being significantly larger than that of the two branch channels, which is related to the larger pressure fluctuations in the main channel. When the flow split ratio is relatively large, the positions of the attractors in the phase space of the three channels can be clearly observed: branch 2 is located between the main channel and Branch 1, with a lower flow rate in Branch 2. However, as the flow split ratio decreases, the phase space trajectory of branch 2 gradually approaches that of branch 1, and when the flow split ratio is around 0.5, the two trajectories completely overlap. The figure also shows that when the flow split ratio is 0.46, the attractors of the main channel and the two branch channels are all affected to some extent, but they all return to their previous shapes at smaller flow split ratios.

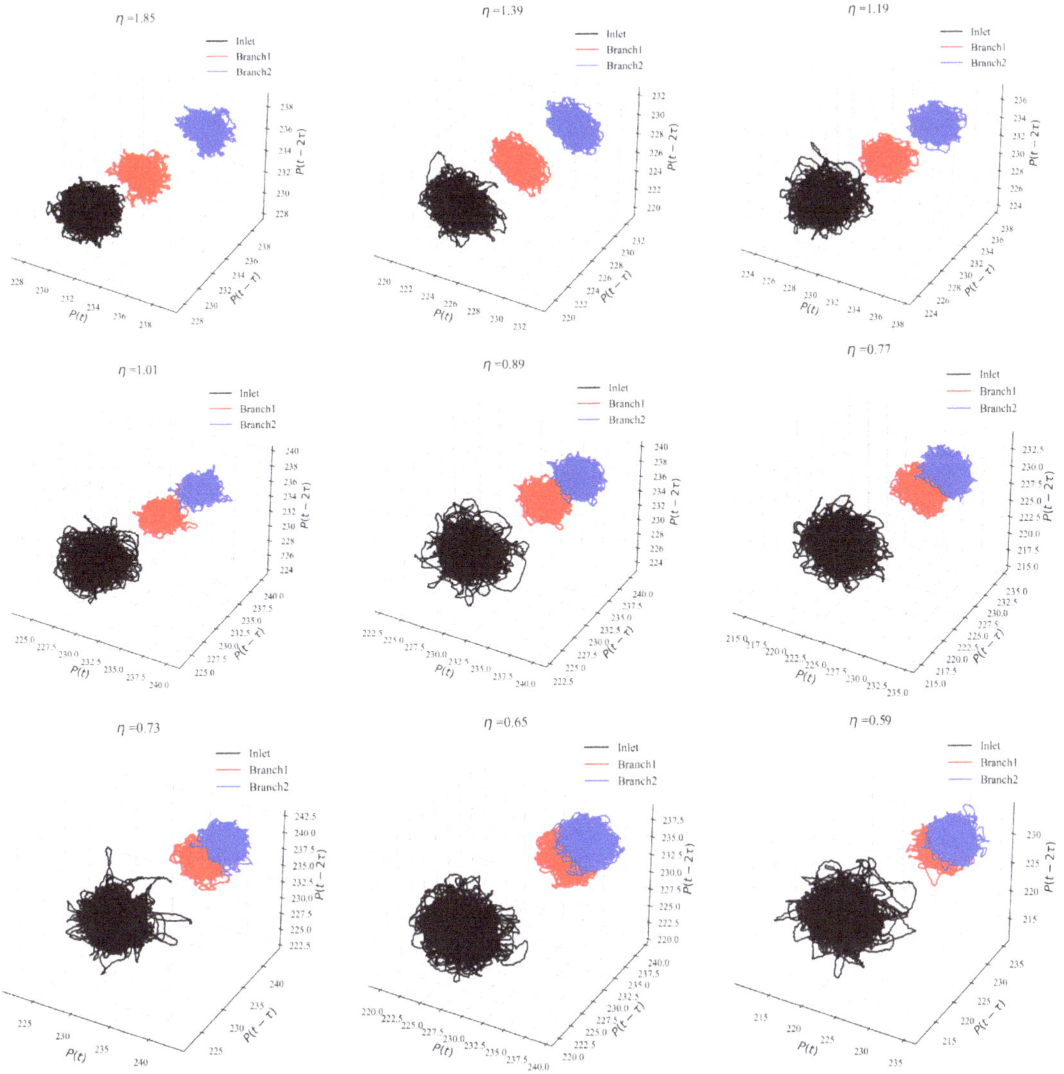

Figure 7. *Cont.*

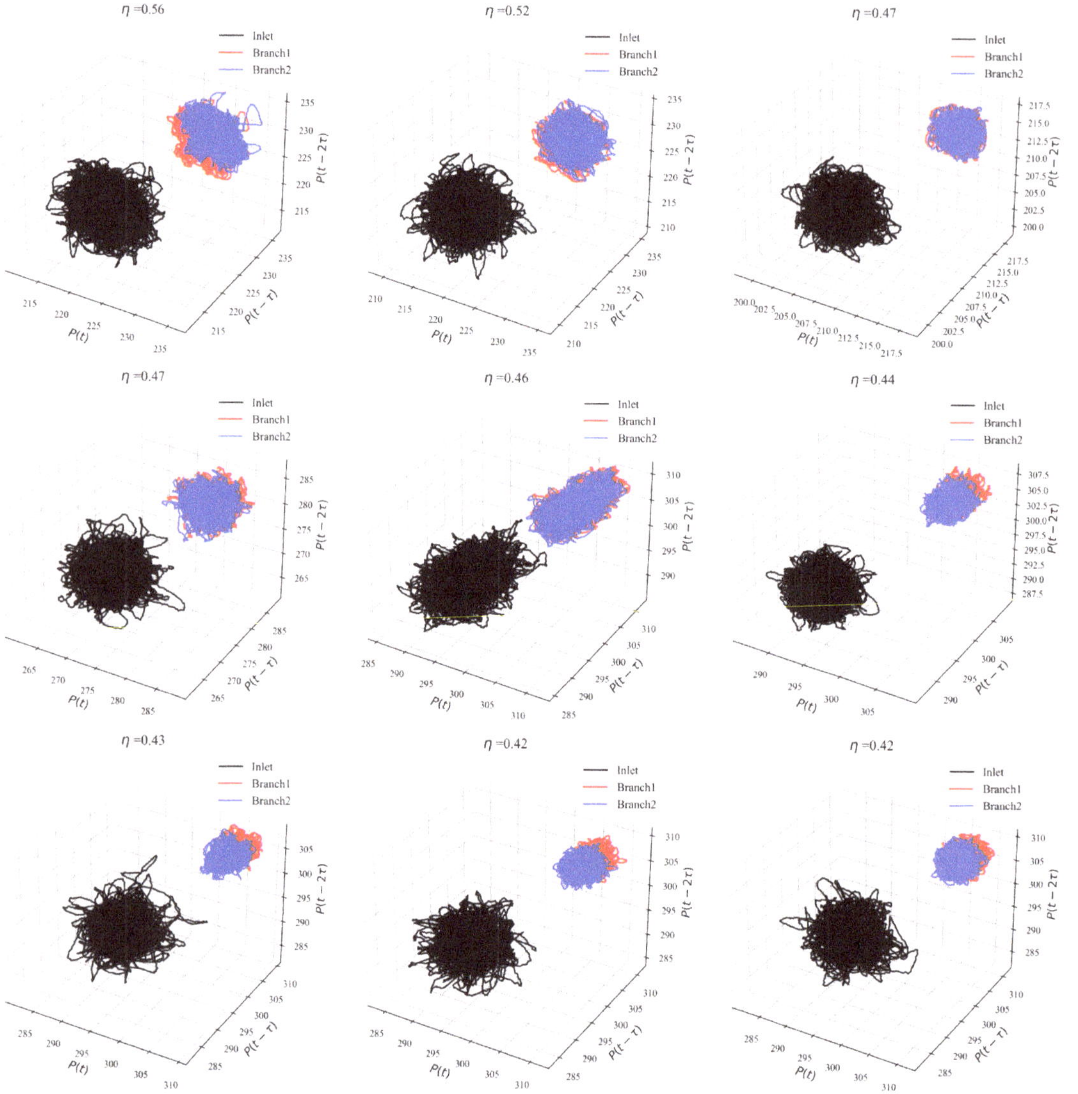

Figure 7. Phase space trajectories at different splitting ratio of Workpiece 1.

In Figure 8, it can be observed that the arrangement of attractors in the phase space of Workpiece 2 is similar to that of Workpiece 1 when the bifurcation ratio is relatively large: the attractor volume of the main channel is the largest, and the attractor of branch 2 is located between the two. Similarly, as the bifurcation ratio decreases, the attractor of branch 2 will move closer to branch 1. However, when the bifurcation ratio decreases to around 0.45, it can be observed that the shapes of all three attractors are stretched, and the shape of the attractor of branch 2 is stretched to span across the main channel and branch 1. To verify whether the above phenomenon actually occurs, the method of constructing a Poincaré section can be adopted to study the trajectory of the high-dimensional dynamical system in a low-dimensional space. Following the description by Kantz et al. in their monograph [22], the phase space trajectories and Poincaré sections complement each other as high-dimensional and low-dimensional features of the dynamical system, respectively, providing a more comprehensive representation of the system's characteristics. Hence, they are jointly presented in the paper. The Poincaré sections of the pressure signals in each channel of Workpiece 1 and Workpiece 2 under different bifurcation ratios are shown in Figures 9 and 10.

The results of the Poincaré sections in Figures 9 and 10 confirm the phenomena observed in the phase space trajectories. The attractors in both workpieces 1 and 2 do exhibit changes with the variation of the flow ratio. This is related to the different angles of the bifurcation in the two workpieces. In the flow process, the fluid is subject to different inertial forces in different directions, leading to different oscillation phenomena within the bifurcation, and causing differences in the pressure signals.

To visualize the vortex structures in spatially orthogonal bifurcations and planar orthogonal bifurcations, the PIV technique was used to observe the cross-sections. The colormap represents the magnitude of vorticity, while the arrows indicate velocity vectors. The results show that in POB, at certain flow split ratios, the formation and breakdown of larger vortex structures, as shown in Figure 11a, can be observed. However, in SOB, only flow behavior with non-uniform vorticity distribution, as shown in Figure 11b, can be observed in both branches, but the formation and breakdown of vortex structures are absent.

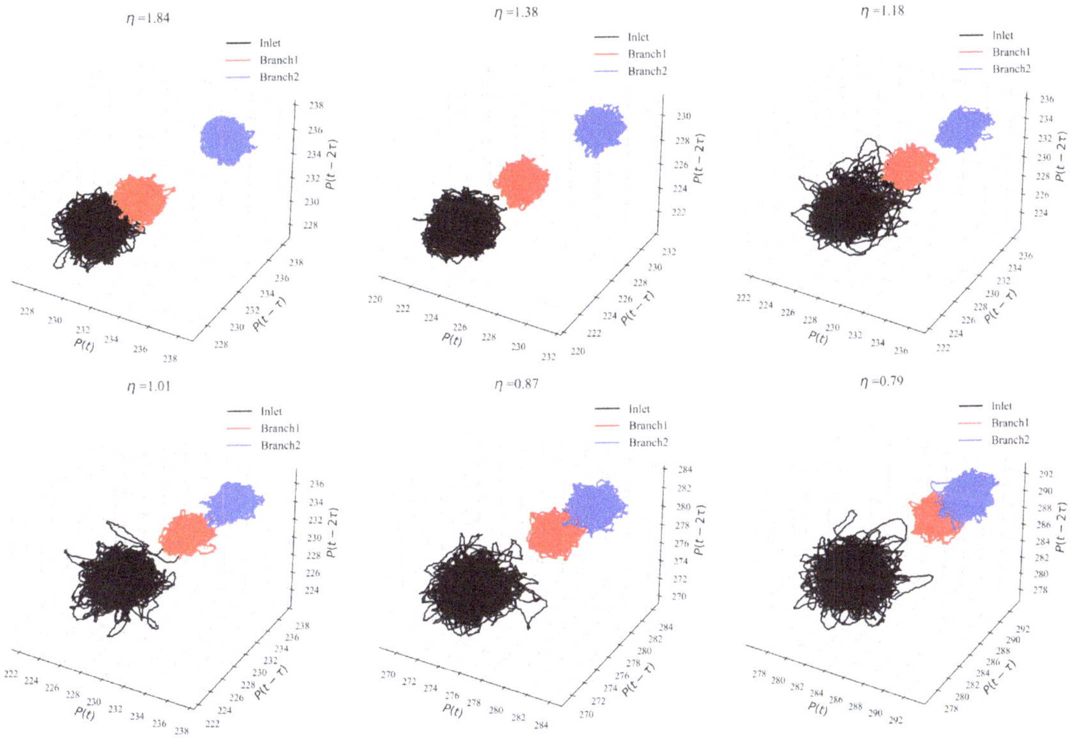

Figure 8. *Cont.*

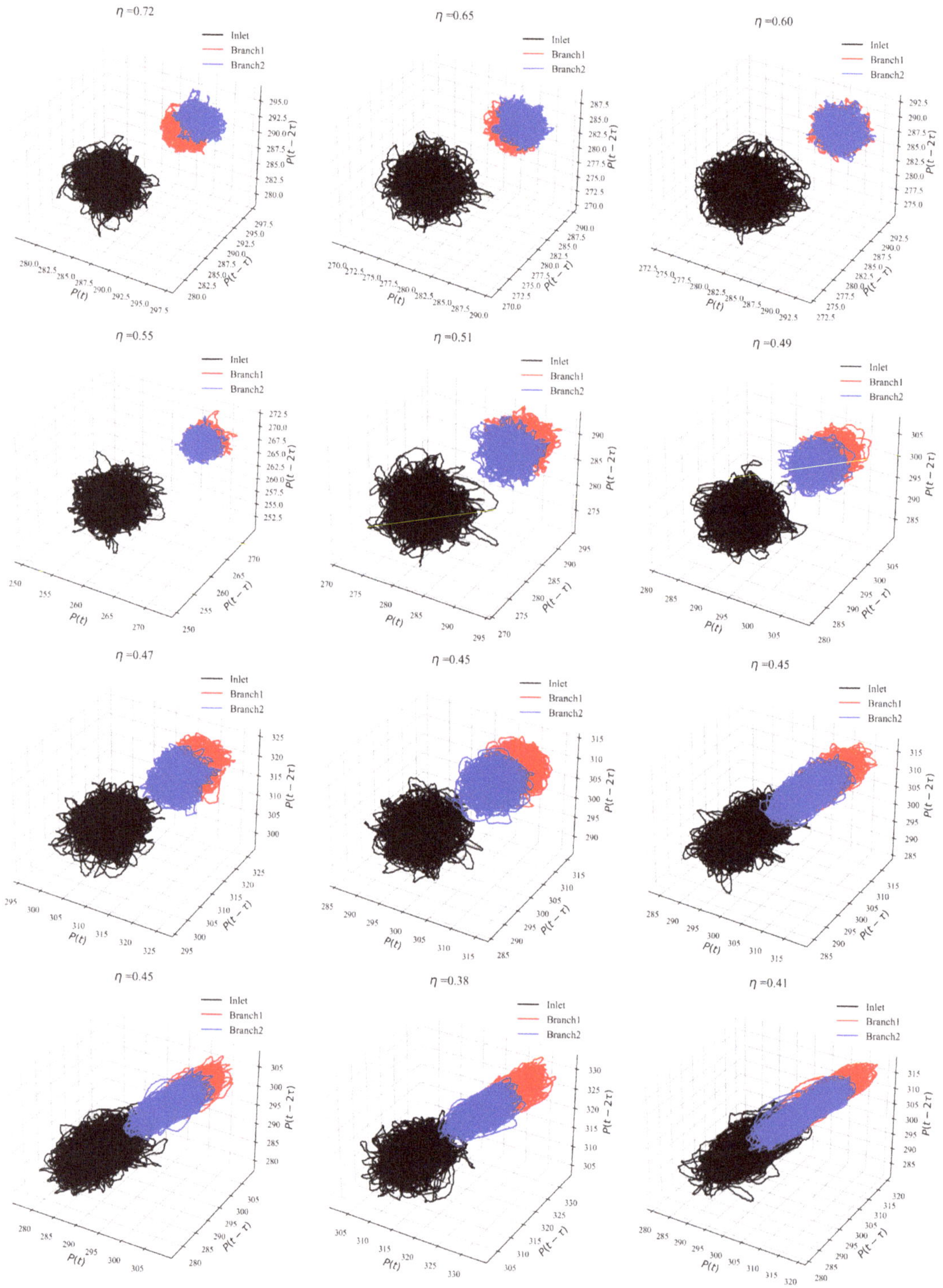

Figure 8. Phase space trajectories at different splitting ratio of Workpiece 2.

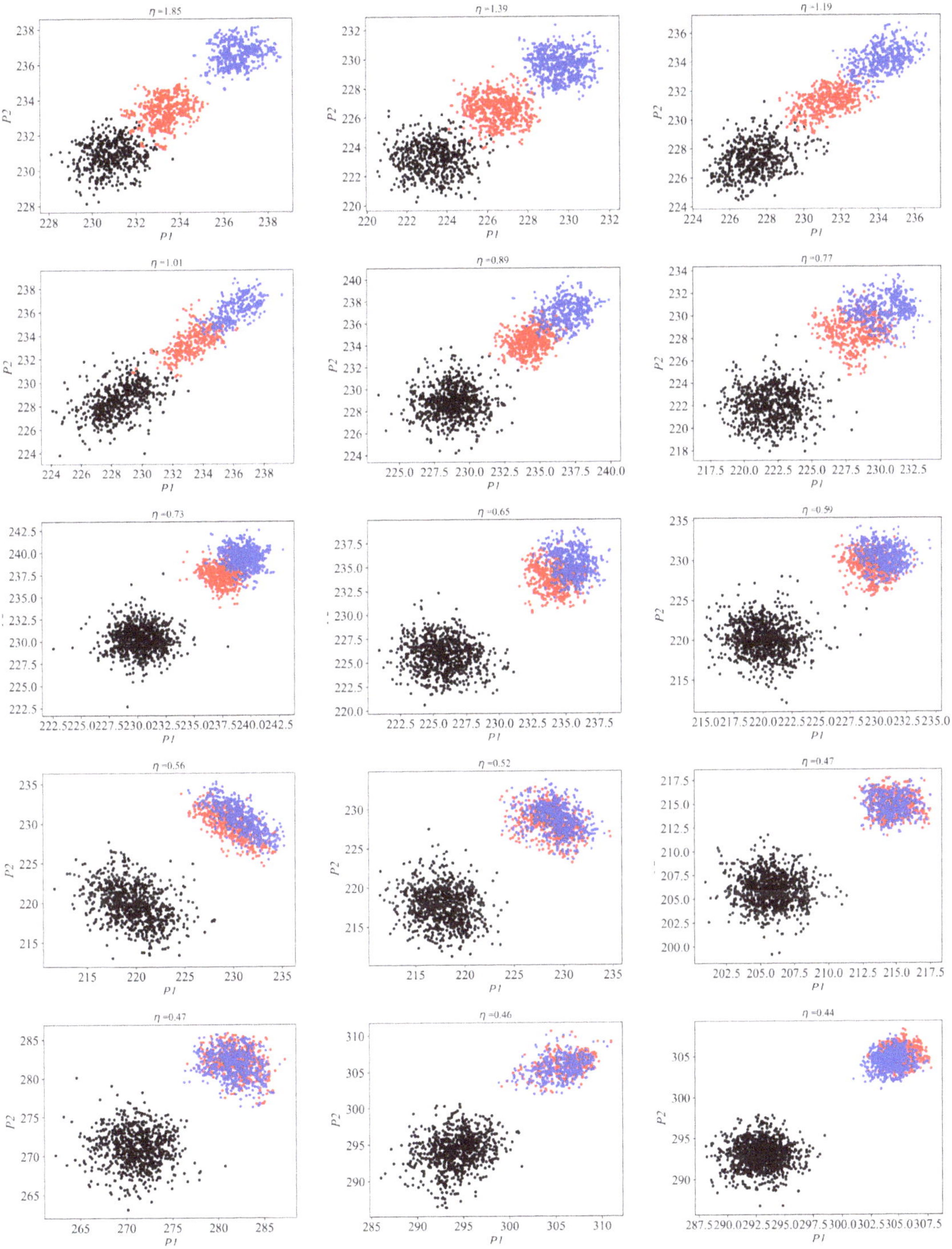

Figure 9. *Cont.*

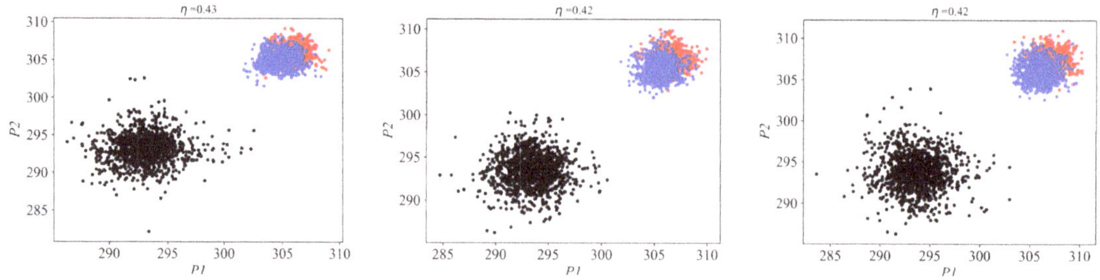

Figure 9. Poincaré sections at different splitting ratio of workpiece 1.

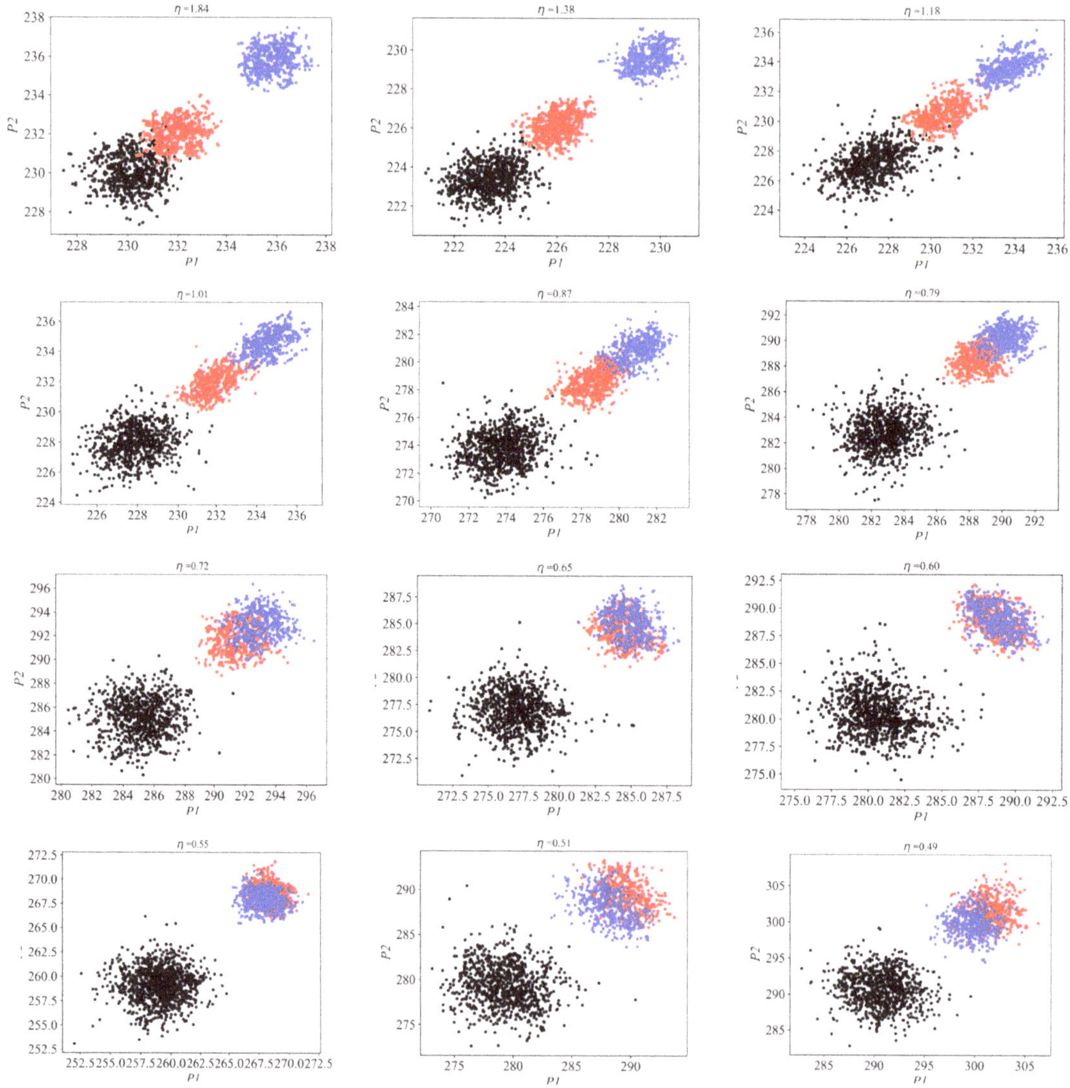

Figure 10. *Cont.*

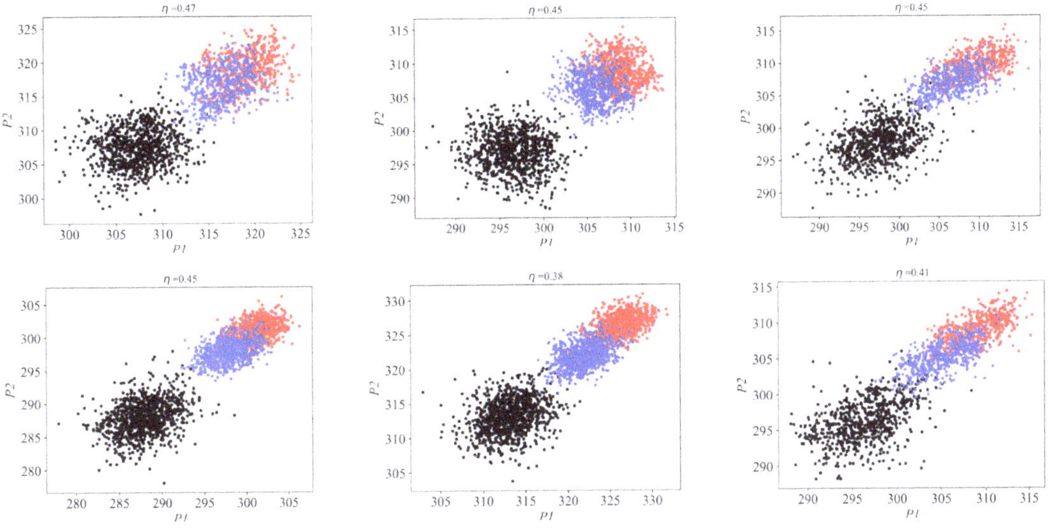

Figure 10. Poincaré sections at different splitting ratio of workpiece 2.

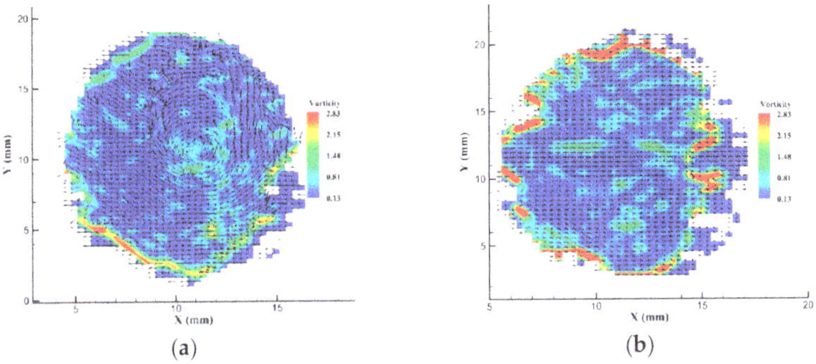

Figure 11. Cross-sectional (**a**) with vortex structures (T-shaped) and (**b**) without vortex structures (SOB).

To simplify the observation process and verify the results obtained from PIV, a string-tracing method was used to observe the number of times the string wraps around in the branch. The basic principle of this technique can be found in Fang et al. [10]. The biggest advantage of using this technique is that it makes it very easy to observe the generation of vortex structures inside the branches. The study showed that in the POB, the strings in branch 1 and branch 2 exhibit a wrapping phenomenon under specific flow conditions, as shown in Figure 12a. However, in the SOB, the behavior of the strings in branch 1 and branch 2 is as shown in Figure 12b.

Based on the PIV and string-tracing results, it can be concluded that the vortex structures in the downstream branches only occur in the POB and are not affected by chirality. This suggests that, compared to the POB, the space orthogonal bifurcation can suppress the formation of vortex structures in the downstream branches. However, the specific mechanism still requires further investigation.

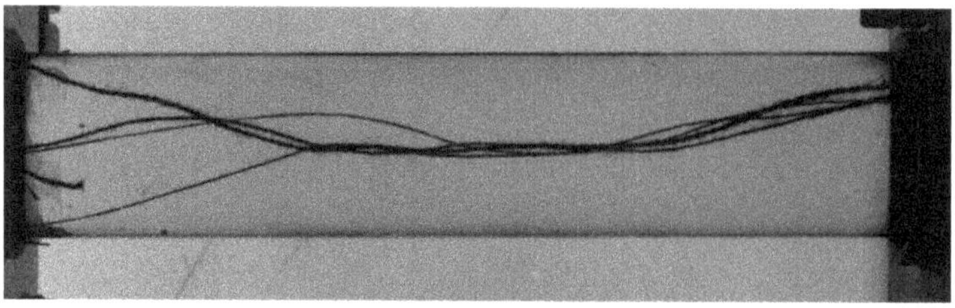

(a)

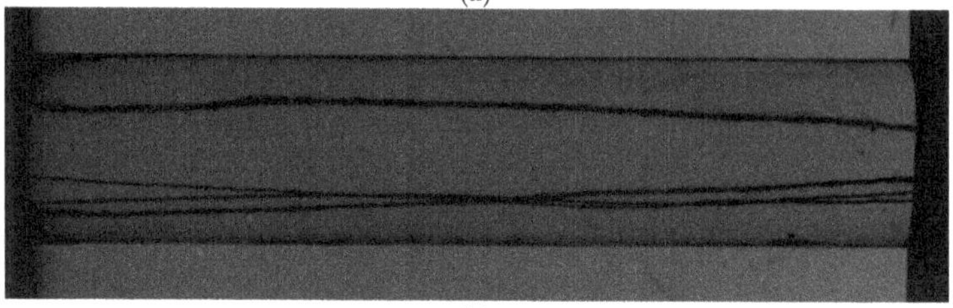

(b)

Figure 12. String-tracing results (**a**) with vortex structure and (**b**) without vortex structure.

4. Conclusions

We constructed a split-type experimental system to investigate the flow behavior inside a chiral symmetric spatial orthogonal bifurcation and measured the pressure signals and vortex formation patterns at a certain cross-section in each channel. Our study revealed the following findings:

1. Chirality structure affects the pressure signal characteristics inside the spatial orthogonal bifurcation, especially when the bifurcation ratio is smaller than 0.5. This effect can be observed in the frequency spectrum, autocorrelation function, phase space trajectory, and Poincaré section. This phenomenon may be related to the different directions of inertial forces.
2. The SOB structure eliminates the large vortex structure in the downstream branch. Through PIV technology and fine wire tracing technology, the vortex structure observed in the planar orthogonal bifurcation cannot be observed in the spatial orthogonal bifurcation. The mechanism behind this phenomenon is not yet fully understood and requires further research at the mechanism level.

This study first investigates the flow behavior of chiral symmetric SOB under different bypass ratios, taking into account the practical engineering background. It provides a comprehensive examination and serves as a reference for future research.

Author Contributions: Conceptualization, Z.F. and Q.F.; methodology, D.Z.; software, X.L.; validation, Z.F., D.Z. and X.L.; formal analysis, X.L.; investigation, Z.F.; resources, D.Z.; data curation, X.L.; writing—original draft preparation, Z.F.; writing—review and editing, Z.F.; visualization, D.Z.; supervision, Q.F.; project administration, Q.F. and L.Y.; funding acquisition, J.L. All authors have read and agreed to the published version of the manuscript.

Funding: This research was supported by the National Natural Science Foundation of China (grant No. 11272026, 11922201, 11927802).

Data Availability Statement: The datasets generated and/or analyzed during the current study are not publicly available due to the need to protect intellectual property rights. However, the datasets are available from the corresponding author on reasonable request. Access to the data will be granted in accordance with any institutional or ethical requirements.

Acknowledgments: The authors would also like to thank Penghui Li, Wentong Qiao, Hu Sun, and Qixiang Gao for their assistance.

Conflicts of Interest: The authors declare no conflict of interest.

References

1. McDonald, C.J.; Devon, M.J. Hollow latex particles: Synthesis and applications. *Adv. Colloid Interface Sci.* **2002**, *99*, 181–213. [CrossRef]
2. Seymour, R.B. The role of fillers and reinforcements in plastics technology. *Polym. Plast. Technol. Eng.* **1976**, *7*, 49–79. [CrossRef]
3. Dreher, S.; Kockmann, N.; Woias, P. Characterization of laminar transient flow regimes and mixing in T-shaped micromixers. *Heat Transf. Eng.* **2009**, *30*, 91–100. [CrossRef]
4. Senn, S.M.; Poulikakos, D. Laminar mixing, heat transfer and pressure drop in tree-like microchannel nets and their application for thermal management in polymer electrolyte fuel cells. *J. Power Sources* **2004**, *130*, 178–191. [CrossRef]
5. Kockmann, N.; Engler, M.; Haller, D.; Woias, P. Fluid Dynamics and Transfer Processes in Bended Microchannels. *Heat Transf. Eng.* **2005**, *26*, 71–78. [CrossRef]
6. Meng, H.; Wang, Z.; Hoi, Y.; Gao, L.; Metaxa, E.; Swartz, D.; Kolega, J. Complex Hemodynamics at the Apex of an Arterial Bifurcation Induces Vascular Remodeling Resembling Cerebral Aneurysm Initiation. *Stroke* **2007**, *38*, 1924–1931. [CrossRef]
7. Casiano, M.J.; Hulka, J.R.; Yang, V. Liquid-Propellant Rocket Engine Throttling: A Comprehensive Review. *J. Propuls. Power* **2010**, *26*, 897–923. [CrossRef]
8. Wang, S.; Shoji, M. Fluctuation characteristics of two-phase flow splitting at a vertical impacting T-junction. *Int. J. Multiph. Flow* **2002**, *28*, 2007–2016. [CrossRef]
9. Wang, S.; Mosdorf, R.; Shoji, M. Nonlinear analysis on fluctuation feature of two-phase flow through a T-junction. *Int. J. Heat Mass Transf.* **2003**, *46*, 1519–1528. [CrossRef]
10. Fang, Z.X.; Li, Y.J.; Zhang, D.W.; Liu, X.K.; Li, J.X.; Yang, L.J.; Fu, Q.F. Pressure Oscillation and Vortex in an Asymmetric T-junction. *Acta* **2023**. *Submitted*.
11. Liu, H.R.; Li, H.P.; He, Y.; Cheng, P.; Zhang, Y.Q.; Feng, B.; Li, H.; Wu, K.; Chen, L. Condensation and asymmetric amplification of chirality in achiral molecules adsorbed on an achiral surface. *Nat. Commun.* **2023**, *14*, 2100. [CrossRef]
12. Hentschel, M.; Schäferling, M.; Duan, X.Y.; Liu, N. Chiral plasmonics. *Sci. Adv.* **2017**, *3*, e1602735. [CrossRef] [PubMed]
13. Fukushima, K.; Kharzeev, D.E.; Warringa, H.J. Chiral magnetic effect. *Phys. Rev. D* **2008**, *78*, 074033. [CrossRef]
14. Diakonov, D.; Petrov, V.; Pobylitsa, P. A chiral theory of nucleons. *Nucl. Phys. B* **1988**, *306*, 809–848. [CrossRef]
15. Pisarski, R.D.; Wilczek, F. Remarks on the chiral phase transition in chromodynamics. *Phys. Rev. D* **1984**, *29*, 338–341. [CrossRef]
16. Vigolo, D.; Radl, S.; Stone, H.A. Unexpected trapping of particles at a T junction. *Proc. Natl. Acad. Sci. USA* **2014**, *111*, 4770–4775. [CrossRef] [PubMed]
17. Chen, K.K.; Rowley, C.W.; Stone, H.A. Vortex dynamics in a pipe T-junction: Recirculation and sensitivity. *Phys. Fluids* **2015**, *27*, 034107. [CrossRef]
18. Ault, J.T.; Fani, A.; Chen, K.K.; Shin, S.; Gallaire, F.; Stone, H.A. Vortex-breakdown-induced particle capture in branching junctions. *Phys. Rev. Lett.* **2016**, *117*, 084501. [CrossRef] [PubMed]
19. Chen, K.K.; Rowley, C.W.; Stone, H.A. Vortex breakdown, linear global instability and sensitivity of pipe bifurcation flows. *J. Fluid Mech.* **2017**, *815*, 257–294. [CrossRef]
20. Xiong, A.; Wei, Q. Experiment study on a sort of strong swirling flow. *Exp. Meas. Fluid Mech.* **1999**, *13*, 8–13. (In Chinese)
21. Su, B.; Huang, K.; Lin, M.; Wang, Q. Experimental investigation on the flow patterns in a T-junction under steady and rolling motion conditions using PLIF. *Chem. Eng. J.* **2023**, *465*, 142772. [CrossRef]
22. Kantz, H.; Schreiber, T. *Nonlinear Time Series Analysis*, 2nd ed.; Cambridge University Press: Cambridge, UK, 2004.

aerospace

MDPI

Article

Heat Transfer and Flow Structure Characteristics of Regenerative Cooling in a Rectangular Channel Using Supercritical CO_2

Jian Liu, Mengyao Xu, Pengchao Liu and Wenxiong Xi *

Research Institute of Aerospace Technology, Central South University, Changsha 410012, China; jian.liu@csu.edu.cn (J.L.); 15130715317@163.com (M.X.); 15538571052@163.com (P.L.)
* Correspondence: 13739076081@163.com

Abstract: At an extremely high Mach number, the regenerative cooling of traditional kerosene cannot meet the requirement of the heat sink caused by aerodynamic heating and internal combustion in a scramjet propulsion system. As a supplement of traditional regenerative cooling, supercritical CO_2 is regarded as an effective coolant in severe heating environments due to its excellent properties of heat and mass transportation. In this paper, the heat transfer and flow structure characteristics of regenerative cooling in a rectangular channel using supercritical CO_2 are analyzed numerically using a validated model. The effect of heat flux magnitude, nonuniform heat flux, acceleration and buoyancy and flow pattern are considered to reveal the regenerative cooling mechanism of supercritical CO_2 in the engine condition of a scramjet. The results indicate that the heat transfer deterioration phenomenon becomes obvious in the cooling channel loaded with relatively high heat flux. Compared with the cooling channels loaded with increased heat flux distribution, the maximum temperature increased for the channel loaded with decreased heat flux distributions. When larger acceleration is applied, a relatively lower wall temperature distribution and higher heat transfer coefficients are obtained. The wall temperature distribution becomes more uniform and the high-temperature region is weakened when the coolants in adjacent channels are arranged as a reversed flow pattern. Overall, the paper provides some references for the utilization of supercritical CO_2 in regenerative cooling at an extremely high Mach number in a scramjet.

Keywords: regenerative cooling; supercritical CO_2; heat flux; acceleration; flow pattern

check for updates

Citation: Liu, J.; Xu, M.; Liu, P.; Xi, W. Heat Transfer and Flow Structure Characteristics of Regenerative Cooling in a Rectangular Channel Using Supercritical CO_2. *Aerospace* **2023**, *10*, 564. https://doi.org/10.3390/aerospace10060564

Academic Editor: Hirotaka Sakaue

Received: 12 May 2023
Revised: 3 June 2023
Accepted: 14 June 2023
Published: 16 June 2023

1. Introduction

Due to its sharply varied thermophysical properties including density, thermal capacity, thermal conductivity and viscosity near the critical point, supercritical CO_2 has superior heat and mass transfer capabilities [1–3]. In addition, as a type of nontoxic and harmless working medium, its accessible and low-cost characteristics make it widely used in power plants and nuclear reactors, as well as in many industrial and engineering applications [4,5]. In recent years, the complex thermal and hydraulic phenomena for supercritical CO_2 flowing in the channels have attracted a lot of attention [6,7]. The pressure–temperature diagram of supercritical CO_2 is shown in Figure 1.

Many researchers have carried out fundamental investigations in simplified tubes to analyze the flow and heat transfer characteristics of supercritical CO_2 [8–18]. Yan et al. [8] numerically investigated the convective heat transfer characteristics of supercritical CO_2 in vertical tubes loaded with uniform and nonuniform heat flux. They found that by adopting nonuniform heat flux, the wall temperature is reduced due to smaller thermal resistance which can be explained by pseudo-phase transition theory. Khalesi et al. [9] numerically studied the flow and heat transfer characteristics of supercritical CO_2 in a rectangular microchannel loaded with uniform heat flux. They pointed out that the Nusselt number was not affected by the Reynold number in the laminar regime at high operating pressures.

Pandey et al. [10] improved the two-layer model for the heat transfer of supercritical CO_2 in a heated pipe by taking acceleration and buoyancy effects into consideration. Additionally, the results were refined and validated using direct numerical simulation data and experimental data. It was found that thermal conductivity and viscosity have a limiting effect on heat transfer. Zhang et al. [11] numerically investigated the mechanisms for the heat transfer behaviors of supercritical CO_2. They analyzed the formation mechanism of the heat-transfer-enhanced phenomenon for low mass flux cases and heat transfer deterioration (HTD) phenomenon for normal mass flux cases. The results indicated that the heat conduction process of the boundary layer played an important role in the overall heat transfer of supercritical CO_2.

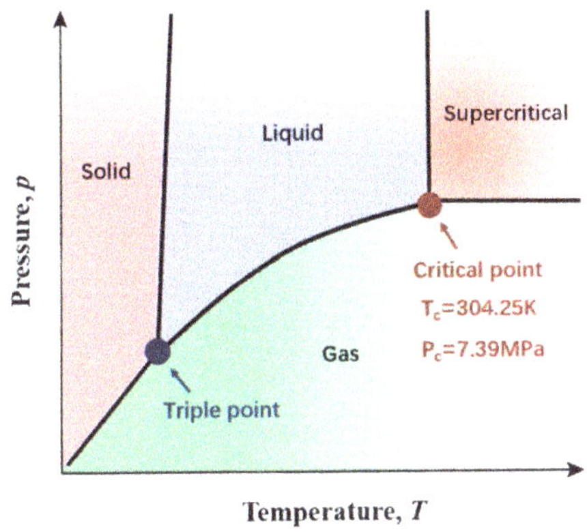

Figure 1. Pressure–temperature diagram of supercritical CO_2.

In addition to fundamental research, many researchers also performed investigations of supercritical CO_2 applied in heat exchangers [19–21], heat sinks [21–23] and other industrial devices [24–26]. Chu et al. [21] carried out experiments based on a supercritical CO_2–water experiment platform to test the thermohydraulic performance of a printed circuit heat exchanger. The results indicated that the supercritical CO_2 had better heat transfer ability than water and the printed circuit heat exchanger had better heat transfer performance under a higher pressure condition. Awais et al. [23] employed supercritical CO_2 as a coolant to enhance the heat transfer performance of minichannel heat sink instead of water fluid. The results revealed that using supercritical CO_2 as a coolant brought about better performance and lower pressure loss compared with water. Cheng et al. [24] experimentally investigated the heat transfer performance of a precooler for the supercritical CO_2 Brayton cycle. The effects of the inlet Reynold numbers and temperature on thermal–hydraulic characteristics and pressure loss were analyzed. The effectiveness of the precooler was also attained. Muto and Kato [26] optimized a cycle scheme of a direct cycle using a dual-expansion supercritical CO_2 gas turbine cycle for nuclear power generation. Evan P. et al. [25] conducted a numerical study of a high-temperature regenerator within a supercritical CO_2 recompression Brayton cycle.

As the main component of a hypersonic vehicle, the scramjet operates in severe thermal environments [27]. Using fuel as the coolant, regenerative cooling is regarded as being an effective and reliable cooling method with a lot of channels arranged inside the combustor wall [28,29], as shown in Figure 2. The fuel flows through the channels absorbing excess heat from combustion chambers and then enters the chamber to proceed with the combustion process [30]. Many works have been conducted to investigate the cooling performance of regenerative cooling using supercritical hydrocarbon fuel in terms of effects of nonuniform heat flux [30–32], flow pattern [33] and acceleration and buoyancy [34–38]. Liu et al. [30] investigated the heat transfer performance and flow structure of supercritical n-decane in a rectangular channel under nonuniform heat flux. The results revealed that the wall temperature distribution was sensitive to the heat flux distribution while the outlet temperature was not affected by nonuniform heat flux. Zhang et al. [33] proposed a bi-directional flow pattern in order to optimize the cooling performance of supercritical hydrocarbon fuel. They compared the temperature distribution of the bi-directional flow pattern with that of the traditional co-directional flow pattern and found that the adoption of a bi-directional flow pattern can improve the cooling performance whilst maintaining simple structures. Jia et al. [35] investigated the effects of gravity on the flow characteristics of supercritical n-decane in a rectangular channel. By observing the streamlines perpendicular to the flow direction, they found that the induced small recirculation zones which indicated the buoyancy played an important role in the mixing of hot and cold fluids.

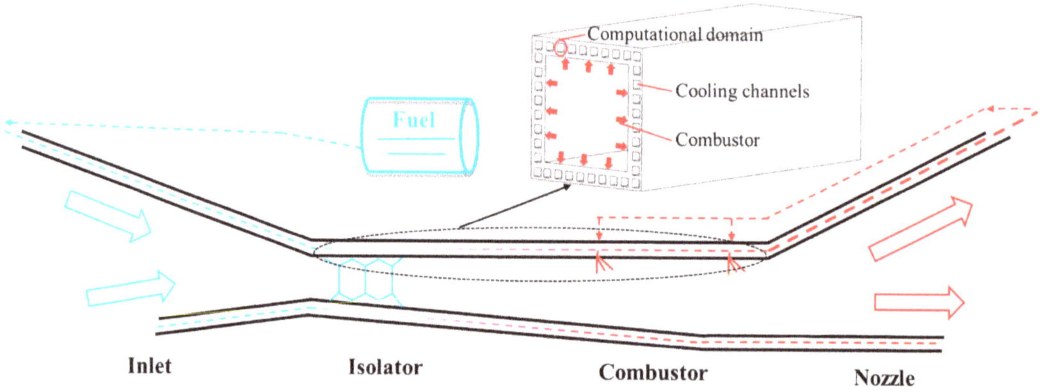

Figure 2. Schematic diagram of regenerative cooling in a scramjet.

However, at extremely high Mach numbers, high-efficiency active cooling technology is urgently needed for a scramjet propulsion system [39–41]. Traditional hydrocarbon fuel cannot supply enough heat sink for the heat load produced by Mach numbers > 8, and supercritical CO_2 is promising to act as an additional coolant for regenerative cooling.

In this work, the flow structure and heat transfer characteristics of supercritical CO_2 under 8 MPa in a regenerative cooling channel with a rectangular cross-section are analyzed. To study the effects of uniform and nonuniform heat flux distribution on flow structure and heat transfer characteristics, different kinds of heat flux magnitudes ranging from 1.5 to 6 MW/m^2 and linearly changed (increased and decreased) heat flux distributions with the same average value of 1.5 MW/m^2 are imposed on the bottom wall for a single channel. Considering the different real flight states of regenerative cooling channels, the effects of acceleration and buoyancy on flow structure and heat transfer characteristics are considered at different magnitudes and direction with values ranging from −3 g to 3 g. In addition, the effects of flow patterns on heat transfer characteristics are analyzed by adopting two adjacent channels with the same and reversed flow directions.

2. Model Description

2.1. Geometry Description and Grid Conditions

For a typical regenerative cooling process, there are a lot of cooling channels set inside the combustor wall to absorb excess heat from the combustion chambers. Considering that each cooling channel is heated equally, a single cooling channel with a rectangular cross-section was selected as the computational domain for convenience. Uniform and nonuniform heat flux boundaries were imposed on the bottom wall of the channel to study the effect of axial heat flux on flow structures and heat transfer performance in the cooling channel. The cross-section size of the fluid region was 2 mm × 2 mm. The length of the heated section was 300 mm. Additionally, two 50 mm extended sections without heat flux loaded were, respectively, set at the beginning and the end of the channel to ensure a fully developed flow process. Taking the influence of fluid–solid coupling and the actual geometry of the regenerative cooling channel into account, there were four solid walls placed around the fluid region, with a height of 1 mm (z-axis) and width of 0.5 mm (y-axis), as shown in Figure 3.

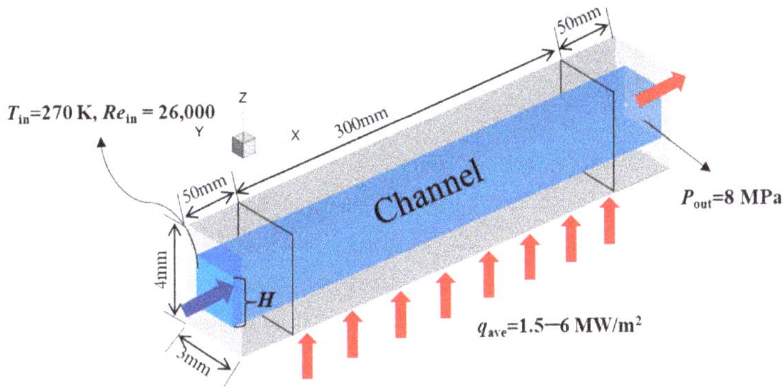

Figure 3. Schematic diagram of regenerative cooling channel used in this paper.

Structural grids were used to discretize the computational domain, and the cross-section is shown in Figure 4. To meet the accurate and reliable requirement, the size of the first layer and the growth rate of the system were set as 0.01 mm and 1.1, respectively. Additionally, the corresponding wall y+ of the simulation was below 1 to meet the requirement of the k-ω SST turbulence model.

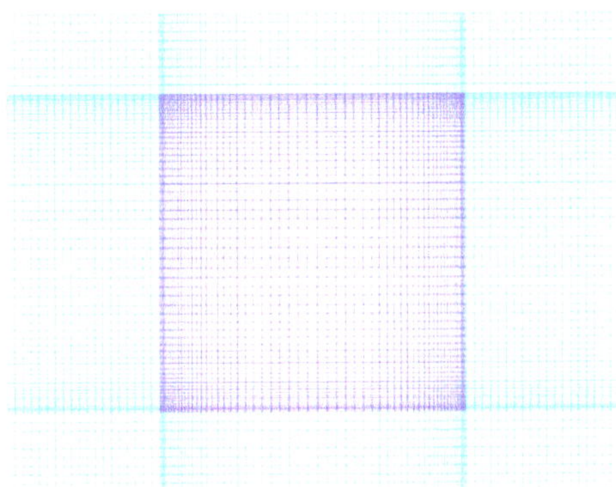

Figure 4. The cross-section of the grids' systems.

2.2. Solution Methods and Convergence Criterion

The commercial software ANSYS FLUENT 19.2 was employed to carry out the simulations. The governing equations included mass, momentum and energy conservation equations. The finite volume method was adopted to discretize the governing equations. A double-precision pressure-based steady solver was applied to obtain the flow and heat transfer performance of supercritical CO_2 in the cooling channel. In addition, the second-order upwind scheme was adopted for spatial discretization. The SIMPLEC algorithm was used to couple the velocity and pressure.

The simulations were regarded as being converged when the residuals of the continuity equation, velocity components, turbulent kinetic energy and specific energy dissipation rate items were less than 10^{-5} and the residual of the energy equation was less than 10^{-8}. It took about 10,000–20,000 iterations before reaching convergence.

2.3. Boundary Conditions and Thermophysical Properties

Three-dimensional flow and heat transfer was performed in the rectangular channel introduced in Section 2.1. The mass flow rate of the inlet was set as 0.0032 kg/s (Re_{inlet} = 26,000) with an initial temperature of 270 K. The gauge pressure of the outlet was set as 8 MPa and the effect of pressure variations on thermophysical properties was neglected. The average heat flux was imposed on the bottom heated with a value of 1.5 MW/m^2–6.0 MW/m^2 and the other walls were set as adiabatic walls. To better analyze the flow structure in the passage, gravity and acceleration were considered in the streamwise–normal section (*x*-*z* section). Additionally, the basic case was named Case A1. Four kinds of influence factors including heat flux magnitude, nonuniform heat flux distribution, acceleration and buoyancy effects and flow patterns under engine-like conditions were studied in this paper.

To study the effect of heat flux magnitude and distribution on flow structures and the heat transfer characteristics of supercritical CO_2 in regenerative cooling channels, different kinds of heat flux magnitudes (Cases A1–A3), increased heat flux distribution (Cases B1–B3) and decreased heat flux distribution (Cases B4–B6) were imposed on the bottom wall for a single channel. The heat flux magnitude ranged from 1.5 to 6 MW/m^2 for Case A1 to Case A3. For nonuniform heat flux distributions, with the same average heat flux of 1.5 MW/m^2, the linear variation ratios for the heat flux distribution function were different. Under the consideration of different real flight environments, the effects of acceleration and buoyancy were considered at different magnitudes and directions with values ranging from −3 g to 3 g, named as Cases C1 to C7, respectively. Finally, the effects of flow patterns on the temperature distribution and heat transfer characteristics were analyzed by adopting two adjacent channels with the same (Case D1) or reversed (Case D2) flow directions. The details of the heat flux distribution, acceleration values and flow pattern of all of the cases are listed in Table 1.

Table 1. All of the cases considered in the paper. A: Effect of heat flux magnitude; B: effect of heat flux distributions; C: effect of acceleration and buoyancy; D: effect of flow pattern.

Study Variables	Case Name	Heat Flux Function/ Correlation (MW/m^2)	Channel Form and Flow Pattern	Acceleration Value (m/s^2)
Uniform heat flux distribution	A1	Uniform, q = 1.5	Single	$a_z = -9.81$
	A2	Uniform, q = 3	Single	$a_z = -9.81$
	A3	Uniform, q = 6	Single	$a_z = -9.81$
Nonuniform heat flux distribution	B1	Increased, q = 0.75 + 3.75x	Single	$a_z = -9.81$
	B2	Increased, q = 1 + 2.5x		
	B3	Increased, q = 1.25 + 1.25x		
	B4	Decreased, q = 2.25 − 3.75x		
	B5	Decreased, q = 2 − 2.5x		
	B6	Decreased, q = 1.75 − 1.25x		
Acceleration and buoyancy effects	C1	Uniform, q = 1.5	Single	$a_z = 0$
	C2			$a_z = -9.81$
	C3			$a_z = -19.62$
	C4			$a_z = -29.43$
	C5			$a_z = 29.43$
	C6			$a_x = 14.715$, $a_z = 25.487$ (30°)
	C7			$a_x = 25.487$, $a_z = -14.715$ (60°)
Flow pattern	D1	Uniform, q = 1.5	Adjacent, same direction	$a_z = -9.81$
	D2		Adjacent, reversed direction	

The thermophysical properties of supercritical CO_2 including density, thermal capacity, thermal conductivity and viscosity near the supercritical point change sharply, which may cause complex flow and heat transfer performance. The corresponding property data were from SUPERTRAPP software developed by the National Institute of Standards and Technology, the United States [42]. The piecewise linear interpolation method was applied to input the thermophysical properties, including thermal capacity, thermal conductivity, density and viscosity. More than 100 piecewise linear functions were used for interpolations in the range of the temperature variations. Additionally, the details of the thermophysical properties of supercritical CO_2 under 8 MPa are shown in Figure 5.

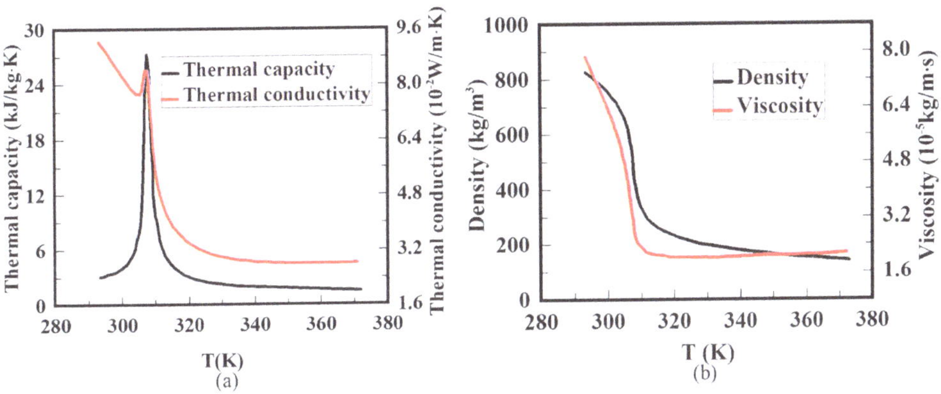

Figure 5. Thermophysical properties of supercritical CO_2 at 8 MPa. (**a**) Thermal capacity and thermal conductivity; (**b**) density and viscosity.

The material of the solid walls was Steel 304. The density and the thermal capacity are 7930 kg/m^3 and 500 J/(kg·K). The linear correction of the thermal conductivity of Steel 304 is 16.3 W/(m·K) at 373 K and 21.5 W/(m·K) at 773 K.

2.4. Model Validation and Mesh Independence Study

It is significant to verify the reliability of the turbulence model and determine the sensitivity of the mesh systems for a turbulence flow calculation. To verify the reliability of the turbulence model, the experimental results in the previous work [43] were used for comparisons, which used supercritical CO_2 as the working fluid in a circular tube. A circle tube with a diameter of 2 mm and a heated section of 290 mm was adopted to perform verification, as shown in Figure 6a. The inlet temperature and Re were 298.15 K and 9000, respectively. The heat flux was uniform with a value of 13,626 W/m^2. Three kinds of turbulence models, i.e., k-ε RNG, k-ω SST and Transition SST, were selected to carry out numerical simulations. Additionally, the values of the wall temperature along the streamwise direction were obtained as shown in Figure 6b.

As seen in Figure 6b, the wall temperature predicted by the *k-ω* SST turbulence model has good agreement with the experimental results obtained by Li et al. [43] compared with the other turbulence models. The prediction error of the *k-ω* SST model is within ±0.13%. Larger error is found at the beginning of the heated region due to the different boundary effects in the experiment and simulation. In addition, the *k-ω* SST turbulence model has shown great advantages in dealing with the transmission of turbulent shear force and wall bounded flow. Overall, the *k-ω* SST turbulence model provided enough accuracy in the simulations. Therefore, *k-ω* SST turbulence was selected for the simulations below.

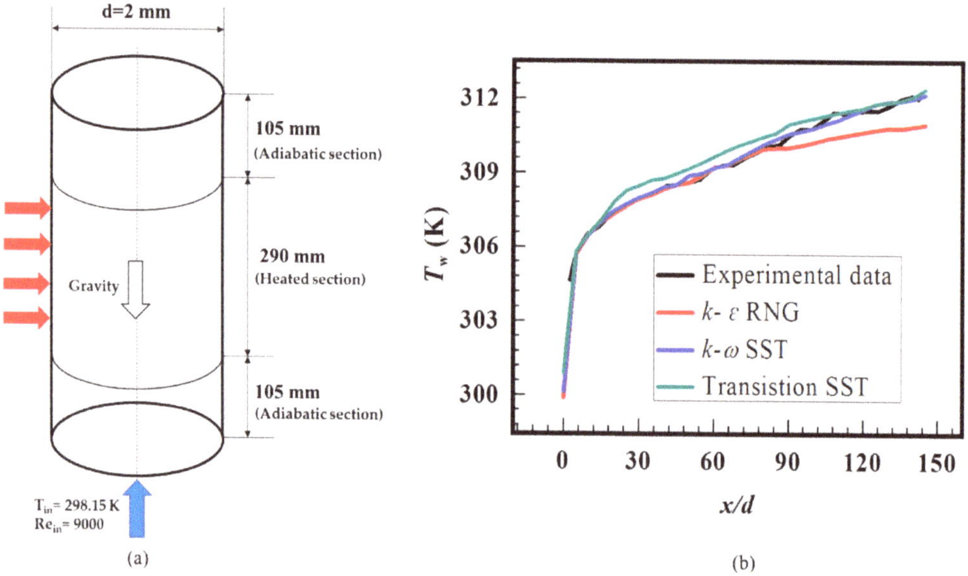

Figure 6. (**a**) The circular tube used for simulation; (**b**) Verification of the turbulence model.

Mesh independence was performed using the k-ω SST turbulence model, and Case A was selected to evaluate the mesh sensitivity. Four kinds of mesh regimes were built, respectively, with grid numbers of 2.38, 3.30, 3.67 and 5.35 million. They were named Mesh 1, Mesh 2, Mesh 3 and Mesh 4, respectively. Thirty points were built on the center line of the heated surface along the streamwise direction and the area-weighted average temperature distribution was obtained, as shown in Figure 7a. As shown in Figure 7a, the center line temperature distributions overlap together. Therefore, the maximum temperatures on the center line for all of the meshes are extracted and compared in Figure 7b, i.e., Point A, Point B, Point C, Point D. They have relatively close values especially for Mesh 3 and Mesh 4. When the grid numbers increase from 3.67 M to 5.35 M, the maximum temperature has no changes. Comprehensively considering the simulation accuracy and computational effort, Mesh 3 was selected.

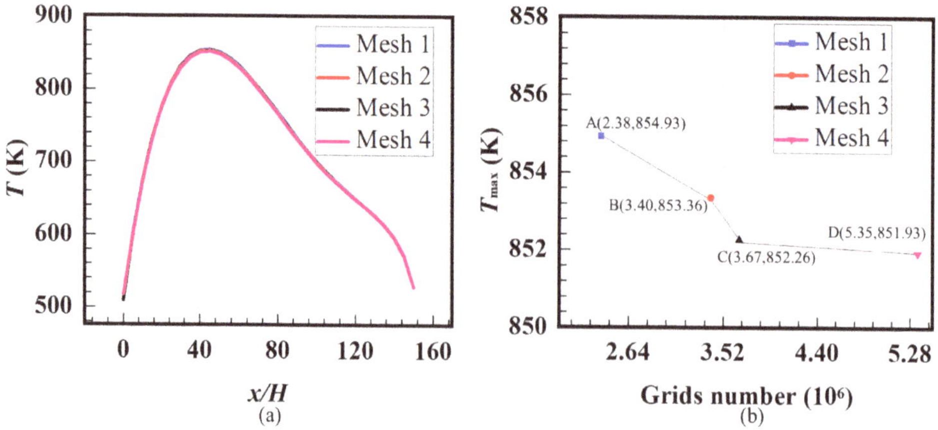

Figure 7. Comparison of center line temperature for the heated wall calculated by four grid systems: (**a**) center line wall temperature distributions; (**b**) maximum wall temperature.

3. Results and Discussions

For the convenience of analysis, the HTC is defined as follows:

$$h = \frac{q}{\left(T_w - T_f\right)} \tag{1}$$

where T_f is the bulk fluid temperature and is calculated as follows:

$$T_f = \frac{\int_A \rho u C_p T d_A}{\int_A \rho u C_p d_A} \tag{2}$$

where A is the area of the fluid cross-section, and ρ, u, C_p and T are the local density, velocity, thermal capacity and temperature of supercritical CO_2 in the same region, respectively.

3.1. Effect of Heat Flux Magnitude

Figure 8 shows the temperature and HTC distributions along the x-axis for the cases loaded with different heat flux magnitudes. To display the distributions, a total of 61 y-z sections of solid and fluid regions were created along the streamwise direction from $x/H = 0$ to $x/H = 150$, respectively, and each data point was obtained by area-weighted averaging all of the results in the section. As shown in Figure 8a, the bulk fluid temperature almost always rises with flow developing in the channel. It is noted that with an increased heat flux value, the increased trend of fluid temperature becomes strong in the downstream region. There is a relatively "weakened" region in Figure 8a in the region of $40 < x/H < 80$, where the fluid temperature increases much more slowly along the cooling channel. The "weakened" region is caused by the core part of the mainstream which is close to the critical point, although the average temperature is much larger than the critical temperature which is mainly determined by the high-temperature regions close to the heated bottom wall. As shown in Figure 5a, when fluid temperature is near the critical point, supercritical CO_2 possesses a high thermal capacity. It takes more heat for the core part of the fluids to increase the temperature; so, the temperature of the whole cross-section increases much more slowly. The phenomenon is much clearer for the cases with the increased heat flux magnitude, from Case A1 to Case A3. As seen in Figure 8b, a phenomenon of HTD is observed with the increased wall temperature in the region of $80 < x/H < 120$. The HTD becomes obvious for the case with the largest heat flux magnitude, i.e., Case A3. The HTC firstly decreases and then increases, as shown in Figure 8c. Additionally, when the flow is near the end of the heated section, the HTC increases strongly. In a certain range of heat flux from 1.5 MW/m^2 to 3 MW/m^2, the HTC increased with the increased heat flux. However, with the heat flux increased to 6 MW/m^2, the HTC decreases. The phenomenon of HTD is also clearly presented in Figure 8c, especially in the case with the largest heat flux (Case A3).

3.2. Effect of Nonuniform Heat Flux Distributions

To present the nonuniform distribution more clearly, a diagram of nonuniform heat flux distribution about x/H is shown in Figure 9, which is consistent with Table 1. Among them, Cases B1–B3 are loaded with increased heat flux distributions with different increase rates, while Cases B4–B6 are loaded with decreased heat flux distributions.

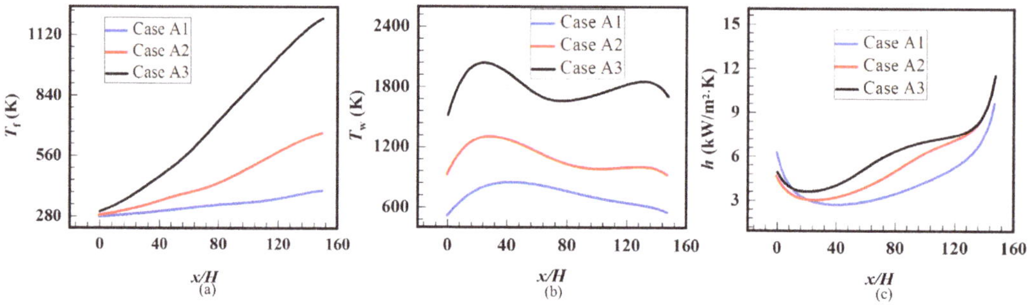

Figure 8. Temperature and HTC distributions for cases with uniform heat flux distributions. (**a**) Fluid temperature; (**b**) average wall temperature; (**c**) HTC.

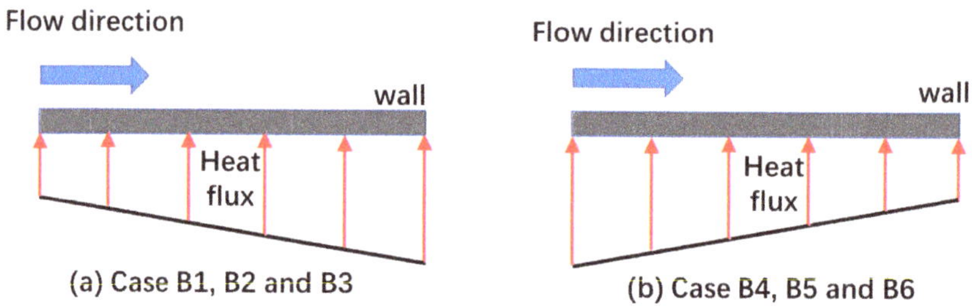

Figure 9. Diagram of nonuniform heat flux distribution. (**a**) For Case B1–B3 (loaded with increased heat flux distributions); (**b**) For Case B4–B6 (loaded with decreased heat flux distributions).

Figure 10 displays the temperature and HTC distributions along the streamwise direction for the cases loaded with nonuniform heat flux (Cases B1–B6). From Cases B1 to B3, the increased heat transfer flux distributions are loaded with the gradually weakened variations while keeping the same average heat flux. Additionally, for Cases B4 to B6, the decreased heat transfer distributions are loaded with the gradually weakened variations. As shown in Figure 10(a1), the bulk fluid temperature displays an overall increased trend except for a "weakened" region of $80 < x/H < 120$ where the core part of the fluid approaches the critical point. If the inclination ratio of the increased heat flux distribution decreases, i.e., the variations weaken, the bulk fluid temperature in the upstream region increases and that in the downstream region decreases. The cases with increased heat flux distribution reduce the "weakened" region which decrease the overall heat transfer. The wall temperature distributions increase sharply and then decrease slowly after reaching the maximum value, as shown in Figure 10(a2). With the inclination ratio of heat flux decreasing, the maximum wall temperature increases and occurs earlier in the channel. For the cases loaded with the decreased heat flux distributions, the bulk fluid temperature demonstrates a relatively fast increase and then a relatively slow decrease followed by a relatively slow increase, as presented in Figure 10(b1), which enlarges the effect where the fluid approaches the critical point. Comparing Figure 10(a2,b2), the maximum wall temperature for the cases with decreased heat flux distributions is higher than that for the cases with increased heat flux distributions.

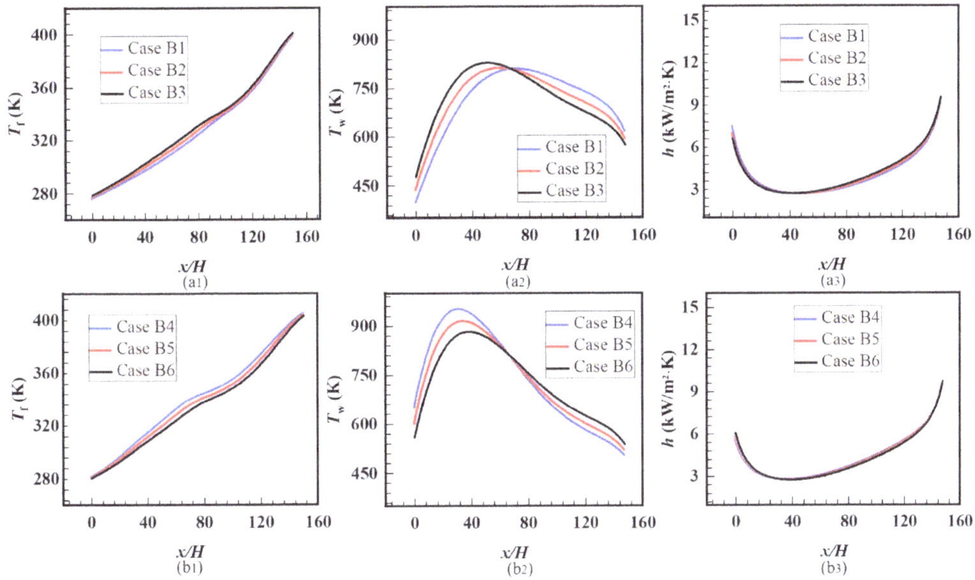

Figure 10. Temperature and HTC distributions for cases with nonuniform heat flux distributions. (**a1**) Fluid temperature for cases with increased heat flux distributions; (**a2**) average wall temperature for cases with increased heat flux distributions; (**a3**) HTC for cases with increased heat flux distributions; (**b1**) fluid temperature for cases with decreased heat flux distributions; (**b2**) wall temperature for cases with decreased heat flux distributions; (**b3**) HTC for cases with increased heat flux distributions.

According to the energy principle of energy conservation, the fluid temperature at the outlet remains the same regardless of the different heat flux distributions for Case B1–Case B6. As shown in Figure 10(a3), the HTC firstly decreases quickly and then increases relatively slowly for the cases loaded with increased heat flux distributions. HTC increases quickly in the region of $120 < x/H < 150$, which is more obvious in cases with decreased heat flux, as shown in Figure 10(b3). It is also indicated that the cases with the decreased heat flux distributions can improve the heat transfer in the downstream region.

Figure 11 exhibits the temperature contours of the bottom heated wall loaded with nonuniform heat flux distributions. Obviously, all of the cases present a firstly increased and then decreased temperature distribution, which is consistent with Figure 10(a2,b2). Overall, the temperature distributions for Case B1–Case B3 are relatively more uniform than those for Case B4–Case B6. For Cases B1–B3, the gradually increased heat flux distributions delay the occurrence of the high-temperature regions. However, a much stronger high-temperature region is found in the upstream region for the cases loaded with the decreased heat flux distributions, due to the larger heat flux and the core part being far from the critical point.

The streamlines and temperature contours on the y-z sections for Case B1 (loaded with increased heat flux distribution) and Case B4 (loaded with decreased distribution) are shown in Figure 12a and Figure 12b, respectively. The sections range from $x/H = 25$ to 150. As shown in Figure 12, the low-temperature regions are located in the core part of the mainstream and the temperature close to the walls is very high due to the high heat conduction from the solid materials. When the fluid in the core part of the mainstream approaches the critical point, large heat absorption happens to weaken the increased trend of the fluid temperature which corresponds with the "weakened" region in Figure 10(a1,b1). From the figure, when the flow develops along the channel at the section of $x/H = 75$, the core part is close to the critical point. A pair of counter-rotating vortices is found in most contours which are formed from the buoyancy effects. As seen in Figure 5, the density of the supercritical fluids decreases with increased temperature. In the presence of density difference, the vortices are driven by the gravity. It is observed that the mainstream is downward flow at the beginning, and the flow near the sidewalls is upward.

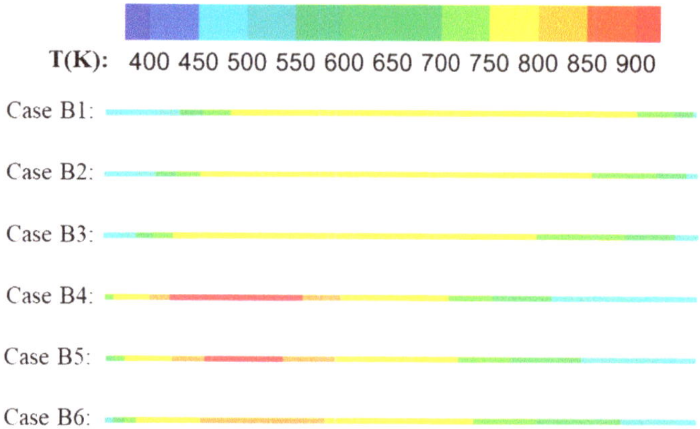

Figure 11. Temperature contours of the bottom wall for cases with increased and decreased heat flux distributions.

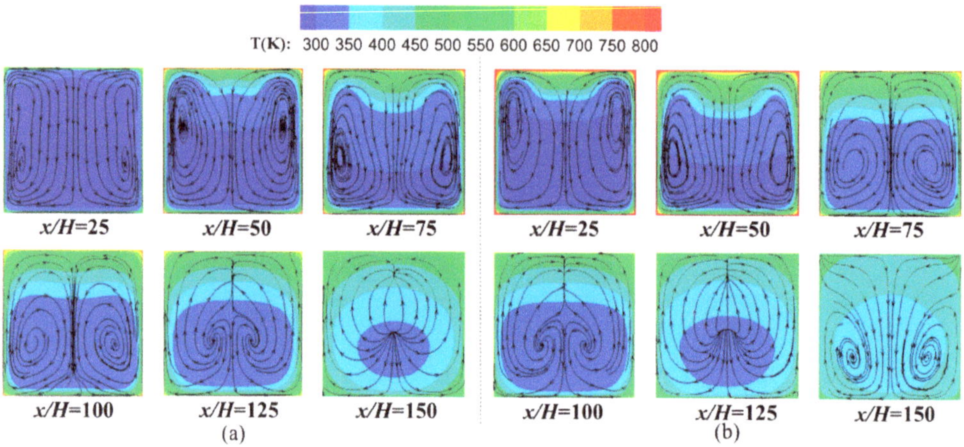

Figure 12. Streamlines and temperature contours on *y*-*z* sections for the cases with nonuniform heat flux distributions. (**a**) Case B1; (**b**) Case B4.

As shown in Figure 12a, the fluids near the solid walls are heated fast upstream which is related to the high wall temperature. However, the fluid temperature near the solid wall decreases downstream along the *x*-axis and the overall distribution becomes uniform. It is concluded that the wall temperature gradually decreases downstream but is always higher than that of the mainstream. The distribution in Case B4 (shown in Figure 12b) is roughly the same as that in Case B1, but the fluid is heated more quickly upstream due to the higher temperature difference between the mainstream and solid walls. Additionally, compared with Case B1, the temperature distribution for Case B4 is more uniform downstream. It is also noted that the scale of the vortices is related to the scale of the low-temperature region. The core of the vortices is more obvious in the section with a larger low-temperature region.

3.3. Effect of Acceleration and Buoyancy Effects

Figure 13 displays the temperature and HTC distributions for the cases loaded with accelerations at different magnitudes and directions. For Case C1 to Case C5, the magnitudes of the acceleration range from −3 g to 3 g along the normal direction including the states of overweight and weightlessness, respectively, 0 (Case C1), −g (Case C2), −2 g (Case C3), −3 g (Case C4) and +3 g (Case C5). As shown in Figure 13(a1), the average temperature of the fluid and solid decreases during the fluid approaching the critical point with increased accelerations. The phenomenon is more obvious in the distribution of the wall temperature distribution in Figure 13(a2). It is noted that the direction of accelerations along the z-axis have no significant influence on the fluid and wall temperature distributions by comparing the results in Case C4 and Case C5. The transverse accelerations along the streamwise direction are also considered in the work shown in Figure 13b. From the z-axis, the accelerations take effect in the directions of 0° (Case C4), 30° (Case C6) and 60° (Case C7), respectively, with the gradually increased transverse accelerations (a_x). The transverse accelerations have no obvious influence on the fluid temperature distributions, as seen in Figure 13(b1). However, the wall temperature decreases with increased transverse accelerations in the upstream region, which has benefits for the heat transfer shown in Figure 13(b2). For HTC distribution, the variation becomes gentle when the uniform heat flux is loaded. Larger HTC distributions are obtained via the case with the larger acceleration, regardless of the acceleration direction. The transverse accelerations can reduce the wall temperature in the upstream region and increase the corresponding HTCs.

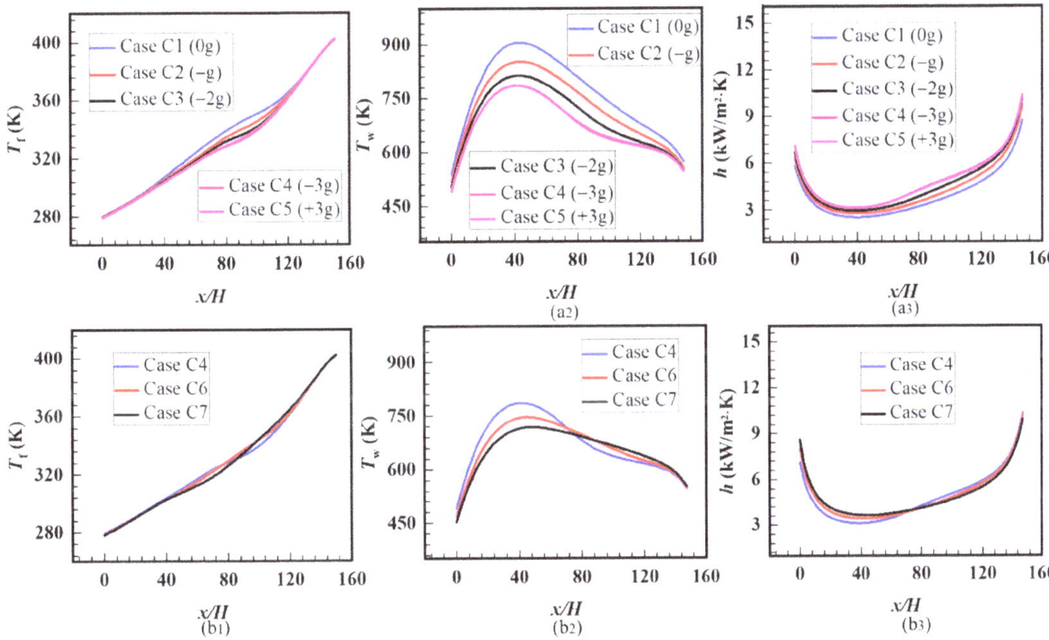

Figure 13. Comparison of temperature and HTC distributions for cases with different kinds of accelerations. (**a1**) Average fluid temperature for cases with different acceleration magnitudes; (**a2**) average wall temperature for cases with different acceleration magnitudes; (**a3**) HTC for cases with different acceleration magnitudes; (**b1**) fluid temperature for cases with different acceleration directions; (**b2**) wall temperature for cases with different acceleration directions; (**b3**) HTC for cases with different acceleration directions.

Figure 14a,b present the streamlines and temperature contours on the *y-z* sections for cases with accelerations in the states of overweight (Case C4) and weightlessness (Case C5). It is found that the temperature and streamline distributions for Case C4 and Case C5 are almost symmetrical along the acceleration directions, although the loaded accelerations show the reversed directions. This also indicates that the heating and temperature condition are similar for the bottom heated wall and the other three walls due to the quick heat conduction inside the solid material. Different from the streamlines in Figure 12, the second pair of vortices are formed from x/H = 75 and then gradually disappear at x/H = 125 due to the strengthened buoyancy effect at the larger accelerations. It is speculated that the mixing of cold and hot fluids becomes stronger at the larger accelerations.

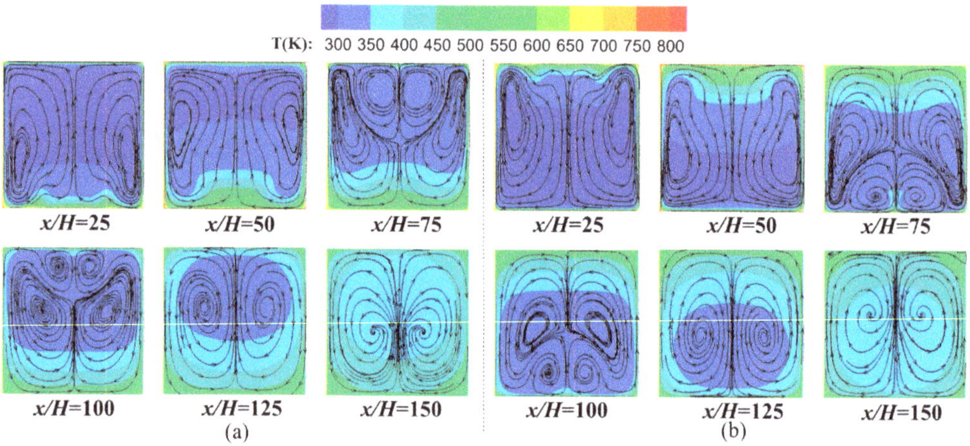

Figure 14. Streamlines and temperature contours on *y-z* sections for the cases with ±3 g accelerations. (**a**) Case C4 (−3 g); (**b**) Case C5 (+3 g).

3.4. *Effect of Flow Pattern*

To investigate the effect of flow pattern on heat transfer performance, adjacent channels with the same and reversed flow directions were designed to obtain the flow and heat transfer characteristics, i.e., Case D1 and Case D2. The computation domain contained two rectangular channels, as shown above. The total width was 6 mm and the height was 4 mm. Because the existence of extended sections may affect the temperature variations under the pattern of the two channels, the length of the channels was designed without upstream and downstream extended channels. The other boundary conditions were consistent with Case A1, including the mass flow rate, heat flux, operation pressure and inlet temperature. The gravity effect was also considered.

The fluid temperature contours in the spanwise–normal (*y-z*) sections for the cases with the same (Case D1) and reversed (Case D2) flow directions are shown in Figure 15 at an interval of x/H = 15 along the streamwise direction. Two identical fluid temperature distributions were obtained in Case D1, as shown in Figure 15a. From the fluid temperature distribution near the wall, it is indicated that the wall temperature firstly increases and then decreases, which also has agreement with Figure 12. It is observed that the fluid temperature approaching the bottom heated wall is higher than the fluid near the other sidewalls in the upstream region, and the difference gradually decreases in the downstream region. With flow developing along the channel, the fluid temperature becomes more uniform and a maximum average temperature is obtained at the outlet. Slightly different from Figure 15a, the fluid temperature distributions in Channel 1 (CH1) and Channel 2 (CH2) are symmetrical along the streamwise direction, as seen in Figure 15b. Compared with Figure 15a, the fluid temperature near the walls for the case with the reversed flow direction is lower, which indicates the decreased wall temperature for Case D2.

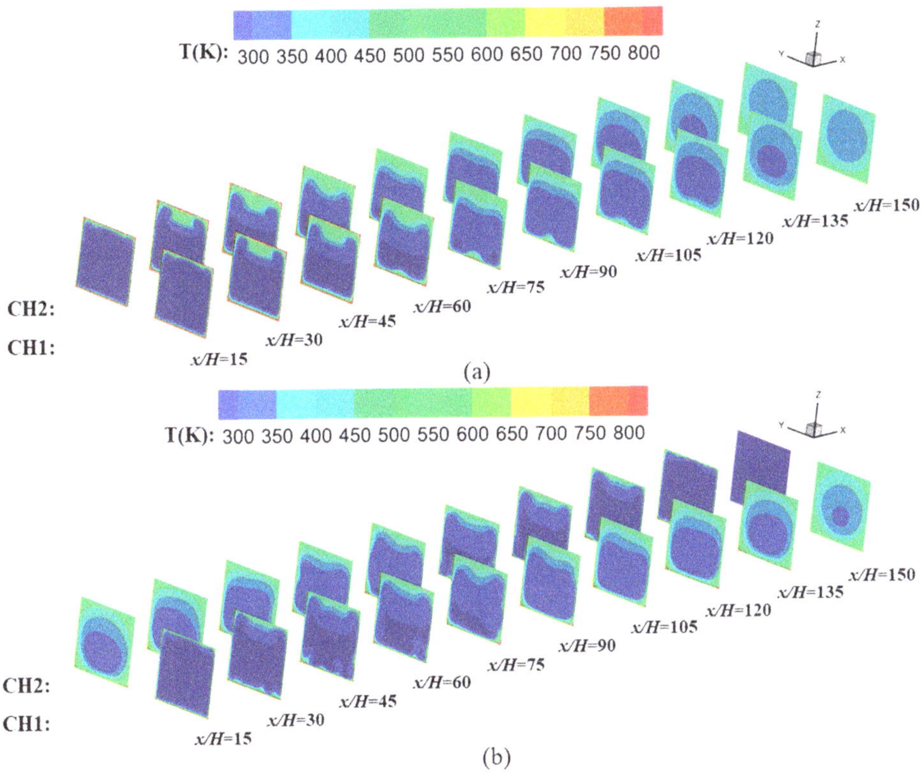

Figure 15. Bulk fluid temperature contours along x-axials in adjacent cooling channels. (**a**) Case D1; (**b**) Case D2.

The temperature and HTC distributions along the x-axis in adjacent cooling channels with the same (Case D1) and reversed (Case D2) flow directions are presented in Figure 16. Due to the symmetrical distribution of Case D1, the results for one channel are plotted and shown in Figure 16a. Using the flow pattern of the reversed flow directions, the "weakened" region disappears due to the transverse heat conduction of two adjacent channels and the fluid increases along the streamwise direction more smoothly. The wall temperature distributions exhibit great differences in Figure 16b. For Case D1, the wall temperature first increases to the maximum value and then gradually decreases. However, for Case D2, the maximum temperature is much lower and the distributions become more uniform. It is indicated that heat conduction inside solid materials is more dominant in determining the wall temperature compared with the convective heat transfer from the fluid due to the excellent thermal conductivity of the steel. Overall, the pattern of reversed flow directions greatly promotes the heat transfer between the two adjacent solid channels and avoids local high-temperature regions in Case D1. For HTC, Case D1 exhibits a trend of firstly decreasing and then quickly increasing, as shown in Figure 16c. For the pattern with reversed flow directions, the fluctuations in HTC are relatively gentle, despite showing the same trend.

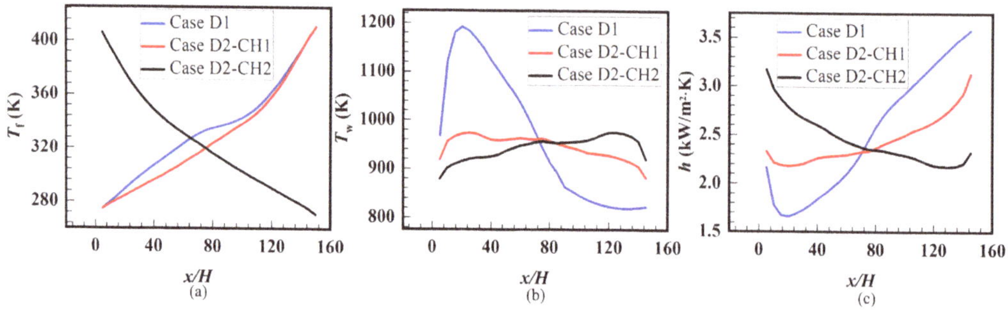

Figure 16. Temperature and HTC distributions along *x*-axials in adjacent cooling channels with the same and reversed flow directions: (**a**) fluid temperature; (**b**) average wall temperature; (**c**) HTC.

To further analyze the bulk fluid and average wall temperature distributions for Case D1 and Case D2, the maximum temperature, average temperature and distribution standard deviations of each channel were extracted from Figure 16 and compared in Table 2.

Table 2. Maximum temperature, average temperature and distribution standard deviations of *y*-*z* sections in Case D1 and Case D2.

Temperature Type	Case Name	Maximum Temperature (K)	Average Temperature (K)	Distribution Standard Deviation (K)
Bulk fluid temperature	Case D1 Case D2	411 411 (0)	335 328 (−2.09%)	36 39 (+8.33%)
Average wall temperature	Case D1 Case D2	1192 973 (−18.37%)	971 946 (−2.57%)	137 23 (−83.21%)

For bulk fluid temperature, the maximum temperatures in Case D1 and Case D2 are relatively close. The average temperature in Case D2 is lower than that in Case D1 by 2.09%, but the standard deviation is slightly higher than that in Case D1 by 8.33%. For average wall temperature, the maximum temperature in Case D2 reduces by 18.37% compared with Case D1. The standard deviation for Case D2 is much lower than that in Case D1 by 83.21%, which indicates the great improvements in wall temperature uniformity. It is concluded that the adoption of an adjacent channel with reversed flow direction can improve the uniformity of the wall temperature associated with the reduced maximum temperature.

4. Conclusions

In this paper, flow structures and heat transfer characteristics of supercritical CO_2 in a rectangular cross-section regenerative cooling channel were investigated under engine-like conditions. The effects of heat flux magnitude, nonuniform heat flux, acceleration and buoyance and flow patterns were considered. The mechanism of heat transfer and three-dimensional flow structure were comprehensively analyzed. The paper provides a good insight into supercritical CO_2 used as a coolant for regenerative cooling in a scramjet at an extremely high Mach number. Some significant conclusions from this study are listed as follows:

(1) With the effect of sharply varied thermophysical properties of supercritical CO_2, the wall temperature displays obvious fluctuations in the temperature variation range, especially regarding the core part of the fluid temperature close to the critical point. The phenomenon is more obvious in the case with a higher heat flux magnitude associated with the phenomenon of HTD in the downstream region of $80 < x/H < 120$.

(2) The cases with the linear decreased heat flux distributions enlarge the "weakened" region, which has benefits for heat transfer in the region of $70 < x/H < 90$. HTC

increases with the increased heat flux in the considered range of 0.75 MW/m^2 to 2.25 MW/m^2. Compared with channels loaded with increased heat flux, the maximum temperature increases for the cases loaded with decreased heat flux.

(3) The average temperatures of the fluid and solid both decreased during the period of the fluid approaching the critical point with increased accelerations. Larger HTC distributions were obtained via the case with the larger acceleration, regardless of the acceleration direction. Transverse accelerations can reduce the wall temperature in the upstream region and increase the corresponding HTCs.

(4) Compared with the case of adjacent channels arranged in the same direction, the wall temperature distribution becomes more uniform for the case arranged with reversed flow directions. It is indicated that heat conduction inside solid materials is more dominant in determining the wall temperature compared with convective heat transfer from the fluid, due to the excellent thermal conductivity of the steel. The maximum temperature decreased by 18.37% and the uniformity of the wall temperature field improved by 83.21%.

Author Contributions: Methodology, J.L. and W.X.; software, M.X. and P.L.; investigation, M.X.; original draft preparation, J.L. and M.X.; review and editing, P.L. and W.X. All authors have read and agreed to the published version of the manuscript.

Funding: This work was supported by the Key Laboratory Continuously Supporting Project (grant no. WDZC6142703202216), the National Natural Science Foundation of China (grant no. 12272133) and the Natural Science Foundation of Hunan Province (grant no. 2023JJ40733).

Data Availability Statement: All the results presented in this paper are available.

Conflicts of Interest: The authors declare no conflict of interest.

Nomenclature

Latin characters	
a	acceleration (m/s^2)
Cp	fluid thermal capacity (J/kg·K)
d	diameter of the tube (m)
g	gravity (m/s^2)
h	heat transfer coefficient (W/m^2·K)
H	height of the channel (m)
k	turbulent kinetic energy (m^2/s^2)
P	pressure (Pa)
q	heat flux (W/m^2)
T	temperature (K)
x	streamwise direction
y	spanwise direction
z	normal direction
Greek symbols	
β	thermal expansion coefficient (W/m·K)
λ	thermal conductivity (W/m·K)
μ	fluid dynamic viscosity (Pa·s)
ρ	fluid density (kg/m^3)
ω	specific energy dissipation rate (s^{-1})
Subscripts	
f	fluid
s	solid
Abbreviations	
HTC	heat transfer coefficient
HTD	heat transfer deterioration
Re	Reynolds number

References

1. Ehsan, M.M.; Guan, Z.; Klimenko, A.Y. A comprehensive review on heat transfer and pressure drop characteristics and correlations with supercritical CO_2 under heating and cooling applications. *Renew. Sustain. Energy Rev.* **2018**, *92*, 658–675. [CrossRef]
2. Cabeza, L.F.; de Gracia, A.; Fernández, A.I.; Farid, M.M. Supercritical CO_2 as heat transfer fluid: A review. *Appl. Therm. Eng.* **2017**, *125*, 799–810. [CrossRef]
3. Liu, G.; Huang, Y.; Wang, J.; Liu, R. A review on the thermal-hydraulic performance and optimization of printed circuit heat exchangers for supercritical CO_2 in advanced nuclear power systems. *Renew. Sustain. Energy Rev.* **2020**, *133*, 110290. [CrossRef]
4. Lecompte, S.; Ntavou, E.; Tchanche, B.; Kosmadakis, G.; Pillai, A.; Manolakos, D.; De Paepe, M. Review of Experimental Research on Supercritical and Transcritical Thermodynamic Cycles Designed for Heat Recovery Application. *Appl. Sci.* **2019**, *9*, 2571. [CrossRef]
5. Xie, J.; Liu, D.; Yan, H.; Xie, G.; Boetcher, S.K. A review of heat transfer deterioration of supercritical carbon dioxide flowing in vertical tubes: Heat transfer behaviors, identification methods, critical heat fluxes, and heat transfer correlations. *Int. J. Heat Mass Transf.* **2020**, *149*, 119233. [CrossRef]
6. Xu, W.; Li, M.; Li, Y.; Zhao, J. A comprehensive review of the effect of lubricant on the flow characteristics of supercritical CO_2 during cooling. *Int. J. Refrig.* **2022**, *133*, 145–156. [CrossRef]
7. Fan, Y.; Tang, G.; Li, X.; Yang, D. General and unique issues at multiple scales for supercritical carbon dioxide power system: A review on recent advances. *Energy Convers. Manag.* **2022**, *268*, 115993. [CrossRef]
8. Yan, C.; Xu, J.; Wang, S.; Liu, G. Numerical study of convective heat transfer to supercritical CO_2 in vertical heated tubes. *Int. Commun. Heat Mass Transf.* **2022**, *137*, 106242. [CrossRef]
9. Khalesi, J.; Sarunac, N.; Razzaghpanah, Z. Supercritical CO_2 conjugate heat transfer and flow analysis in a rectangular microchannel subject to uniformly heated substrate wall. *Therm. Sci. Eng. Prog.* **2020**, *19*, 100596. [CrossRef]
10. Pandey, S.; Laurien, E.; Chu, X. A modified convective heat transfer model for heated pipe flow of supercritical carbon dioxide. *Int. J. Therm. Sci.* **2017**, *117*, 227–238. [CrossRef]
11. Zhang, Q.; Li, H.; Liu, J.; Lei, X.; Wu, C. Numerical investigation of different heat transfer behaviors of supercritical CO_2 in a large vertical tube. *Int. J. Heat Mass Transf.* **2020**, *147*, 118944. [CrossRef]
12. Li, W.; Yu, Z.; Wang, Y.; Li, Y. Heat transfer of supercritical carbon dioxide in a tube-in-tube heat exchanger—A CFD study. *J. Supercrit. Fluids* **2022**, *181*, 105493. [CrossRef]
13. Ye, K.; Zhang, Y.; Yang, L.; Zhao, Y.; Li, N.; Xie, C. Modeling convective heat transfer of supercritical carbon dioxide using an artificial neural network. *Appl. Therm. Eng.* **2018**, *150*, 686–695. [CrossRef]
14. Liu, Z.; He, Y.; Li, Y.; Qu, Z.; Tao, W. Heat transfer characteristics of supercritical CO_2 flow in metal foam tubes. *J. Supercrit. Fluids* **2015**, *101*, 36–47. [CrossRef]
15. Wang, J.; Guan, Z.; Gurgenci, H.; Hooman, K.; Veeraragavan, A.; Kang, X. Computational investigations of heat transfer to supercritical CO_2 in a high horizontal tube. *Energy Convers. Manag.* **2018**, *157*, 536–548. [CrossRef]
16. Zhang, G.; Hu, P.; Chen, L.; Liu, M. Experimental and simulation investigation on heat transfer characteristics of in-tube supercritical CO_2 cooling flow. *Appl. Therm. Eng.* **2018**, *143*, 1101–1113. [CrossRef]
17. Xu, W.; Li, Y.; Wang, Y.; Li, M.; Zhao, J.; Li, M.; Tian, H. Experimental investigations on cooling heat transfer of CO_2-lubricant mixtures in horizontal tubes at supercritical pressure: A review. *Int. J. Refrig.* **2022**, *139*, 168–179. [CrossRef]
18. Yuan, N.; Bi, D.; Zhai, Y.; Jin, Y.; Li, Z.; Wang, H. Flow and heat transfer performance of supercritical pressure carbon dioxide in pipes with discrete double inclined ribs. *Int. J. Heat Mass Transf.* **2020**, *149*, 119175. [CrossRef]
19. Dong, X.; Zhang, C.; Wu, Y.; Lu, Y.; Ma, C. Thermodynamic Analysis and Optimization Design of a Molten Salt–Supercritical CO_2 Heat Exchanger. *Energies* **2022**, *15*, 7398. [CrossRef]
20. Wang, X.; Xu, W.; Xu, B.; Xiong, C.; Guo, S.; Chen, Z. Theoretical and numerical analysis of conjugate heat transfer for supercritical CO_2 flowing in printed circuit heat exchanger. *J. Supercrit. Fluids* **2022**, *189*, 105717. [CrossRef]
21. Chu, W.; Li, X.; Ma, T.; Chen, Y.; Wang, Q. Experimental investigation on SCO_2-water heat transfer characteristics in a printed circuit heat exchanger with straight channels. *Int. J. Heat Mass Transf.* **2017**, *113*, 184–194. [CrossRef]
22. Sarkar, J. Improving thermal performance of micro-channel electronic heat sink using supercritical CO_2 as coolant. *Therm. Sci.* **2019**, *23*, 243–253. [CrossRef]
23. Awais, A.A.; Saeed, M.; Kim, M.H. Performance enhancement in minichannel heat sinks using supercritical carbon dioxide (sCO_2) as a coolant. *Int. J. Heat Mass Transf.* **2021**, *177*, 121539. [CrossRef]
24. Cheng, K.; Zhou, J.; Zhang, H.; Huai, X.; Guo, J. Experimental investigation of thermal-hydraulic characteristics of a printed circuit heat exchanger used as a pre-cooler for the supercritical CO_2 Brayton cycle. *Appl. Therm. Eng.* **2020**, *171*, 115116. [CrossRef]
25. Reznicek, E.P.; Hinze, J.F.; Nellis, G.F.; Anderson, M.H.; Braun, R.J. Simulation of the supercritical CO_2 recompression Brayton power cycle with a high-temperature regenerator. *Energy Convers. Manag.* **2021**, *229*, 113678. [CrossRef]
26. Muto, Y.; Ksto, Y. Optimal Cycle Scheme of Direct Cycle Supercritical CO_2 Gas Turbine for Nuclear Power Generation Systems. *J. Power Energy Syst.* **2008**, *2*, 1060–1073. [CrossRef]
27. Jiang, J.; Zhang, R.; Le, J.; Liu, W.; Yang, Y.; Zhang, L.; Zhao, G. Regeneratively cooled scramjet heat transfer calculation and comparison with experimental data. *Proc. Inst. Mech. Eng. Part G J. Aerosp. Eng.* **2014**, *228*, 1227–1234. [CrossRef]
28. Lee, S.H.; Lee, J.; Kim, D.; Kim, S.; Yang, I. Heat transfer characteristics of kerosene phase change based on fuel flow rates in the regenerative cooling heat exchanger of scramjet engines. *J. Mech. Sci. Technol.* **2021**, *35*, 2733–2741. [CrossRef]

29. He, N.; Liu, C.; Pan, Y.; Liu, J. Progress of Coupled Heat Transfer Mechanisms of Regenerative Cooling System in a Scramjet. *Energies* **2023**, *16*, 1025. [CrossRef]
30. Liu, J.; Xu, M.; Ma, K.; Liu, C.; Xi, W. Heat transfer and flow structures of supercritical n-decane in a regenerative cooling channel loaded with non-uniform heat flux. *Front. Energy Res.* **2022**, *126*, 16648714. [CrossRef]
31. Zhang, J.; Zhou, Q.; Zhao, X.; Jiang, Y.; Fan, W. Numerical Investigation on the Heat Transfer of n-Decane in a Horizontal Channel with Axially Nonuniform Heat Flux under Supercritical Pressure. *Aerospace* **2022**, *9*, 326. [CrossRef]
32. Zhang, T.; Jing, T.; Qin, F.; Sun, X.; Li, W.; He, G. Topology optimization of regenerative cooling channel in non-uniform thermal environment of hypersonic engine. *Appl. Therm. Eng.* **2023**, *219*, 119384. [CrossRef]
33. Zhang, C.; Gao, H.; Zhao, J.; Zhou, J.; Wen, D. Cooling performance optimization of supercritical hydrocarbon fuel via bi-directional flow pattern. *Appl. Therm. Eng.* **2022**, *212*, 118563. [CrossRef]
34. Yu, J.; Huang, L.; Sun, S.; Yu, J.; Li, Y. Flow and heat transfer of RP-3 aviation kerosene in horizontal tube at supercritical pressure under altered gravity levels. *Appl. Therm. Eng.* **2022**, *207*, 118162. [CrossRef]
35. Jia, D.; Wang, N.; Pan, Y.; Liu, C.; Wang, S.; Yang, K.; Liu, J. Flow and Heat Transfer Characteristics of Supercritical N-Decane in Adjacent Cooling Channels with Opposite Flow Directions. *Energies* **2021**, *14*, 1071. [CrossRef]
36. Wen, J.; Gu, X.; Liu, Y.; Wang, S.; Li, Y. Effect of surface tension, gravity and turbulence on condensation patterns of R1234ze(E) in horizontal mini/macro-channels. *Int. J. Heat Mass Transf.* **2018**, *125*, 153–170. [CrossRef]
37. Tian, K.; Yang, P.; Tang, Z.; Wang, J.; Zeng, M.; Wang, Q. Effect of pyrolytic reaction of supercritical aviation kerosene RP-3 on heat and mass transfer in the near-wall region. *Appl. Therm. Eng.* **2021**, *197*, 117401. [CrossRef]
38. Sun, F.; Li, Y.; Sunden, B.; Xie, G. The transport and thermodynamic characteristics of thermally oscillating phenomena in a buoyancy-driven supercritical fuel flow. *Int. J. Therm. Sci.* **2021**, *159*, 106550. [CrossRef]
39. Gou, J.; Yan, Z.; Hu, J.; Gao, G.; Gong, C. The heat dissipation, transport and reuse management for hypersonic vehicles based on regenerative cooling and thermoelectric conversion. *Aerosp. Sci. Technol.* **2021**, *108*, 106373. [CrossRef]
40. Zhu, Y.; Peng, W.; Xu, R.; Jiang, P. Review on active thermal protection and its heat transfer for airbreathing hypersonic vehicles. *Chin. J. Aeronaut.* **2018**, *31*, 1929–1953. [CrossRef]
41. Zhang, S.; Li, X.; Zuo, J.; Qin, J.; Cheng, K.; Feng, Y.; Bao, W. Research progress on active thermal protection for hypersonic vehicles. *Prog. Aerosp. Sci.* **2020**, *119*, 100646. [CrossRef]
42. Harvey, A.; Bell, I.; Huber, M.; Laesecke, A.; Lemmon, E.; Meyer, C.; Muzny, C.; Perkins, R. Thermophysical Properties of Carbon Dioxide and CO_2-Rich Mixtures, Report to US Department of Energy. 2016. Available online: https://tsapps.nist.gov/publication/get_pdf.cfm?pub_id=919336 (accessed on 8 May 2023).
43. Li, Z.; Jiang, P.; Zhao, C.; Zhang, Y. Experimental investigation of convection heat transfer of CO_2 at supercritical pressures in a vertical circular tube. *Exp. Therm. Fluid Sci.* **2010**, *34*, 1162–1171. [CrossRef]

aerospace

MDPI

Article

Experimental Study on the Dynamic Characteristics of Gas-Centered Swirl Coaxial Injector under Varying Ambient Pressure

Xiaoguang Zhang [1], Wentong Qiao [2,*], Qixiang Gao [2,*], Dingwei Zhang [2], Lijun Yang [2,3] and Qingfei Fu [2,3]

1 Xi'an Aerospace Propulsion Institute, Xi'an 710100, China
2 School of Astronautics, Beihang University, Beijing 100191, China
3 Ningbo Institute of Technology, Beihang University, Ningbo 315100, China
* Correspondence: qiaowentong@buaa.edu.cn (W.Q.); buaagqx@buaa.edu.cn (Q.G.)

Abstract: To determine the dynamic characteristics of a gas-centered swirl coaxial injector under backpressure, an experimental system of dynamic injection in a backpressure chamber was constructed. Filtered water and nitrogen were used as simulant media for rocket propellants, which are typically used with this kind of injector. An inertial flow pulsator was manufactured to generate the pulsation of the flows that feed to the liquid injector. The electric conductance method was adopted to measure liquid film thickness. After the pulsation of incoming flow in the feedline was tested, and the operating conditions for the injector to start pulsating were validated, the effects of the chamber backpressure and the recess length of the injector on the dynamic characteristics of spray, such as liquid film thickness, breakup length, and amplitude of pulsation, have been investigated in detail. Experimental results demonstrated that the increase in chamber backpressure prompts the liquid sheet to rupture earlier with a shorter breakup length, which results from the increased density of the ambient gas. Chamber backpressure suppresses the pulsation of the outlet flow, especially for a longer recess length. Moreover, a decrease in the recess length results in a reduction in breakup length due to an intense gas–liquid shearing in a narrower recess section. For a lower backpressure, the amplitude of outlet flow generally increases when the recess length increases. However, this phenomenon is not obvious for the conditions of higher backpressure and lower pulsation frequency.

Keywords: gas-centered swirl coaxial injector; chamber backpressure; dynamic characteristics; pulsation flow

Citation: Zhang, X.; Qiao, W.; Gao, Q.; Zhang, D.; Yang, L.; Fu, Q. Experimental Study on the Dynamic Characteristics of Gas-Centered Swirl Coaxial Injector under Varying Ambient Pressure. *Aerospace* **2023**, *10*, 257. https://doi.org/10.3390/aerospace10030257

Academic Editor: Kevin Lyons

Received: 10 January 2023
Revised: 24 February 2023
Accepted: 6 March 2023
Published: 8 March 2023

1. Introduction

The gas-centered swirl coaxial (GCSC) injectors are widely used in the combustion chamber of liquid rocket engines, such as NK-33, RD180, YF-100, and YF-115 [1]. In the combustion chamber, the GCSC injector ejects the liquid fuel that encloses the oxygen-rich gas in the middle [2–4]. The use case of this is typically the staged combustion cycle engines. One propellant is burned with a small amount of the other in the preburner, producing a hot gas mixture. Then, this gas mixture (oxygen-rich gas) passes through the turbine and is injected into the thrust chamber [2]. Under the action of gas–liquid shear force and centrifugal force, the conical liquid sheet breaks up and is atomized into liquid filaments and droplets. Further, the propellant burns in the chamber. However, liquid rocket engines often encounter the challenge of unstable combustion, which can be devastating to the combustion chamber. With a better understanding of the unstable combustion problem, it has been found that the dynamic characteristics of the injector play an important role in the stable operation of liquid rocket engines [5]. The injector not only performs the task of spraying and mixing, but also acts as an amplifier, phase regulator, exciter, and oscillator in the overall dynamic system of the engine. Whether it is the pressure/flow pulsation from the supply system or the backpressure pulsation in the combustion chamber, the dynamic characteristics of the injector can be adjusted to cut off the instability coupling between them [6]. Therefore, it is important to study the dynamic response characteristics of the injector to the incoming pressure/flow oscillation when the chamber backpressure varies.

In recent years, there have been many experimental and theoretical studies on injection dynamics, especially on swirl injectors, single-component injectors, and dual-component coaxial injectors [2,7–10]. Bazarov [11] pioneered the work on injector dynamics. The injector frequency characteristic equation was derived analytically by using the inviscid hydrodynamic equation that controls the internal flow of the injector. After that, a detailed study of different injectors and propellants was carried out to derive the relationship between the amplitude and phase difference of each output parameter, and to describe in detail the engineering calculation method of injector dynamic characteristics [12–16]. More relevant to this paper is the study of the dynamic characteristics of GCSC injectors. Park et al. [17] performed a study on the dynamic characteristics of the GCSC injectors by inducing excitation on the liquid flow through a mechanical pulsator. The influence of the geometry structure (tangential inlet diameter and gap thickness) and the momentum flux ratio on the dynamic gains and the spray performance were investigated in detail. In addition to adding perturbation to the liquid flow, they also added acoustic excitation in the gas flow of the GCSC injector by a speaker [18]. The excitation effect is enhanced by the increment of the momentum flux ratio. Additionally, the spray patterns have different shapes depending on the excitation frequency. Similarly, Sahoo and Gadgil [19] experimentally studied the self-pulsation dynamics in a GCSC injector by varying the momentum flux ratio for different values of swirl number, recess length, and lip thickness. They pointed out that the constrained shear layer instability between the liquid and the gas streams inside the recess region, the spread rate of the gas jet, and its interaction location with the liquid stream dictate the presence of dominant pulsation frequencies and even their higher harmonics in many recessed configurations. The dynamic characteristics of GCSC injectors with different tangential inlet diameters were investigated by Oh et al. [20]. A mechanical pulsating flow generator was used to generate excitation in the supply line. They concluded that as the tangential inlet diameter increases, the mass flow rate increases and the spray cone angle decrease. As the tangential inlet diameter decreases, the injector becomes less effective as an amplifier, confirming that a sufficiently small tangential inlet can be used as a damper. The above studies on the dynamic characteristics and pulsating spray of the GCSC injector were conducted under the atmospheric pressure, and hardly involved the case of GCSC injectors in the combustion chamber under high backpressure conditions.

Up to now, a number of scholars have attempted to study the steady-state sprays and their mechanisms using various types of injectors under ambient pressure conditions [21–26]. To investigate the spray under high backpressure similar to a real rocket engine, the most critical thing is to build a backpressure chamber. Kenny et al. [27] fabricated a backpressure chamber that could reach 9.65 MPa. The spray experiments using a liquid swirl injector showed that an increase in the backpressure of the chamber increased the average liquid film thickness of the nozzle and made the spray cone angle smaller when the mass flow rate was constant. Kim [23] used a liquid–liquid swirl coaxial injector to perform spray experiments under high backpressure conditions and concluded that the aerodynamic force of ambient gas significantly affects the breakup of a swirl spray. However, the breakup mechanism of liquid–liquid swirl coaxial spray is very much controlled by the impact force by the interaction of propellants. Cho et al. [28] used the backpressure chamber to investigate the characteristics of a cryogenic swirl flow in a liquid swirl injector at subcritical to supercritical conditions. It is shown that the cryogenic flow behavior changes significantly when the environmental conditions change from subcritical to supercritical conditions. The interface initially produces spiral waves and then develops into a coiled vortex structure. However, the frequency of flow instability does not change dramatically during the transition from subcritical to supercritical conditions. For a liquid-centered swirl coaxial injector, Bai et al. [21] explored the self-pulsation characteristics under various chamber backpressures. As the backpressure increased, the flow in injector changed from the outer mixing flow to the critical mixing flow, and finally developed into the inner mixing flow. Moreover, for a gas–liquid swirl coaxial under backpressure, Chen et al. [29] summarized the effects of the gas–liquid ratio and ambient pressure on the spatial distribution of the spray. They concluded that the Sauter mean diameter increased with an increase in backpressure. An increase in the gas–liquid ratio caused a significant increase in the axial velocity near the axis of the spray filed. The above studies on the spray of various single-component and dual-component injectors under backpressure mainly focus on the effects of ambient backpressure, operating conditions, and geometry structure on the steady-state characteristics of the spray and its mechanism. However, few have dealt with the dynamic characteristics of the injectors under backpressure conditions.

In summary, for the various types of injectors used in liquid rocket engines, researchers have mainly focused on two aspects of the spray: the injector dynamic characteristics at atmospheric pressure and the steady-state characteristics under backpressure conditions. However, for the GCSC injector, few studies have been conducted on the dynamic characteristics under backpressure conditions. This can provide technical guidance for suppressing the combustion instability in the combustion chamber. Therefore, the present paper aimed to experimentally study the spray dynamic characteristics of the GCSC injector under various chamber backpressures. An inertial flow pulsator was manufactured to generate the dynamic pulsation of the coming flow in the liquid feedline. The liquid film thickness was measured by electric conductance method. An experimental system of dynamic injection with a backpressure chamber was constructed. The effects of the chamber backpressure and the recess length of the injector on the dynamic characteristics of spray, such as liquid film thickness, breakup length, and amplitude characteristics of pulsation, have been investigated in detail. The experimental results can help researchers to obtain some insights into the mechanism of the dynamic characteristics of the spray on the combustion instability.

2. Experimental Methods

2.1. Experimental Facilities

As depicted in Figure 1, the experimental apparatus of injector dynamics characteristics in this study consists of a supply system of simulated propellant, an excitation system of pulsating flow, a measuring system, and a backpressure environment system. When using a pump to supply liquid to the injector, a vibration along the feedline will be generated and interfere with the flow/pressure pulsation that feeds to the injector, so the pressurized water supply method is adopted. The nitrogen gas in the high-pressure nitrogen tank 2 is supplied to the working medium storage tank 1 after decompression through relief valve 17. Filtered water enters the main feedline under the squeeze of nitrogen and passes through the liquid flow meter 5. After the water flows through the flow pulsation generator 6, the flow and pressure oscillations propagating downstream will be generated. The gas feedline of injector uses nitrogen as the supply medium, which is provided by the high-pressure nitrogen tank 3 in Figure 1. A gas flow controller 4 is used to control the flow rate of the nitrogen gas.

The spray field information is captured using a high-speed camera 9 through the observation window of backpressure chamber 8. The pressure and flow signals in injector are measured using a dynamic pressure transducer and a homemade liquid film thickness sensor. The signals are displayed in real time through the data acquisitor 14 and stored in the computer 19. The high-pressure nitrogen tank 10 is used to inflate the backpressure chamber 8 to provide a certain ambient pressure. The mist droplets generated during the spraying process will splash on the surface of the observation window, affecting the photographic performance of the high-speed camera 9. Therefore, the observation window is purged with an air curtain using the high-pressure nitrogen tank 10.

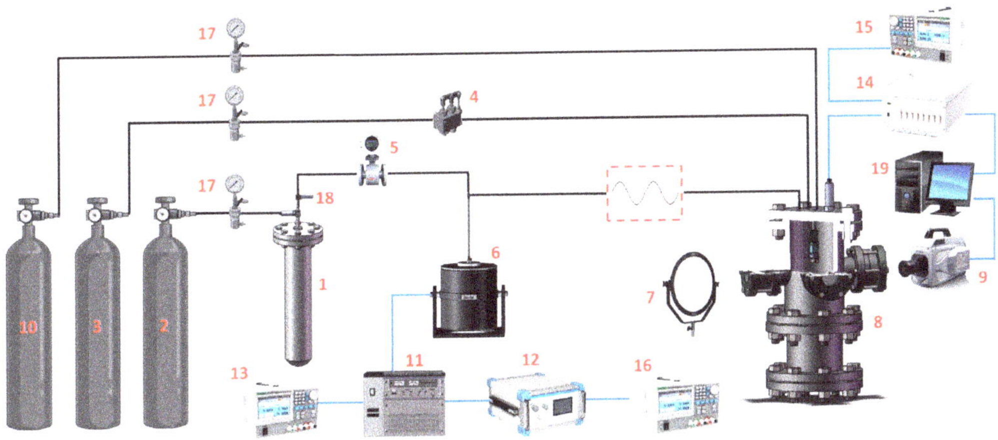

Figure 1. Experimental facilities and setup: 1—working medium storage tank; 2, 3, 10—high-pressure nitrogen tank; 4—gas flow controller; 5—liquid flow meter; 6—flow pulsation generator; 7—LED lamp; 8—backpressure chamber; 9—high-speed camera; 11—power amplifier; 12—signal generator; 13, 15, 16—DC power supply; 14—data acquisitor; 17—relief valve; 18—stop valve; 19—computer.

The schematic of the GCSC injector is presented in Figure 2. The gas is injected from the injector housing into the cylindrical passage of the gas injector. The outer liquid injector is designed as an open-end swirl structure, which adopts two tangential passages that are located every 180°. The liquid in the cavity between the liquid injector and the injector housing is injected into the annular aperture between the liquid injector and the gas injector through the tangential passages. In this way, annular liquid film with swirl motion is formed. The conical liquid sheet at the injector exit is broken and atomized by centrifugal effect and aerodynamic force. The injector recess length ΔL represents the distance between the gas injector outlet and the liquid injector outlet. To investigate the effect of recess length ΔL on the injector dynamic characteristics, three GCSC injectors with different geometrical parameters were manufactured, in which the recess lengths ΔL are 13 mm, 8 mm, and 3 mm, respectively. The other structure and parameters of the injectors were determined by referring to a previous study [10] and considering the injection conditions in present experiments. The key geometrical parameters are listed in Table 1.

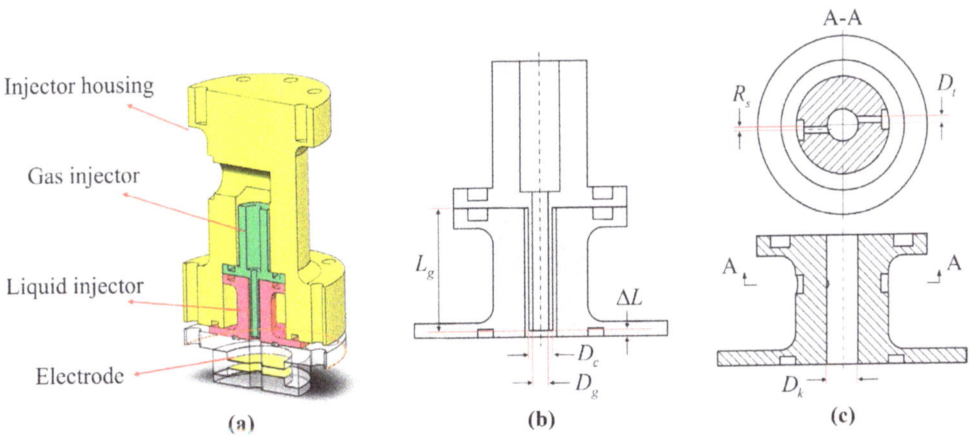

Figure 2. Schematic of GCSC injector. (**a**) assembly diagram with measurement electrodes; (**b**) gas injector; (**c**) liquid injector.

Table 1. Geometrical parameters of the GCSC injector.

No.	D_t (mm)	D_k (mm)	D_c (mm)	D_g (mm)	R_s (mm)	L_g (mm)	n	A	ΔL (mm)
1	0.8	4	3	2	0.5	10	2	20	13
2	0.8	4	3	2	0.5	15	2	20	8
3	0.8	4	3	2	0.5	20	2	20	3

Note: n is the number of tangential passages; A is the geometry characteristics constant of the injector; ΔL is the recess length.

The core equipment of the backpressure environment system is the backpressure chamber that is shown in Figure 3. The material of this backpressure chamber is 0Cr18Ni9 (austenitic stainless steel). It has a height of 726 mm and an inner diameter of 154 mm. The maximum ambient pressure that this chamber can withstand is 3 MPa, and the backpressure fluctuation range is ±5%. It has three optical observation windows in two perpendicular directions, which have gas purge functions. A regulator valve is installed at the bottom for regulating the chamber pressure and exhausting. It can provide backpressure experimental conditions for gas and liquid multi-component injectors.

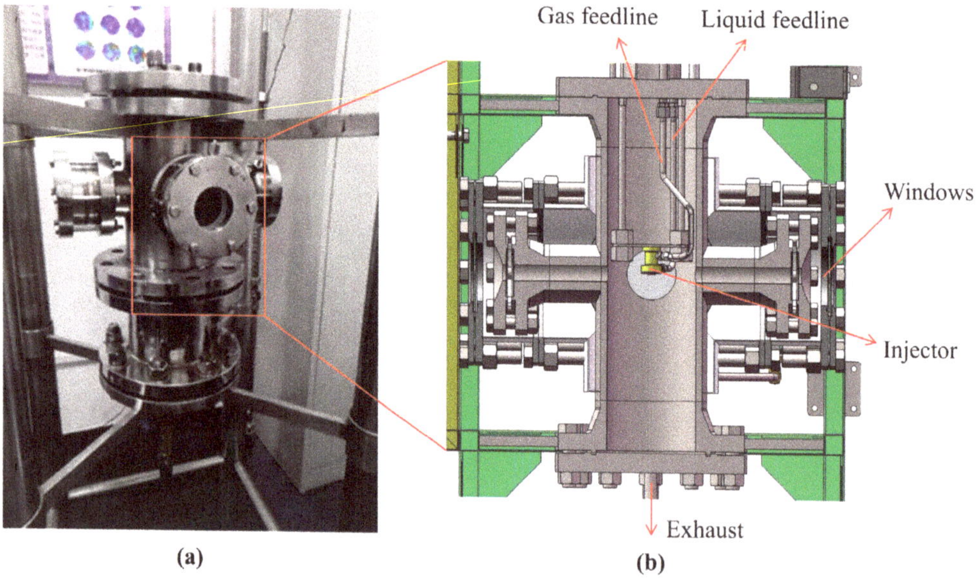

Figure 3. Backpressure chamber: (**a**) physical picture; (**b**) internal schematic.

2.2. Experimental Conditions

The experimental conditions are summarized in Table 2. Because of the relatively simple structure and the stable operating performance, liquid swirl (LS) injector is used to calibrate the measuring system of liquid film thickness (Test 1) and test the inertial flow pulsator (Test 2–4). Test 5 is used to explore the appropriate operating condition for GCSC injector when a flow perturbation is applied into the liquid feedline by turning on the inertial flow pulsator. Exp 1–3 are designed to study the effect of the backpressure value P_b and the recess length ΔL on the dynamic characteristics of GCSC injectors, in which the pressure drop ΔP in liquid feedline and the volume flow rate Q_g of gaseous nitrogen remain constant. The deviation of the pressure drops in liquid feedline and the volume flow rate of gas in these experiments did not exceed 6% and 8%, respectively.

Table 2. Experimental conditions and parameters.

Type	Injector	ΔP (MPa)	Q_g (L/min)	P_b (MPa)	f (Hz)
Test 1	LS	0.45	–	0.5	500
Test 2	LS	0.65	–	1	100
Test 3	LS	0.3	–	1	100
Test 4	LS	0.45	–	1.5	500
Test 5	GCSC 2	0.25	15	0.1	200
Exp 1	GCSC 1	0.25	15	0.3, 0.6, 0.9	180–270
Exp 2	GCSC 2	0.25	15		
Exp 3	GCSC 3	0.25	15		

Note: ΔP is the pressure drop in liquid feedline; Q_g is the volume flow rate of gaseous nitrogen in gas feedline; P_b is the backpressure value of chamber; f is the pulsation frequency from signal generator.

2.3. Experimental Techniques

The dynamic spray patterns were captured instantaneously by a backlighting photography technique. As depicted in Figure 1, a high-speed camera (Photron Fastcam SA-Z, Tokyo, Japan) and a surface LED lamp (Falcon Eyes SO-48TD, Hong Kong, China) were placed opposite to each other on both sides of the observation window of the backpressure chamber. The camera exposure time was set to 20 µs, and the instantaneous gray spray images with 1024 × 688 pixels were obtained. The frame rate was set to be 20,000 frames/s as the best compromise for all experiments.

Considering that the frequency of the dynamic characteristics of the injector is 10–1000 Hz, and the pressure amplitude is 10^3–2×10^6 Pa, a high-frequency dynamic transducer was chosen to measure the pressure in the liquid cavity of the injector. The accuracy of this transducer is 0.5% and the range is 0–4 MPa. A power amplifier was used to amplify the pulsation signal generated by the signal generator, which drives the flow pulsation generator to excitation and meet its frequency range (5–2000 Hz). The data acquisitor adopted a dynamic signal test and analysis system (DH5922D, Donghua Calibration & Testing Co., Taizhou, China) with a sampling frequency of 256 Hz. The system can simultaneously perform 16 signal acquisitions, convert digital–analog signals, and transfer the acquired real-time signals to a computer.

In addition to the methods described above, the following two techniques used in this experimental study are crucial.

2.3.1. Inertial Flow Pulsator

The flow pulsation generator is a key device in the experimental system of the injector's dynamic characteristic, which is used to generate the flow/pressure oscillation in the liquid feedline. The working principle of the existing hydro-mechanical pulsator (Figure 4a) is that the flow pulsation generator is powered by a motor, and a rotating multi-orifice disk rotates at high speed. The orifices on the rotating disk connect to the flow pipe in some predetermined period, so the flow fluctuates at a certain frequency [30]. However, this pulsator will be subject to leaks due to the poor sealing of complex mechanical structures. In particular, when the inlet pressure in the manifold is high, this issue is more serious. Therefore, this pulsator is not suitable for the dynamic spray experiments in a backpressure chamber. In this paper, a new type of inertial flow pulsator (Figure 4b) was designed and manufactured, which refers to the comparison of different types of flow pulsation generators by Bazarov [31].

The principle of the inertial flow pulsator is that the oscillation in liquid flow is excited by the excitation of mechanical vibrations of a part of the pressurized manifold (i.e., oscillation tube). The pulsation is generated by the rapid displacement of a rigid section of tubing through which the liquid flows. The liquid will generate pressure pulsations determined by the periodic displacement of its mass instead of moving parts. Thus, the sealing performance is good, even if the inlet pressure in the manifold is high. First, the sinusoidal signals of different frequencies generated by the signal generator are amplified by a power amplifier. Then, the amplified signal is transmitted to the pulse exciter. Finally, the oscillation tube of the pulse exciter will make a periodic reciprocating motion corresponding to the frequency of the given signal, driving the liquid in the oscillation tube to flow periodically. In this way, a pulsating flow at a specific frequency can be output into the liquid feedline.

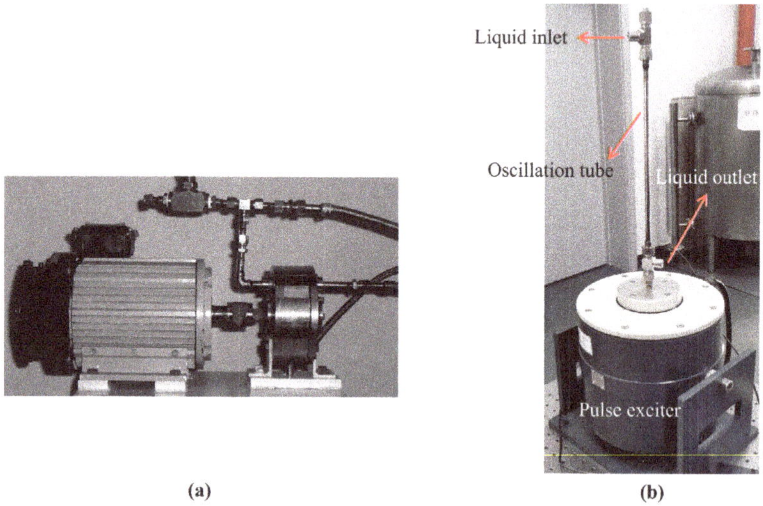

Figure 4. Flow pulsation generator: (**a**) hydro-mechanical pulsator; (**b**) inertial flow pulsator.

To improve the pulsating performance, Bazarov [31] argued that the velocity of liquid in the oscillation tube should not be greater than 12 m/s and the length of this tube should be less than 1/10 of the length of the pressure wave of the excited frequency in the liquid. Therefore, the size of the oscillation tube can be designed according to the flow rate and the pulsation frequency required for the experiment, as shown in Figure 4b.

2.3.2. Electric Conductance Method

The flow pulsation of liquid at injector outlet manifests as the change of annular liquid film thickness, whereas the liquid film thickness can be measured via the electric conductivity of liquid between two electrodes in injector chamber. Therefore, the purpose is to calibrate the relation between the voltage of two electrodes and the liquid film thickness. The measurement principle of the electric conductance method (ECM) is elaborated by taking an example of an LS injector with a simple structure. The schematic and structure parameters of LS injector are shown in Figure 5 and Table 3, respectively. Two ring electrodes composed of porous titanium are placed near the outlet of LS injector. The conductivity between the two electrodes varies with the liquid film thickness. The expression of the resistance of annular liquid column R_{ring} is

$$R_{ring} = \frac{\rho_r L_r}{S} \quad (1)$$

where ρ_r is the resistivity of conductive liquid, L_r is the length of annular liquid column, and S is the cross-sectional area of annular liquid column, which is expressed as

$$S = \pi\left[r^2 - (r-h)^2\right] \quad (2)$$

where r is the inner radius of injector outlet, and h is the liquid film thickness. Substituting Equation (2) into Equation (1), the theoretical expression between the annular liquid film thickness h and the resistance of annular liquid column R_{ring} can be obtained:

$$h = r - \sqrt{r^2 - \frac{\rho_r L_r}{\pi R_{ring}}} \tag{3}$$

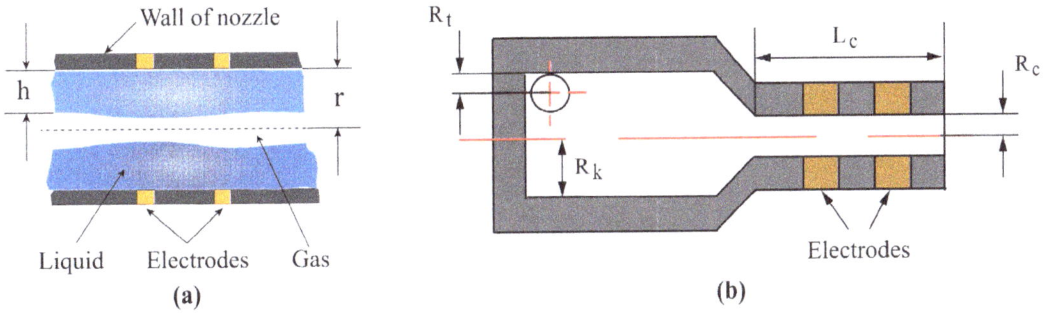

Figure 5. Principle of ECM: (**a**) schematic of ECM; (**b**) schematic of LS injector.

Table 3. Geometrical parameters of the LS injector.

R_t (mm)	R_k (mm)	R_c (mm)	L_c (mm)	n_s	A_s
1	3.2	1	10	2	2.2

Note: n_s is the number of tangential passages of LS injector; A_s is the geometry characteristics constant of LS injector.

Cylindrical ceramic rods of different diameters were placed in the swirl chamber to calibrate the measuring system of liquid film thickness. When the liquid flowed through the swirl chamber, liquid films of different thicknesses were generated in the annular space between the inner wall of injector and the ceramic rods of different diameters. By recording the corresponding voltage signal value, the calibration curve of liquid film thickness and voltage could be obtained as drawn in Figure 6. The same measurement principle of liquid film thickness and the installation method of electrodes were also adopted for the experiment of the dynamic characteristics of GCSC injectors.

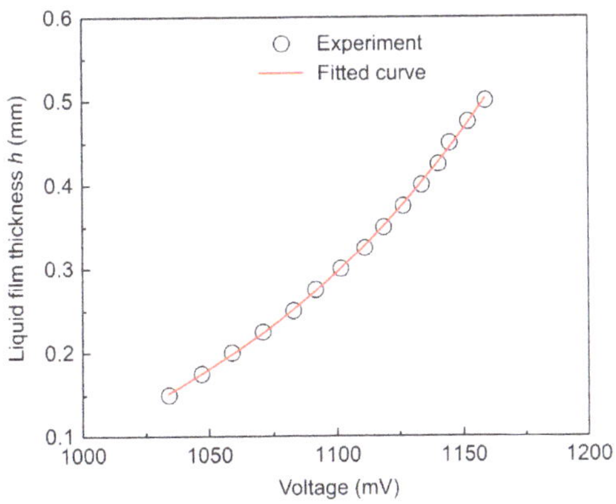

Figure 6. Calibration curve of the LS injector: voltage vs. film thickness.

The small fluctuations of liquid film thickness in injector and the considerable noise associated with the flow process make it difficult to measure the conductivity of the liquid column between the electrodes. In order to solve this problem, a homemade liquid film thickness sensor based on a lock-in amplifier circuit was used, which can meet the requirements of liquid film thickness acquisition at frequency up to 500 Hz. The raw data of the measured liquid film thickness are shown in Figure 7.

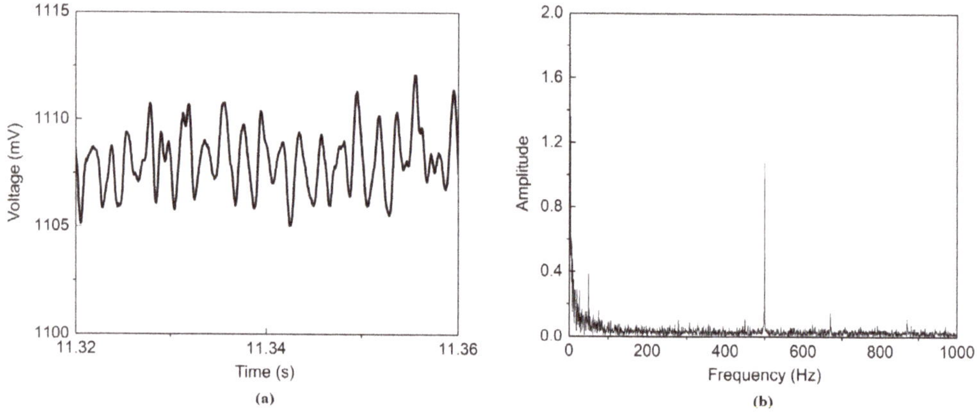

Figure 7. Raw data of LS injector when $P_b = 0.5$ MPa, $\Delta P = 0.45$ MPa, and $f = 500$ Hz: (**a**) raw data of liquid film thickness; (**b**) fast Fourier transform (FFT).

There are several sources of uncertainty in the electric conductance method to measure liquid film thickness [13]. First, the machining of the calibration ceramic rods caused uncertainty. A series of rods with different diameters correspond to different liquid film thicknesses. The rods were fabricated with a tolerance of 0.001 mm, and it is assumed that the machining error complied with the even distribution. The second source of uncertainty was the change of environmental conditions, such as the variation of the specific conductivity of the liquid. To reduce this uncertainty, the test medium for calibrations and experiments adopted a sodium chloride solution (1 g sodium chloride per 10 L distilled water). Third, less important sources of uncertainty include the limitations of this measuring system, which produces an average conductivity signal. This method measures the average liquid film thickness between the two electrodes, whereas the liquid film thickness varies axially. However, this error was omitted because the distance between the electrodes was much smaller than the wavelength of the liquid film surface [30]. Moreover, type A standard uncertainty caused by measurement repeatability is another source of uncertainty. In steady condition, the measurement was repeated several times. The standard deviation of the measured film thickness was calculated using the Bessel formula. The maximum standard deviation is denoted as the type A standard uncertainty. Table 4 summarizes the above sources of uncertainty. The combined standard uncertainty $u_c = \sqrt{u_A^2 + u_B^2} = 0.017$ mm.

Table 4. Assessment of uncertainty of measured liquid film thickness.

Source of Uncertainty	Component of Standard Uncertainty			
	Semibreadth of the Range (a)	**Confidence Coefficient (k_c)**	**Type A Standard Uncertainty (u_A)**	**Type B Standard Uncertainty ($u_B = a/k_c$)**
Calibration rod	0.001 mm	$\sqrt{3}$	–	0.0006 mm
Environmental effect	–	–	–	0
Measurement repeatability	–	–	0.017 mm	–
Combined uncertainty (u_c)	0.017 mm			

3. Results and Discussion

3.1. The Pulsation Test of Inertial Flow Pulsator and the Validation of Operating Conditions

A series of tests were conducted on the excitation system of pulsating flow under backpressure conditions to determine the feasibility of the inertial flow pulsator that was manufactured in this study. The LS injector was mounted in the backpressure chamber. The outflow exit of the liquid in the inertial flow pulsator was connected to the injector using a metal hose. The pressure signal in the liquid cavity of the injector is collected as the output signal for the test. As shown in Figure 8, the inertial flow pulsator can generate sinusoidal pressure signals with sufficiently large amplitudes under a backpressure environment, where the amplitude varies by more than 15%. Moreover, the spectrum analysis (Figure 8b) shows that the pulsation frequency of the output pressure signal is consistent with the one excited by the upstream signal generator (100 Hz). For the dynamic characteristics experiment under high backpressure, the inlet pressure in the manifold increases with the increase in the chamber backpressure when the pressure drop in the manifold remains constant. Comparing the pressure pulsation curves obtained under different operating conditions (Figure 8c), it is found that a larger input pressure leads to a larger amplitude of oscillation. Therefore, the inertial flow pulsator is a viable solution to excite flow pulsations for the injectors in dynamic spray experiments under high backpressure.

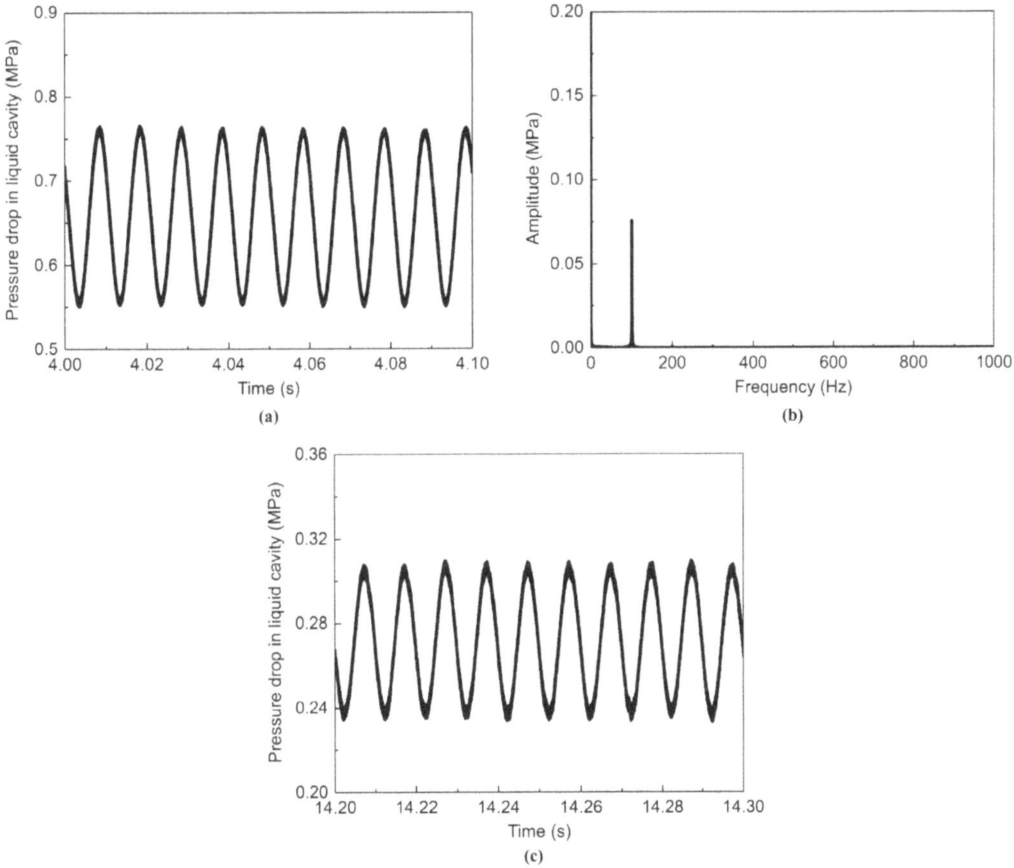

Figure 8. Pressure signal of LS injector when P_b = 1 MPa and f = 100 Hz: (**a**) ΔP = 0.65 MPa; (**b**) FFT; (**c**) ΔP = 0.3 MPa.

In the experiments that simulate the spray process of real rocket engines, it is difficult to produce high-frequency pulsations in supply pressure [30]. This is because the increase in frequency makes it difficult for the flow pulsation generator to generate pressure pulsation of incoming flow at high backpressure. The inertial flow pulsator is based on mechanical vibrations causing the pulsation of the pipeline flow. Considering the influence of mechanical vibration on the stable operation of the whole experimental system, the inlet and outlet of the inertial flow pulsator are connected with the main feedline by means of a metal hose. In this way, the vertical displacement of the oscillation tube can be matched to buffer the influence of the system pipeline oscillation on the pulsation signal of the liquid pressure in the tube. Using this inertial flow pulsator, high-frequency pressure pulsations can be generated in a high backpressure environment. The pressure pulsation at 500 Hz from the pulse exciter, after passing through the feedline, causes a pressure signal in the liquid cavity of LS injector as shown in Figure 9. It can be seen that this inertia flow pulsator can still cause a regular pulsation of the incoming flow at a higher frequency.

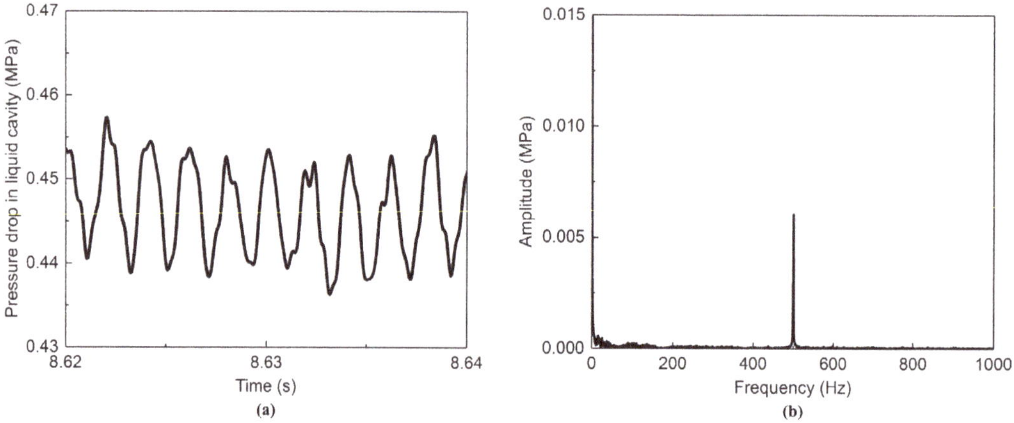

Figure 9. Pressure signal of LS injector when P_b = 1.5 MPa, ΔP = 0.45 MPa, and f = 500 Hz: (**a**) pressure curve; (**b**) FFT.

The liquid injector of the GCSC injector used in the experiments is an open-ended swirl structure. It is more difficult for the open-ended swirl injector to cause the oscillation of the liquid film at the injector outlet because there is no converging section downstream of the swirl chamber [32]. In order to find the proper conditions that are easy to start oscillating, it is necessary to validate the operating conditions for the GCSC injector. In the case of the same power output of the power amplifier and the unchanged working conditions, the experiments were conducted in different supply orders of liquid and gas phases. The test results show that the different supply order of the liquid and gas flow caused the amplitudes of pressure drop in the liquid cavity of the GCSC injector to be different, as shown in Figure 10.

The ambient pressure of the backpressure chamber is atmospheric pressure. The output pulsation frequency of the signal generator (i.e., the pressure pulsation frequency of the inertial flow pulsator) is 200 Hz. The pressure drop ΔP in the liquid feedline is 0.25 MPa (corresponding to the liquid flow rate of 23 g/s). The gas flow rate is regulated to 15 L/min by the gas flow controller. As seen in Figure 10, the amplitude of the dimensionless pulsation of the pressure drop in the liquid cavity caused by supplying liquid flow first is larger than that caused by supplying gas flow first. Moreover, the variation of pressure pulsation of the former is more regular. The possible reason for this is that supplying the gas phase first will cause the nitrogen gas flow to occupy the exit of the liquid channel at the recess section of the injector. Then, opening the liquid feedline, the gas will have a strong shear effect on the liquid film, causing the liquid film on the nozzle wall to pulsate non-harmonically. The pulsation of the liquid film thickness is transmitted to the liquid cavity, resulting in an unsmooth pressure curve.

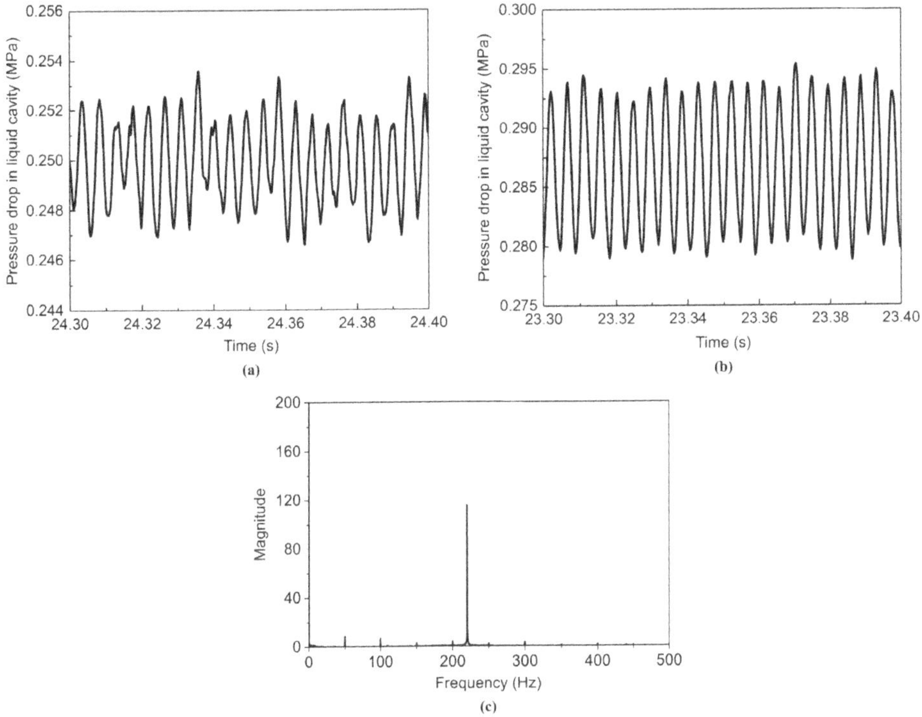

Figure 10. Pressure pulsation curve of liquid cavity of GCSC injector and the FFT graph: (**a**) supply gas first; (**b**) supply liquid first; (**c**) FFT.

3.2. *Steady State*

The steady-state experiments were first conducted using the No. 2 GCSC injector. In this work, the steady-state characteristics of the injector atomization were obtained firstly. Then, the inertial flow pulsator was turned on for the experimental study of dynamic characteristics. In these experiments, the pressure drop ΔP in the liquid feedline including the injector was 0.25 MPa (corresponding to the liquid flow rate of 23 g/s). The gas flow rate Q_g was controlled to 15 L/min by the gas flow controller. The experiments were conducted under the operating conditions of 0.3 MPa, 0.6 MPa, and 0.9 MPa for the chamber backpressures, respectively.

To capture the details of the spray field, the spray field was photographed using a high-speed camera with a sampling frequency of 20,000 frames per second. The resulting instantaneous spray patterns under different chamber backpressures P_b are shown in Figure 11. Several possible criteria for the breakup length are defined in the literature. In this study, the spray breakup length was defined as the vertical distance from the injector exit to the first point where the liquid sheet started to break up into liquid filaments. The breakup length was obtained using 20 images from the backlighting photography. Additionally, the error bars were obtained from the measurement uncertainty of 20 samples, which were analyzed by solving the sample standard deviation. Figure 12 shows the effect of chamber backpressure on the spray breakup length L.

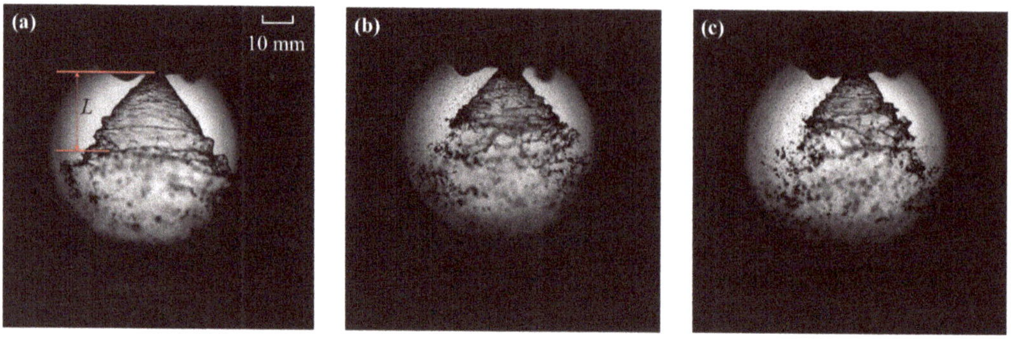

Figure 11. Spray patterns of the GCSC injectors under different chamber backpressures: (**a**) $P_b = 0.3$ MPa; (**b**) $P_b = 0.6$ MPa; (**c**) $P_b = 0.9$ MPa.

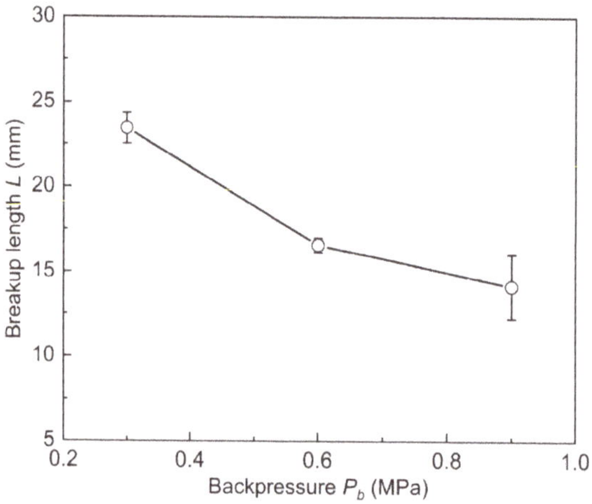

Figure 12. Effect of the chamber backpressure P_b on the breakup length L.

The steady-state spray images under different chamber backpressure conditions (Figure 11) show that the surface of the liquid sheet becomes folded from smooth with increasing ambient pressure. The development of dense instable waves prompts the liquid sheet to rupture earlier. This is demonstrated by the fact that the spray breakup length L becomes shorter with increasing the chamber backpressure P_b (Figure 12). The increase in backpressure P_b essentially leads to an increase in the density of the ambient gas, which strengthens the aerodynamic force of the gas on the annular liquid sheet downstream of the injector outlet [22,33]. Therefore, the larger chamber backpressure drives the liquid sheet surface to be more unstable and wrinkled, making the breakup length L of the liquid sheet in the spray field shorter.

Furthermore, the mechanism by which the backpressure P_b shortens the breakup length L by increasing the ambient gas density can be analyzed from the linear instability theory. For an annular swirling liquid sheet, Kim [33] considered the attenuation of the sheet and corrected the normalized breakup length relation according to the injection conditions and the ambient gas based on the work of Dombrowski [34] and Hagerty [35], as shown in the following:

$$\frac{r_0}{h_0 \tan\theta}\left[\left(\frac{L\tan\theta}{r_0}+1\right)^{3/2}-1\right]=C\left(\frac{\rho_g}{\rho_l}\right)^{-1}We_l^{-1/2} \tag{4}$$

where r_0 is the orifice radius; h_0 is the initial liquid sheet thickness at the injector exit; θ is the half of the spray angle; ρ_g and ρ_l are the gas and the liquid density, respectively; C is a constant; We_l is the liquid Weber number, which is given by

$$We_l = \frac{\rho_l u^2 h_s}{\sigma} \tag{5}$$

where u is the axial velocity of the liquid sheet; h_s is the liquid sheet thickness at the injector exit; σ is the surface tension. In these experiments, the operating conditions (such as pressure drop and the gas flow rate) were fixed except for the chamber backpressure so that the We_l was constant. However, the increased ambient pressure will result in an increment in the ambient gas density ρ_g. As stated by Kim [33], the spray angle θ before breakup is independent of the ambient gas density ρ_g and is only a function of We_l. Therefore, in Equation (4), it is concluded that when the gas density ρ_g increases, the breakup length L will decrease. The above result of theoretical analysis is consistent with the trend of the effect of the backpressure P_b on the breakup length L observed in Figure 12.

Comparison with the previous study [22,33] revealed that the effect of backpressure on the spray cone angle was less pronounced in these experiments in the backpressure chamber. In contrast, the previous study showed that the increase in the backpressure leads to a decrease in the spray cone angle. The possible reasons for this phenomenon in the present experiments are explained as follows. The spraying environment of the backpressure chamber is a confined space. The slight increase in ambient pressure in the chamber during the spraying process hinders the continuous supply of the gas flow in the upstream feedline. Although the increase in the backpressure will make the spray cone angle smaller, the reduction of gas flow in the GCSC injector makes the spray angle larger. Eventually, the chamber backpressure has less effect on the spray cone angle.

3.3. *The Effect of the Chamber Backpressure*

After acquiring the steady-state spray data of the GCSC injector, the dynamic characteristics of the spraying are investigated by turning on the inertial flow pulsator. The raw data (i.e., voltage) of the liquid film pulsation are measured using the liquid film thickness sensor. The liquid film thickness at the GCSC injector outlet is calibrated using a method similar to that in Section 2.3.2. As shown in Figure 13, the raw voltage signal of liquid film is converted into the liquid film thickness value. Accordingly, it is possible to convert a series of raw voltage signals measured by a liquid film thickness sensor into a dynamic value of liquid film thickness over time, as shown in Figure 14. Based on the processing results of the Fourier transform (Figure 14c), it is found that the main frequency of the pulsating signal of liquid film thickness was 220 Hz, which is consistent with the pulsation frequency outputted by the signal generator.

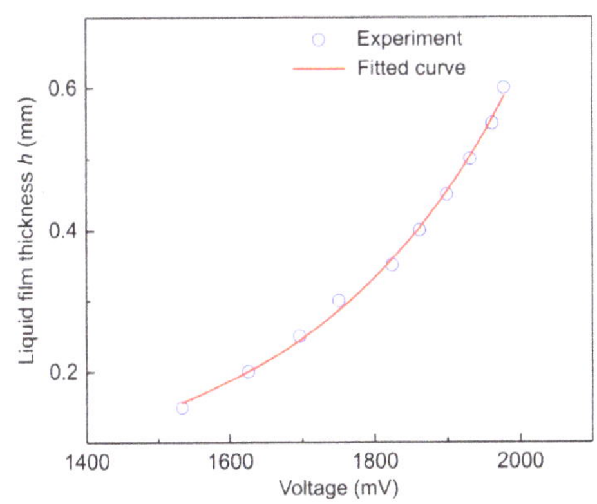

Figure 13. Calibration curve of the GCSC injector: voltage vs. film thickness.

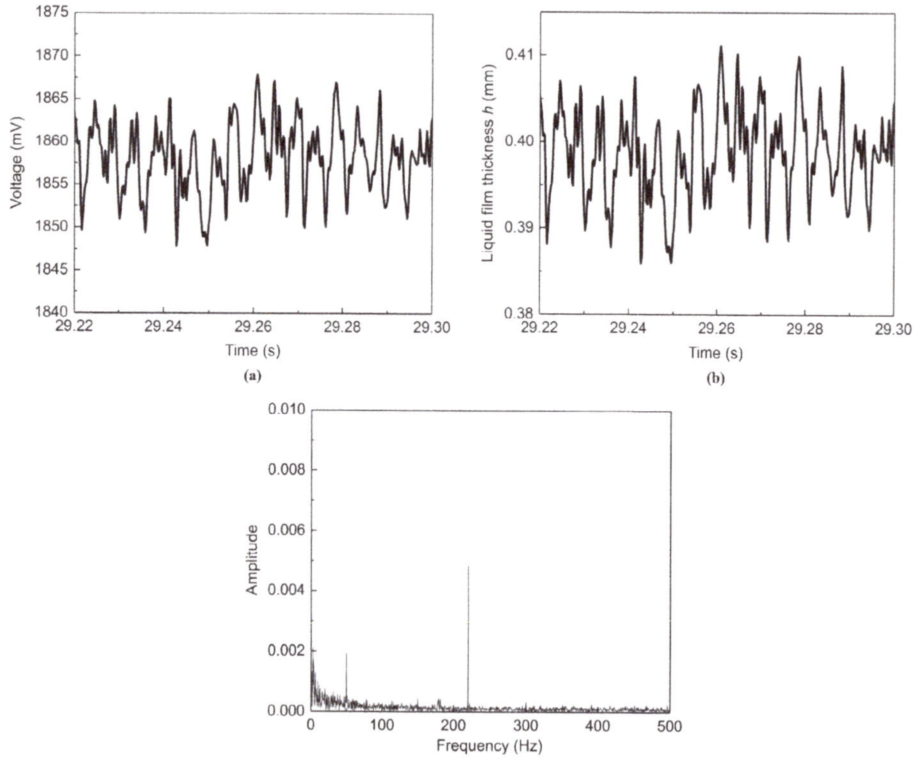

Figure 14. Pulsation curve of liquid film thickness and the FFT graph when f = 220 Hz: (**a**) raw data of liquid film thickness; (**b**) liquid film thickness; (**c**) FFT.

The pressure pulsation signal in the liquid cavity of the injector is measured using a dynamic pressure transducer in the liquid injector. According to the calibration curve of the liquid film thickness of the GCSC injector, the pulsation curve of the liquid film thickness is then plotted against the pressure signal in Figure 15 for comparison. It can be seen that the pressure in the liquid cavity varies steadily and regularly in the form of a sinusoidal signal. This pressure perturbation in the swirl chamber causes the instability of the liquid film, which in turn causes the liquid film thickness to pulsate. The liquid film thickness evolves temporally with irregular periodic variations accompanied by the pressure pulsations.

The dynamic theory of injector addresses the accurate prediction of the dynamics characteristics of a swirl injector when the inflow pressure fluctuates, which was developed by Bazarov [36]. By using the inviscid fluid mechanics equation that controls the internal flow of the injector, the frequency characteristic equation of an injector is derived by an analytical method. The dynamic behavior of an injector is characterized by the response function given by

$$\Pi_{\phi} = \left(\frac{Q'}{\overline{Q}}\right) / \left(\frac{\Delta P'}{\overline{\Delta P}}\right) \tag{6}$$

According to this theory, the pulsations of flow rate or pressure drop bring about the development of surface waves on the gas–liquid interface as well as vortex waves of different circumferential velocities within the swirling liquid itself.

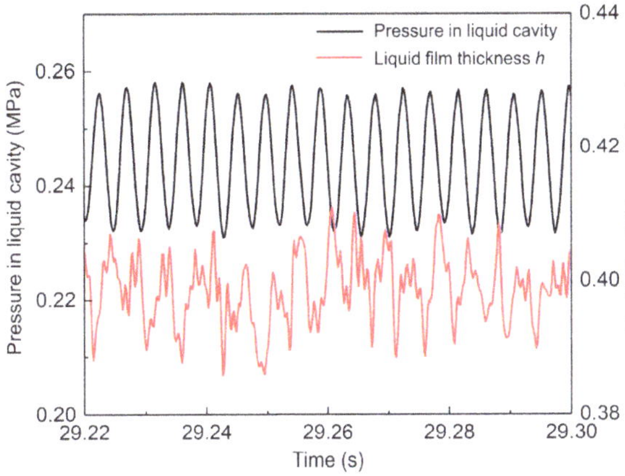

Figure 15. Pulsation curves of the pressure in liquid cavity and the liquid film thickness when f = 220 Hz.

It can be seen from Equation (6) that a vital objective of the experiments on the dynamic characteristics of an injector is to obtain the pulsating flow of the injector. Axial velocity fluctuations and liquid film thickness fluctuations are in the same phase. Therefore, according to the dynamic theory of injector, the instantaneous volume flow rate of the swirl injector can be obtained by multiplying the axial velocity of the flow and the flow area in the swirl chamber, as presented in the following equation:

$$\begin{aligned} Q &= \overline{Q} + Q' \\ &= \pi W_a \left[\left({R_N}^2 - {r_M}^2 \right) + \left(\sqrt{\frac{2{R_N}^4}{{R_N}^2 + {r_M}^2}} + 2r_M \right) \xi \right] \end{aligned} \tag{7}$$

where r_M is the radius of gas vortex, which can be found by the difference between the nozzle radius and the liquid film thickness; R_N is the nozzle radius of the swirl injector; ξ is the amplitude of liquid film thickness, which is obtained by subtracting its average thickness from the liquid film thickness curve in Figure 15; W_a is the steady-state axial velocity of the flow, which is calculated from the steady-state flow rate $\overline{Q}$ by the following equation:

$$W_a = \frac{\overline{Q}}{\pi\left({R_N}^2 - {r_m}^2\right)} \tag{8}$$

Using Equations (7) and (8), the liquid film thickness is converted into the flow rate at the injector outlet, and the pulsation of the flow rate at the injector outlet can be further obtained. According to the definition of the injector dynamic characteristics (Equation (5)), the influence of various backpressure conditions on the dynamic characteristics of GCSC injectors with different recess structures can be summarized as shown in Figure 16. The variation range of pulsation frequency is 180–270 Hz. It can be seen from the amplitude–frequency characteristic curve of each GCSC injector that when the chamber backpressure P_b is constant, the amplitude characteristic of the injector will show a downward trend as the pulsation frequency increases. This indicates that the increase in the pulsation frequency of the pressure in the swirl chamber of the injector can suppress the pulsation amplitude of the liquid film to a certain extent.

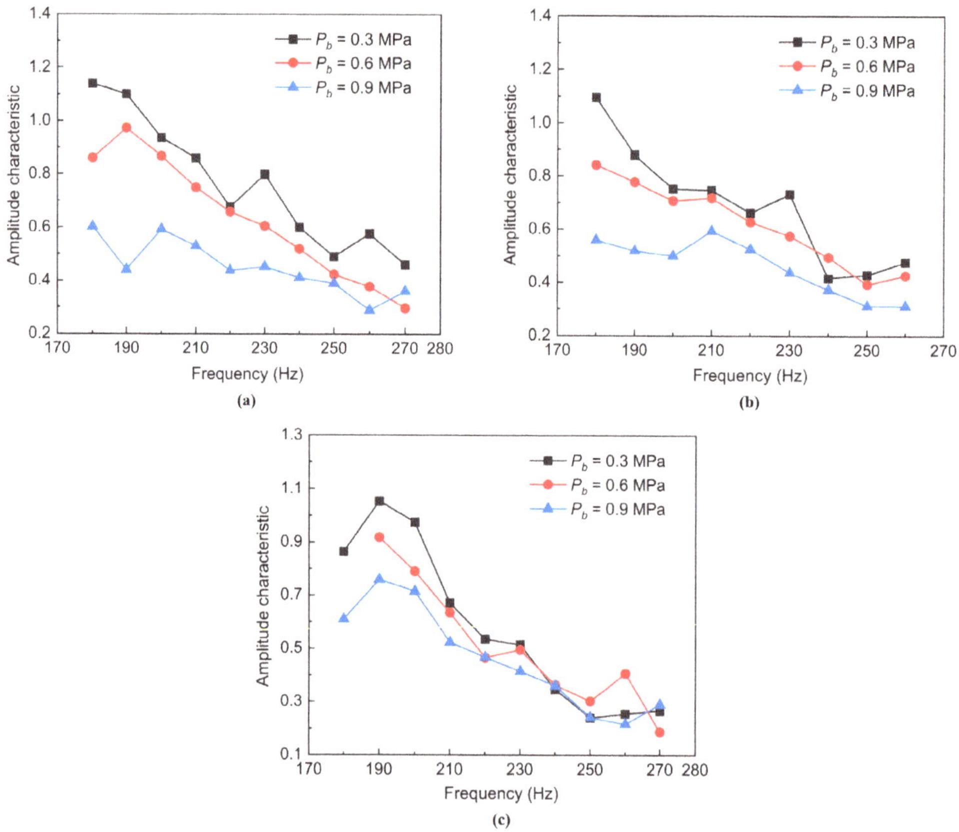

Figure 16. Effect of the chamber backpressure P_b on the dynamic characteristics of different GCSC injectors: (**a**) No.1 injector (ΔL = 13 mm); (**b**) No.2 injector (ΔL = 8 mm); (**c**) No. 3 injector (ΔL = 3 mm).

It can be further seen from Figure 16 that for each GCSC injector, the dimensionless amplitude of the flow rate at the injector outlet tends to decrease as the chamber backpressure P_b increases. That is, the increase in chamber backpressure P_b has a suppressive effect on the pulsation of the outlet flow of the injector; in particular, when the recess length ΔL of the injector is larger, this phenomenon is more obvious. This is because the density of the ambient gas in the chamber increases when the chamber backpressure P_b increases. Then, the liquid sheet ejected from the injector will consume more momentum when it moves the surrounding quiescent gas. Therefore, the momentum transferred from the moving liquid sheet to the quiescent ambient gas is greater, resulting in a reduction in the pulsation amplitude of the flow rate at the injector outlet and a decrease in the transfer function. This is similar to the influence of the chamber backpressure on the dynamic characteristics of an LS injector of the single-component fluid [33].

3.4. *The Effect of the Recess Length*

In order to study the effect of the recess length of the GCSC injector on the dynamic characteristics under chamber backpressure, the recess lengths ΔL of the No.1, No.2, and No.3 injectors used in the experiments were designed to be 13 mm, 8 mm, and 3 mm, respectively. Nitrogen gas was charged into the backpressure chamber to make the ambient backpressure P_b remain 0.6 MPa. The pressure drop ΔP in the liquid feedline was 0.25 MPa.

The gas flow rate Q_g in the gas feedline was 15 L/min. The pulsation frequency range of the signal generator was 180–270 Hz.

The instantaneous spray patterns of the GCSC injectors with different recess lengths are shown in Figure 17. The trend of the effect of the recess length ΔL on the breakup length L of the liquid sheet is obtained by processing the spray images, as shown in Figure 18. The injector with a shorter recess length ΔL has a liquid sheet that ruptures earlier. That is, a decrease in the recess length ΔL leads to a decrease in the spray breakup length L. The possible reason for this is that the reduction in recess length ΔL reduces the distance between the gas injector outlet and the liquid injector outlet. The wall of the shorter recess section has little inhibiting effect on the perturbation of the annular liquid film, compared with the longer recess section. Additionally, the gas flow at the center significantly shears the liquid sheet. As a result, the break-up process of the centrifugal liquid sheet at the injector exit is more intense. Ultimately, it is manifested as a reduction in the spray breakup length L.

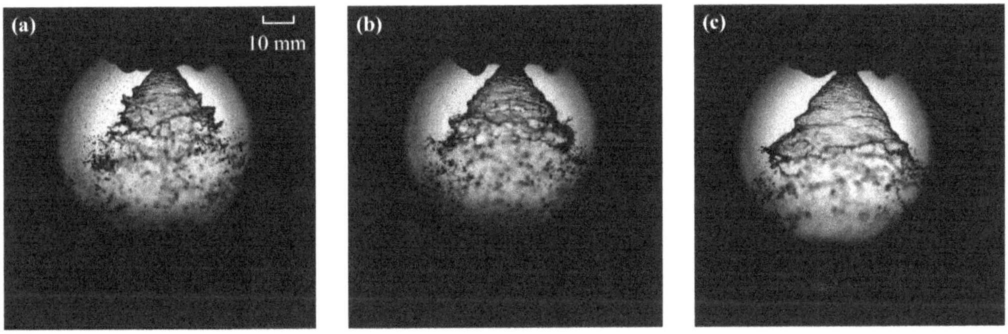

Figure 17. Spray patterns of the GCSC injectors with different recess lengths: (**a**) ΔL = 3 mm; (**b**) ΔL = 8 mm; (**c**) ΔL = 13 mm.

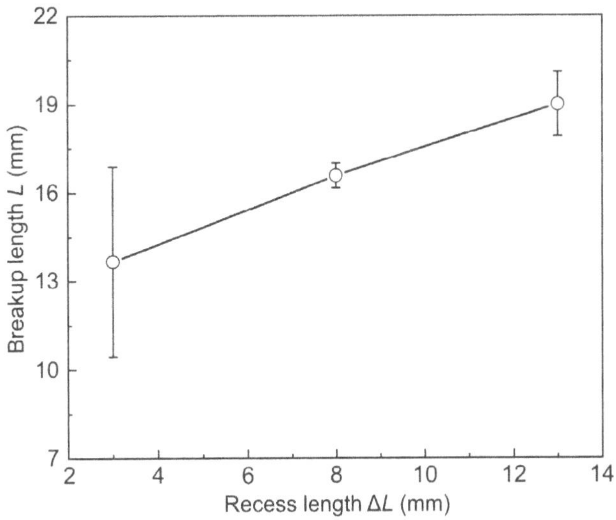

Figure 18. Effect of the recess length ΔL on the breakup length L.

In addition, the influence mechanism of the recess length on the breakup length of the liquid sheet of injection in backpressure can be analyzed by the linear instability theory. For a swirl coaxial injector, the quantitative relationship between the recess length and the disturbance amplitude of the liquid sheet can be expressed as follows [23]:

$$\eta = \eta_0 \exp\left(-k \cdot \Delta L\right) \tag{9}$$

where η is the disturbance amplitude of the liquid sheet at the injector exit; η_0 is the initial disturbance amplitude inside the recess section; k is the disturbance wave number; ΔL is the recess length of the swirl coaxial injector. In these experiments about the effect of the recess length, the operating conditions, such as the pressure drop, the gas flow rate, and the chamber backpressure, were fixed. Additionally, the initial disturbance amplitude η_0 of the liquid sheet inside the recess section was constant. However, the disturbance amplitude η of the liquid sheet at the injector exit would be different with the changed recess length ΔL. From Equation (9), the disturbance amplitude η of the liquid sheet decreases as the recess length ΔL increases. This decreased disturbance amplitude η increases the breakup length L of the liquid sheet according to the theory of Clark and Dombrowski [37], as shown in the following equation:

$$L^{3/2} = \left(\frac{9\rho_l K U^2}{32\left(\rho_g U^2 k - \sigma k^2\right)}\right)^{1/2} \cosh^{-1}\left[8(\eta k)^{-2} + 1\right] \tag{10}$$

where K is the sheet thickness parameter; U is the axial sheet velocity. From the above, it is theoretically explained that an injector with a larger recess length produces a longer breakup length (Figure 18), forming a more stable conical liquid sheet.

Using the similar approach in Section 3.3, the effect of the recess length ΔL of the GCSC injectors on the dynamic characteristics of spraying can be obtained under different chamber backpressure conditions. Figure 19 shows the effect of various recess lengths on the dynamic characteristics of the GCSC injectors in the pulsation frequency ranges of 180–270 Hz. It can be seen from the figures that for lower backpressure conditions, as the recess length ΔL increases, the amplitude–frequency characteristics of the injector generally increase. This is because when the recess section of the GCSC injector becomes longer, the gas from the outlet of the gas injector will enhance the impact effect in the recess section, to tear the liquid film inside the injector tempestuously. This leads to an increase in the oscillation of the liquid flow at the injector outlet, which in turn increases the amplitude–frequency characteristics of the spraying for the injector.

However, it is not difficult to see that the above influence law is not obvious in the lower range of the pulsation frequency of the incoming flow (about 220 Hz or less). In particular, when the chamber backpressure P_b is 0.9 MPa, the influence of the recess length ΔL of the injector on the amplitude–frequency characteristics of the outlet flow rate is not obvious, as shown in Figure 19c. The possible reason is that as the pulsation frequency of the pressure decreases, the oscillation amplitude of the liquid film inside the injector becomes relatively large. Additionally, the GCSC injector is a gas–liquid coaxial structure with a gas center. When the outlet width of the injector is constant, the scouring effect of the gas flow on the liquid film will be strengthened, resulting in irregular disturbance for the larger pulsation of the liquid film. Therefore, the influence of the recess length ΔL on the dynamic characteristics of the GCSC injector was driven not to be obvious at low pulsation frequencies.

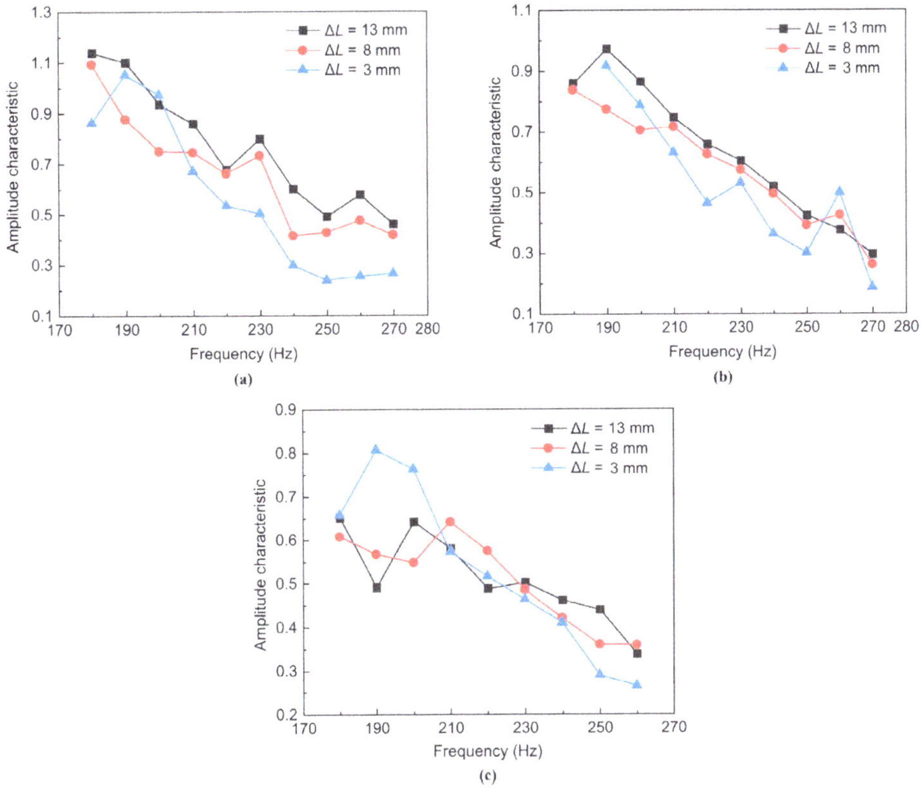

Figure 19. Effect of the recess length ΔL on the dynamic characteristics of GCSC injector under different chamber backpressures: (**a**) P_b = 0.3 MPa; (**b**) P_b = 0.6 MPa; (**c**) P_b = 0.9 MPa.

4. Conclusions

Experimental observations of the dynamic characteristics generated by the GCSC injectors under various ambient pressures have been presented and analyzed. In order to simulate the high backpressure ambient in the combustion chamber of a rocket engine, an experimental system of the dynamic injection in a backpressure chamber is constructed. An inertial flow pulsator is designed and manufactured to generate the dynamic pulsation of the flows that feed to the injector. The electric conductance method is adopted to measure steady and instantaneous liquid film thickness. The steady state of the spraying under backpressure is shown in detail. Emphatically, the effects of the chamber backpressure and the recess length of the injectors on the dynamics characteristics of the spray are investigated. The conclusions are drawn as follows:

(1) The issue of liquid leakage is solved by the inertial flow pulsator, which can generate sinusoidal pressure signals with sufficient amplitudes that vary by more than 15% under backpressure. A larger input pressure leads to a larger amplitude of pulsation. This pulsator meets the pressure pulsation with a frequency of 500 Hz. A sodium chloride solution is used to calibrate the liquid film thickness to improve the accuracy of the electric conductance method. The measurement error of this method is 0.017 mm, satisfying the experimental accuracy of liquid film thickness.

(2) The increase in chamber backpressure leads to an increase in the gas density, which strengthens the aerodynamic force of the gas on the liquid sheet. Then, the surface of the liquid sheet becomes more folded, prompting the liquid sheet to rupture earlier with a shorter breakup length. An increase in backpressure increases the momentum transfer of

the gas–liquid flow, resulting in a reduction in the pulsation amplitude of the liquid film. Therefore, chamber backpressure suppresses the pulsation of the flow at the injector outlet, especially for a longer recess length.

(3) A decrease in the recess length leads to a reduction in the breakup length because of the enhanced shearing effect in a narrower recess section. As the recess length increases, the amplitude characteristics of outlet flow generally increase, especially for a lower backpressure. Moreover, when the recess length and the chamber backpressure are constant, the increase in the pulsation frequency of the pressure can suppress the pulsation amplitude of the outlet flow under backpressure.

Author Contributions: Conceptualization, X.Z. and W.Q.; methodology, Q.G.; investigation, Q.G. and D.Z.; writing—original draft preparation, W.Q.; writing—review and editing, W.Q. and Q.F.; supervision, L.Y.; funding acquisition, L.Y. and Q.F. All authors have read and agreed to the published version of the manuscript.

Funding: This work was supported by the National Natural Science Foundation of China (Grant Nos. 12272026, 11922201 and U1837211).

Data Availability Statement: Not applicable.

Acknowledgments: The authors thank the reviewers for their great help during the review progress.

Conflicts of Interest: The authors declare no conflict of interest.

References

1. Kang, Z.; Wang, Z.; Li, Q.; Cheng, P. Review on pressure swirl injector in liquid rocket engine. *Acta Astronaut.* **2018**, *145*, 174–198.
2. Im, J.H.; Cho, S.; Yoon, Y.; Moon, I. Comparative study of spray characteristics of gas-centered and liquid-centered swirl coaxial injectors. *J. Propul. Power* **2010**, *26*, 1196–1204. [CrossRef]
3. Inoue, C.; Watanabe, T.; Himeno, T.; Uzawa, S. Liquid Jet Dynamics and Primary Breakup Characteristics at Near-Field of Coaxial Type Injector. In Proceedings of the 46th AIAA/ASME/SAE/ASEE Joint Propulsion Conference & Exhibit, Nashville, TN, USA, 25–28 July 2010.
4. Cohn, R.; Strakey, P.; Bates, R.; Talley, D.; Muss, J.; Johnson, C. Swirl Coaxial Injector Development. In Proceedings of the 41st Aerospace Sciences Meeting and Exhibit, Reno, NV, USA, 6–9 January 2003.
5. Bazarov, V. Influence of Propellant Injector Stationary and Dynamic Parameters on High Frequency Combustion Stability. In Proceedings of the 32nd Joint Propulsion Conference and Exhibit, Lake Buena Vista, FL, USA, 1–3 July 1996.
6. Harrje, D.T.; Reardon, F.H. *Liquid Propellant Rocket Combustion Instability*; Scientific and Technical Information Office, National Aeronautics and Space Administration, U.S. Government Printing Office: Washington, DC, USA, 1972; Volume 1.
7. Ismailov, M.; Heister, S.D. Dynamic response of rocket swirl injectors, part II: Nonlinear dynamic response. *J. Propul. Power* **2011**, *27*, 412–421. [CrossRef]
8. Li, P.; Yang, L.; Fu, Q.; Fang, Z. Spray characteristics of the nanoparticle-containing gel propellants by using an improved single-phase nozzle. *Fuel* **2022**, *315*, 122968. [CrossRef]
9. Ismailov, M.M. Modeling of Classical Swirl Injector Dynamics. Ph.D. Thesis, Purdue University, West Lafayette, IN, USA, 2010.
10. Liu, L.; Zhang, S.; Fu, Q. Spray characteristics of gel fuel in a gas-centered coaxial swirl injector with variable recess ratios. *Fuel* **2022**, *319*, 123766. [CrossRef]
11. Bazarov, V.G.; Yang, V. Liquid-propellant rocket engine injector dynamics. *J. Propul. Power* **1998**, *14*, 797–806. [CrossRef]
12. Fu, Q.F.; Yang, L.J.; Wang, X.D. Theoretical and experimental study of the dynamics of a liquid swirl injector. *J. Propul. Power* **2010**, *26*, 94–101. [CrossRef]
13. Fu, Q.F.; Yang, L.; Qu, Y.Y.; Gu, B. Geometrical effects on the fluid dynamics of an open-end swirl injector. *J. Propul. Power* **2011**, *275*, 929–936. [CrossRef]
14. Bazarov, V. Non-Linear Interactions in Liquid-Propellant Rocket Engine Injectors. In Proceedings of the 34th AIAA/ASME/SAE/ASEE Joint Propulsion Conference and Exhibit, Cleveland, OH, USA, 13–15 July 1998.
15. Bazarov, V. Self-Pulsations in Coaxial Injectors with Central Swirl Liquid Stage. In Proceedings of the 31st Joint Propulsion Conference and Exhibit, San Diego, CA, USA, 10–12 July 1995.
16. Yang, V.; Puri, P. Design and Dynamics of Jet and Swirl Injectors. In *Liquid Rocket Thrust Chambers*; Astronautics and Aeronautics Inc.: Reston, VA, USA, 2004; pp. 19–103.
17. Park, G.; Oh, S.; Yoon, Y.; Choi, J.Y. Characteristics of gas-centered swirl-coaxial Injector with liquid flow excitation. *J. Propul. Power* **2019**, *35*, 624–631. [CrossRef]
18. Park, G.; Lee, J.; Oh, S.; Yoon, Y.; Sohn, C.H. Characteristics of gas-centered swirl coaxial injector with acoustic excitation of gas flow. *AIAA J.* **2017**, *55*, 894–901. [CrossRef]

19. Sahoo, S.K.; Gadgil, H. Dynamics of self-pulsation in gas-centered swirl coaxial injector: An experimental study. *J. Propul. Power* **2021**, *373*, 450–462.
20. Oh, S.; Park, G.; Kim, S.; Lee, H.; Yoon, Y.; Choi, J.Y. A Study on dynamic characteristics of gas centered swirl coaxial injector varying tangential inlet diameter with liquid pulsation. *J. ILASS-Korea* **2017**, *22*, 62–68.
21. Bai, X.; Cheng, P.; Sheng, L.; Li, Q.; Zhang, X.; Kang, Z. Effects of backpressure on self-pulsation characteristics of liquid-centered swirl coaxial injectors. *Int. J. Multiphas. Flow* **2019**, *116*, 239–249.
22. He, X.; Chen, C.; Yang, Y.; Yan, Z. Experimental study on the flow field distribution characteristics of an open-end swirl injector under ambient pressure. *Aerosp. Sci. Technol.* **2020**, *98*, 105691.
23. Kim, D.; Jeong, W.; Im, J.; Yoon, Y. The Characteristics of Swirl Coaxial Injector under Varying Geometric and Environmental Conditions. In Proceedings of the 40th AIAA/ASME/SAE/ASEE Joint Propulsion Conference and Exhibit, Fort Lauderdale, FL, USA, 11–14 July 2004.
24. Zhang, Z.; Liu, Y.; Hu, H. Effects of chamber pressure on the kinematic characteristics of spray flows exhausted from an airblast atomizer. *Exp. Therm. Fluid Sci.* **2022**, *130*, 110514.
25. Zhang, L.; Wang, X.; Li, Y.; Yeh, S.T.; Yang, V. Supercritical fluid flow dynamics and mixing in gas-centered liquid-swirl coaxial injectors. *Phys. Fluids* **2018**, *30*, 075106. [CrossRef]
26. Mayer, W.; Tamura, H. Propellant injection in a liquid oxygen/gaseous hydrogen rocket engine. *J. Propul. Power* **1996**, *12*, 1137–1147. [CrossRef]
27. Kenny, R.J.; Hulka, J.R.; Moser, M.D.; Rhys, N.O. Effect of chamber backpressure on swirl injector fluid mechanics. *J. Propul. Power* **2009**, *25*, 902–913.
28. Cho, S.; Park, G.; Chung, Y.; Yoon, Y.; Bazarov, V.G. Surface instability on cryogenic swirl flow at sub-to supercritical conditions. *J. Propul. Power* **2014**, *30*, 1038–1046. [CrossRef]
29. Chen, C.; He, X.; Liu, C.; Yang, Y.; Tang, Z. Experimental study on the flow field distribution characteristics of a gas–liquid swirl coaxial injector under ambient pressure. *Aerosp. Sci. Technol.* **2021**, *114*, 106757.
30. Fu, Q.F.; Yang, L.J.; Qu, Y.Y. Measurement of annular liquid film thickness in an open-end swirl injector. *Aerosp. Sci. Technol.* **2011**, *15*, 117–124. [CrossRef]
31. Bazarov, V.; Lee, E.; Lineberry, D.; Swanner, B.; Frederick, R. Pulsator Designs for Liquid Rocket Injector Research. In Proceedings of the 43rd AIAA/ASME/SAE/ASEE Joint Propulsion Conference & Exhibit, Cincinnati, OH, USA, 8–11 July 2007.
32. Chung, Y.; Kim, H.; Jeong, S.; Yoon, Y. Dynamic characteristics of open-type swirl injector with varying geometry. *J. Propul. Power* **2016**, *32*, 583–591. [CrossRef]
33. Kim, D.; Im, J.H.; Koh, H.; Yoon, Y. Effect of ambient gas density on spray characteristics of swirling liquid sheets. *J. Propul. Power* **2007**, *23*, 603–611. [CrossRef]
34. Dombrowski, N.; Hooper, P.C. The effect of ambient density on drop formation in sprays. *Chem. Eng. Sci.* **1962**, *17*, 291–305. [CrossRef]
35. Hagerty, W.W.; Shea, J.F. A study of the stability of plane fluid sheets. *J. Appl. Fluid Mech.* **1955**, *22*, 509–514. [CrossRef]
36. Bazarov, V. *Dynamics of Liquid Injectors*; Mashinostroenie: Moscow, Russia, 1979; p. 35.
37. Clark, C.J.; Dombrowski, N. Aerodynamic instability and disintegration of inviscid liquid sheets. *Proc. R. Soc. Lond. A. Math. Phys. Sci.* **1972**, *329*, 467–478.

aerospace

MDPI

Review

Progress of Porous/Lattice Structures Applied in Thermal Management Technology of Aerospace Applications

Jian Liu [1], Mengyao Xu [1], Rongdi Zhang [1], Xirui Zhang [1] and Wenxiong Xi [1,2,*]

1 Research Institute of Aerospace Technology, Central South University, Changsha 410012, China
2 Hunan Zhongsheng Hi-Tech Research Institute, Changsha 412007, China
* Correspondence: 13739076081@163.com

Abstract: With lightweight, multifunctional, and designable characteristics, porous/lattice structures have started to be used in aerospace applications. Porous/lattice structures applied in the thermal management technology of aerospace vehicles have attracted much attention. In the past few years, many related numerical and experimental investigations on flow, heat transfer, modelling methodology, and manufacturing technology of porous/lattice structures applied in thermal management systems have been widely conducted. This paper lists the investigations and applications of porous/lattice structures applied in thermal management technology from two aspects, i.e., heat transfer enhancement by porous/lattice structures and transpiration cooling. In addition, future developments and challenges based on the previous investigations are analyzed and summarized. With the higher requirements of thermal protection for aerospace applications in the future, thermal management technology based on porous/lattice structures shows good prospects.

Keywords: porous/lattice structures; thermal management technology; heat transfer enhancement; transpiration cooling

check for updates

Citation: Liu, J.; Xu, M.; Zhang, R.; Zhang, X.; Xi, W. Progress of Porous/Lattice Structures Applied in Thermal Management Technology of Aerospace Applications. *Aerospace* **2022**, *9*, 827. https://doi.org/10.3390/aerospace9120827

Academic Editor: Peng Hao

Received: 1 November 2022
Accepted: 13 December 2022
Published: 15 December 2022

1. Introduction

Porous/lattice structures are one kind of material system, which is composed of a regular arrangement and adjustable pore structures. Due to their excellent mechanical, adsorption, penetration and other characteristics, porous/lattice structures show good prospects in engineering applications [1]. In recent years, many fundamental and engineering applications have both been performed on porous/lattice structures and have become the hotpot in heat and mass transfer disciplines. With lightweight, large contacting area, and excellent transporting characteristics, porous/lattice structures show great benefits in thermal management technology for aerospace applications [2,3].

Propulsion technology with high flight speed and high efficiency is required for future aircraft, and advanced thermal management technology is also urgently needed [4]. Thermal protection has become the most crucial aspect in the design of a high-speed aircraft. For example, hypersonic vehicles have attracted much attention in recent decades in military applications. However, the largest challenge for hypersonic vehicles is the thermal protection and related problems of pneumatics, structure, and propulsion caused by overheating [5–8]. As shown in Figure 1, the largest heat flux regions of a hypersonic vehicle are found in the leading edge, nose zone, and engine interior, where heat flux is usually more than 10^7 W/m^2.

Traditional passive cooling methods are difficult to satisfy the requirements of thermal protection for a hypersonic vehicle because the melting points for structure materials are limited [9]. In addition, the problem caused by extremely high thermal load, such as deformations by thermal stress, also limits the use of traditional passive cooling methods [10]. Therefore, the advanced active cooling method is very significant for a hypersonic vehicle, especially in leading-edge regions and the combustor.

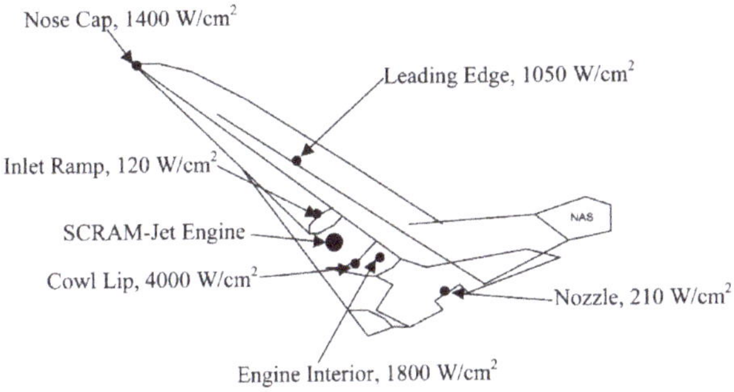

Figure 1. Thermal environment for a hypersonic vehicle. Reprinted/adapted with permission from Ref. [8]. 2005, American Institute of Aeronautics and Astronautics Inc.

With the advantage of lightweight and heat transport characteristics, porous/lattice structures have been introduced to thermal management technology in aerospace applications. According to previous literature, the applications of porous/lattice structures in thermal management technology are mainly grouped into two aspects, i.e., heat transfer enhancement by porous/lattice structures and transpiration cooling.

1.1. Heat Transfer Enhancement by Porous/Lattice Structures

As a new type of lightweight structure, porous/lattice structures can provide large fluid–solid contacting areas and strong flow mixing [11]. The most important applications in thermal management technology are heat transfer enhancement in a cooling channel mounted with porous/lattice structures, such as regenerative cooling in a rocket/scramjet [12,13], turbine blade cooling [14–16], and heat exchangers [17–20].

Many researchers have carried out fundamental investigations in simplified channels and found that, compared with traditional cooling channels, porous/lattice structure-mounted channels offer larger heat transfer enhancement because of the transverse and longitudinal vortices generated [21], which strengthen the mixing of hot and cold flows. Porous/lattice structures with different geometric elements have various heat transfer enhancement and flow characteristics. The geometric structural elements of porous/lattice structures constitute a research interest for researchers in fundamental studies [22]. In addition, the optimization of porous/lattice structure arrangements has also drawn some attention from researchers [23,24].

As an effective active cooling method, cooling channels with porous/lattice structures have started to be used in aerospace devices [25]. For high-speed aircraft in the future, the thermal management of high-temperature components will become a big issue. For the trail edge cooling of a turbine blade, Li et al. [26] numerically investigated the heat transfer of four wedge-shaped channels mounted with different combinations of pin fins, ribs, and Kagome lattices under a strong rotation state. From the results, the cooling efficiency in the trailing edge is greatly improved by Kagome lattice structures. In addition to applications in the active cooling of high-temperature components in engines, the porous/lattice structures are also used in thermal management systems in aerospace devices, such as heat exchangers and heat sinks [18,27].

1.2. Transpiration Cooling

Transpiration cooling is also an important application for porous/lattice structure applications in thermal management technology, which are mainly used in the cooling of aerospace devices, such as turbine blades [28–30], combustion chambers [31], and the leading edge [32] of a hypersonic vehicle. Transpiration cooling is a technique where coolant fluid passes through a number of mini-holes from a porous surface to protect the structure from excessive heat damage from external hot gas [33].

In recent years, many researchers [34–36] have compared the performance of film cooling and transpiration cooling in terms of overall cooling efficiency and the uniformity of temperature fields. Researchers concluded that the transpiration cooling method can produce better cooling effectiveness and more uniform coolant coverage with less coolant compared with traditional film cooling [37].

Aeroengines output thermal energy of high-temperature gas by turbine blades in the form of mechanical work. In a gas turbine, endwalls of the first-stage vane are directly exposed to high-temperature exhaust gas [38]. To achieve high operation efficiency, a high turbine inlet temperature is required, which also brings high thermal protection requirements for blade cooling. As a new and high-efficiency cooling method, transpiration cooling is introduced to the cooling of turbine blades for advanced areoengines [2,39,40].

Transpiration cooling is also applied in combustion chambers of hypersonic vehicles for thermal protection [41,42]. For a hypersonic vehicle, the aerodynamic thermal environment of the leading edge is severe, and active cooling is needed when it operates at a high Mach number [32]. As an efficient active cooling technology, transpiration cooling has great advantages to improve the cooling efficiency of the leading edge of a hypersonic vehicle [43,44], as shown in Figure 2.

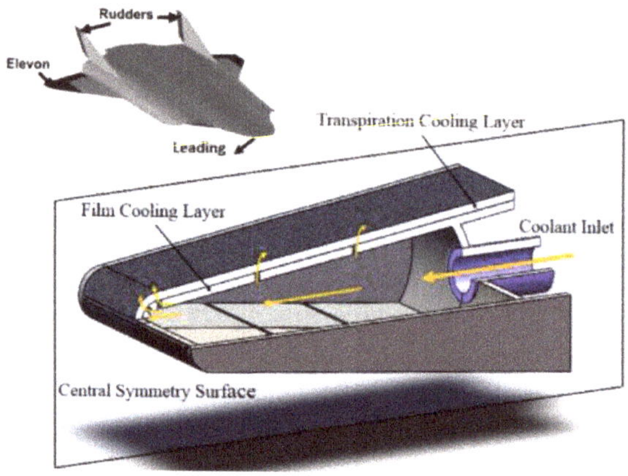

Figure 2. Transpiration cooling of the leading edge for a hypersonic vehicle. Reprinted/adapted with permission from Ref. [44]. 2020, Elsevier Ltd.

In addition to aerospace applications, some fundamental research works have been conducted to reveal the physical mechanisms of transpiration cooling [3,10,45–48]. Some researchers [49,50] aimed at finding the optimal porous/lattice structure and revealing the physical process to experimentally improve the cooling efficiency. Other researchers [51–55] developed numerical models to study the flow and heat transfer characteristics of transpiration cooling in porous/lattice structures. Recently, the research works of transpiration cooling have concentrated on high cooling efficiency, improvement of temperature uniformity, and a quantitatively controlled cooling process [56].

In the following parts, the literature on heat transfer enhancement by porous/lattice structures and transpiration cooling is introduced and summarized in detail. In addition, the existing developments and challenges are provided at the end of the paper.

2. Heat Transfer Enhancement by Porous/Lattice Structures

2.1. *Theory and Mechanisms*

From previous research [57,58], several kinds of complex vortices have been generated, which have benefits in flow mixing when the fluid flows through porous/lattice structures. Convective heat transfer is strengthened with the strengthened mixing of hot and cold flows [59]. For a typical lattice element, there is a horseshoe vortex at the leading edge of the lattice ligaments, a streamwise vortex at the trailing edge, and a counter-rotating longitudinal vortex at the center, as shown in Figure 3. Because of the vortex generation, the local turbulence intensity and flow mixing are both strengthened and then heat transfer is enhanced. However, some research works indicated that the buoyancy effect is weakened because of the porous/lattice structures [60].

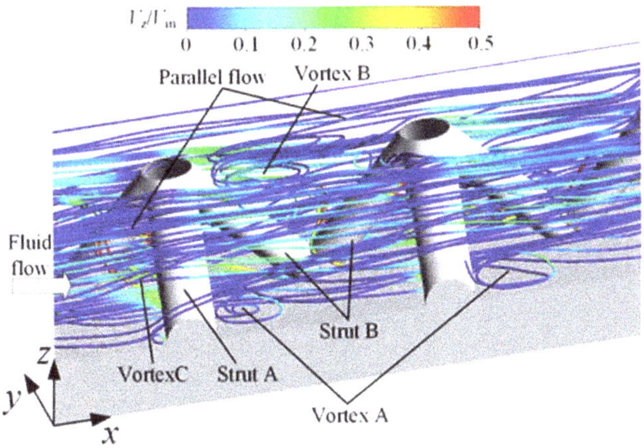

Figure 3. Vortices generated around pyramid lattice struts in a parallel flow regime.

2.2. *Active Cooling for Scramjet Combustors/Turbine Blades*

Aerospace aircraft are developing in the direction of high speed and high efficiency, which also brings higher requirements for thermal management technology [61,62]. Porous/lattice structures have started to be applied in the thermal management of aerospace devices, such as regenerative cooling in a rocket/scramjet [12,63–66] and cooling of a turbine blade [26,66–70].

For rockets or hypersonic vehicles, effective cooling of a combustor is crucial to ensure safe operation under a complicated heat environment and extremely high heat flux [71]. Yu et al. [12] numerically investigated the heat transfer characteristics of a composite Kagome cored porous/lattice structure used for regenerative cooling of a scramjet. The results indicated that the cooling efficiency of the porous/lattice structure channel was better than that of the conventional channel, although with a large pressure drop penalty. Song et al. [63] numerically investigated flow and heat transfer characteristics of the cooling channel with pyramid lattice sandwich structures in a rocket combustor. They concluded that a regenerative cooling channel with pyramid lattice sandwich structures showed excellent cooling performance. Some other references also confirm the generation cooling effect of a rocket engine via porous/lattice structures [65,66].

For turbine blades, operation efficiency is usually raised by increasing the inlet temperature of turbine blades, and then advanced cooling methods are needed [15]. Many researchers design the internal cooling of a turbine blade using porous/lattice structures [68,70] or design a porous/lattice blade [72]. Xu et al. [68] attempted to optimize the internal cooling channel arranged in the turbine blade manufactured with 3D metal printing technology. The Nusselt number is greatly enhanced with the complex flow path in the lattice/porous blade. Du et al. [72] designed latticework with many sub-channels and numerically studied the effects of the jet nozzle location, shapes, and mass flow ratios on the flow and heat transfer characteristics of the turbine blade. The study provided an important reference for the coupling effect of internal cooling with impingement cooling of a porous/lattice blade, as shown in Figure 4 [73].

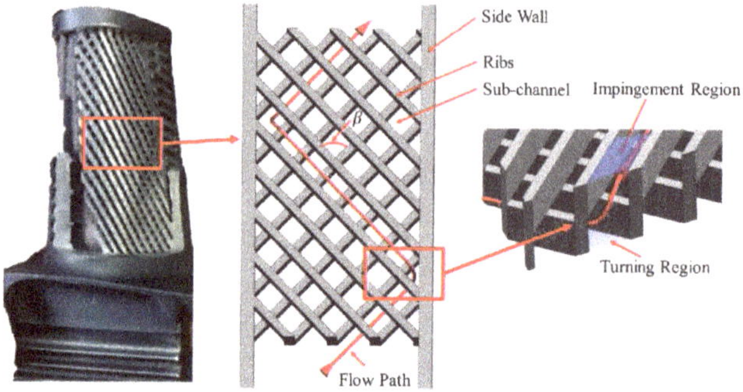

Figure 4. Schematic view of the lattice structure: A typical porous blade and flow network. Reprinted/adapted with permission from Ref. [73]. 2019, Elsevier Ltd.

In addition, porous/lattice structures are introduced to cool the trailing edge of a turbine blade to provide a bigger heat transfer area and strong flow mixing [26,69,74,75]. Shen et al. [69] experimentally investigated the flow and heat transfer performance of a wedge-shaped cooling channel with a Kagome lattice. The results showed increased heat transfer compared with conventional pin-fin channels at the same porosity, as shown in Figure 5. Saha et al. [75] investigated two types of channels with converging porous/lattice structures on the trailing edge of a turbine blade used for internal cooling with different arrangements. Compared with the previous data from a pin-fin channel, the channel with converging porous/lattice structures presented a higher heat transfer with a larger heat transfer area.

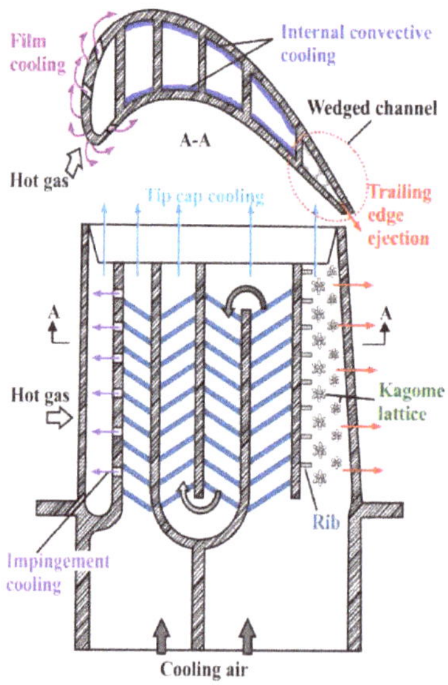

Figure 5. Cooling channel of a gas turbine blade trailing edge mounted with Kagome lattice structures. Reprinted/adapted with permission from Ref. [69]. 2019, Elsevier Ltd.

Other related literature studies on porous/lattice structures applied in the active cooling of scramjet combustors/turbine blades are listed in Table 1.

Table 1. Literature on different applications of porous/lattice structures in aerospace field.

Ref.	Applications	Lattice Type and Geometry	Lattice Material	Coolant	Flow Parameter	Heat Transfer Parameter and Boundary Conditions
[12]	Scramjet combustor	Kagome cored composite lattice structure (d = 2 mm, h = 8 mm, l = 9.8 mm)	C/C composite material	Aviation kerosene	Re = 562–56,234	h = 380–4200 W/(m^2·K); T = 1000 K
[64]	Scramjet combustor	Body-centered cubic lattice (d = 0.28–0.82 mm)	Maraging steel, 17-4 PH, and H13	Water	v = 0.5–2.5 m/s	h = 0–55,000W/(m^2·K); *heat flux* = 0.5 MW/m^2
[63]	Rocket combustor	Pyramid core lattice sandwich (d = 0.4/0.8 mm, l = 12.2 mm)	1Cr18Ni9Ti	Kerosene	q_m = 0.001 kg/s	Nu = 0–110; *heat flux* = 1.5 MW/m^2
[68]	Turbine Blade	Common lattice structure (d = 2 mm)	Nickel-Base superalloy IN718	Air	Re = 6000–12,000	Nu = 110–180; A constant heat flux corresponded the target surface
[72]	Turbine blade	Latticework (l = 224 mm, h = 48 mm)	Aluminium	Cooling air	Re = 44,000	Nu/Nu_0 = 0–5; T = 50 °C
[26]	Trailing edge of turbine blade	Kagome lattice, pin fins, ribs (d = 2.43–5.8 mm, h = 3 mm)	310 stainless steel	Compressible air	Re = 5000	h = 50–170 W/(m^2·K); *heat flux* = 1500 W/m^2
[69]	Trailing edge of turbine blade	Kagome lattice, pin fins, ribs (d = 2.43–5.8 mm, h = 3 mm, l = 3.6 mm)	310 stainless steel (0Cr25Ni20)	Air	Re = 5000–15,000	h = 0–150 W/(m^2·K); T = 335 K
[74]	Trailing edge of turbine blade	Lattice-matrix structures (e/D = 0.02/0.028, p/e = 8–30)	Stainless steel	Air	Re = 10,000–50,000	Nu/Nu_0 = 2.4; T = 120–125 °C
[75]	Trailing edge of turbine blade	Converging lattice structures, pin fins (x/D = 0.6–10)	Low thermal conductivity plastics	Air	Re = 4000–20,000	Nu/Nu =1–4; A time-changing wall temperature

2.3. Applications in Heat Exchangers and Heat Sinks

A heat exchanger is a type of equipment used to realize heat transport from a hot fluid to a cold fluid, which is also widely used in aerospace applications [76]. It is difficult to further improve the efficiency of the traditional fin heat exchanger [77]. The heat exchanger combined with porous/lattice structures has attracted much attention in recent years [78,79], as shown in Figure 6.

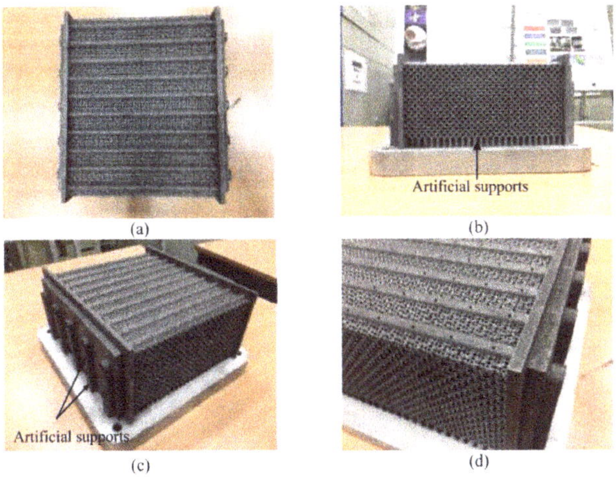

Figure 6. A porous/lattice heat exchanger design. (**a**) top view, (**b**) side view, (**c**) isometric view and (**d**) zoom-in view. Reprinted/adapted with permission from Ref. [80]. 2019, Elsevier Ltd.

The porous/lattice structures used in heat exchangers usually have two types, i.e., an octet truss porous/lattice structure [81–83] or a microporous/lattice structure [84].

The octet truss porous/lattice structure, due to its superior structural characteristics produced by its increased connectivity, has been widely used in multifunctional heat exchangers and electric applications [81–83]. Chaudhari [81] experimentally investigated the effective thermal conductivity and wall Nusselt number in heat exchangers with octet truss lattice porous/lattice structures. The results indicated that the octet truss lattice porous/lattice structure is a good selection to bear compressive loads and dissipate heat. Heat and fluid flow characteristics of octet truss lattice porous/lattice structures with different porosities and pore densities were investigated by Ekade et al. [82]. They found that octet truss lattice porous/lattice structures are more suitable for use in heat exchangers compared with stochastic foams.

Micro-porous/lattice structures draw more attention as a kind of new structural material because of their strengths in lightweight and high structure stability [84,85]. Maloney et al. [85] experimentally studied the heat transfer performance of a heat exchanger with a three-dimensional micro-porous/lattice structure in terms of thermal conductance per unit of volume. The results indicated the multifunctional capability of micro-porous/lattice structures for load bearing and energy absorptions.

Some researchers attempted to apply the porous/lattice structure in heat sinks to improve the cooling efficiency [27,86–89]. Dixit et al. [27] numerically compared several kinds of porous/lattice structures regarding their convection flow at a low Reynolds number and in constant wall temperature conditions. They found that the body-centered-cubic truss, simple-cubic plate, and Kelvin and Octet lattice cells provided better heat transfer than microchannels or channels with open-cell foam elements. Wang et al. [86] numerically investigated the flow and heat transfer characteristics of a cubic body-centered-cubic pin fin heat sink. They found the usage of the combination of conventional pin fin and porous/lattice structure brought a more than 140% improvement in thermal-hydraulic performance.

2.4. Manufacture of Porous/Lattice Structures Cooling Channel

Porous/lattice structure channels are mainly grouped into two types: Metal porous/lattice structures [90] and plastic porous/lattice structures [91]. The selection of lattice/porous material is important in the heat transfer process and, consequently, makes a difference in heat transfer performance. As for the metal lattice, complex metal porous/lattice structures can provide good heat transfer performance because of their high heat conduction coefficient and strong designability [78,92]. Metal porous/lattice structures have been widely applied in aerospace in the main components and related assisting components, such as heat exchangers. Furthermore, some nonmetal porous/lattice materials are also investigated [93], such as composite materials. In addition to metal and nonmetal porous/lattice structures, plastic materials are also used to fabricate porous/lattice structures, which have good capability in shock absorption [91].

The manufacturing methods of porous/lattice structures usually include selective laser melting, additive manufacturing, sheet folding and interweaving, and chemical vapor infiltration. Yun et al. [64] manufactured a series of face-centered cubic lattice channels by additive manufacturing and analyzed their flow and heat transfer characteristics at different porosities and inlet velocities. The effects of different structure materials were compared and it was found that the performance of the H13 steel channel was better than those of maraging steel and 17-4 PH steel channels. Ho et al. [80] fabricated two novel porous/lattice heat exchangers via additive manufacturing, and the thermohydraulic performance of the heat exchangers was experimentally analyzed. The results indicated that with a smaller cell size, the heat exchanger has higher overall thermal conductivity and air-side heat transfer coefficients. Ho et al. [94] produced four structured porous/lattice structures with the same porosity but of different unit element sizes via selective laser melting to study the thermohydraulic properties in terms of the permeability, inertia coefficient, and Nusselt number and provided a theoretical reference for enhancing single-phase forced-convection cooling. Jin et al. [95] fabricated metallic lattices with different punching locations to investigate the effects of a dimensionless punching location shift (s/d) on flow and heat transfer characteristics via the sheet folding method. The results indicated that the heat transfer capability of the pyramidal lattice (s/d = 0) was superior to the lattice with a quarter punching location shift (s/d = 0.25), which was followed by the X-lattice (s/d = 1). Wei et al. [96] manufactured a lightweight C/SiC pyramidal core porous/lattice structure panel via interweaving and the chemical vapor infiltration method. Combining experiments with simulations, the equivalent thermal conductivity of the structure was obtained and indicated the good heat transfer performance of the C/SiC materials.

2.5. Investigations of Porous/Lattice Structure Element

There are various element types of porous/lattice structures (shown in Figure 7). When the fluid flows through porous/lattice structures, different vortices are generated, which affects the flow structures and turbulence intensity [97].

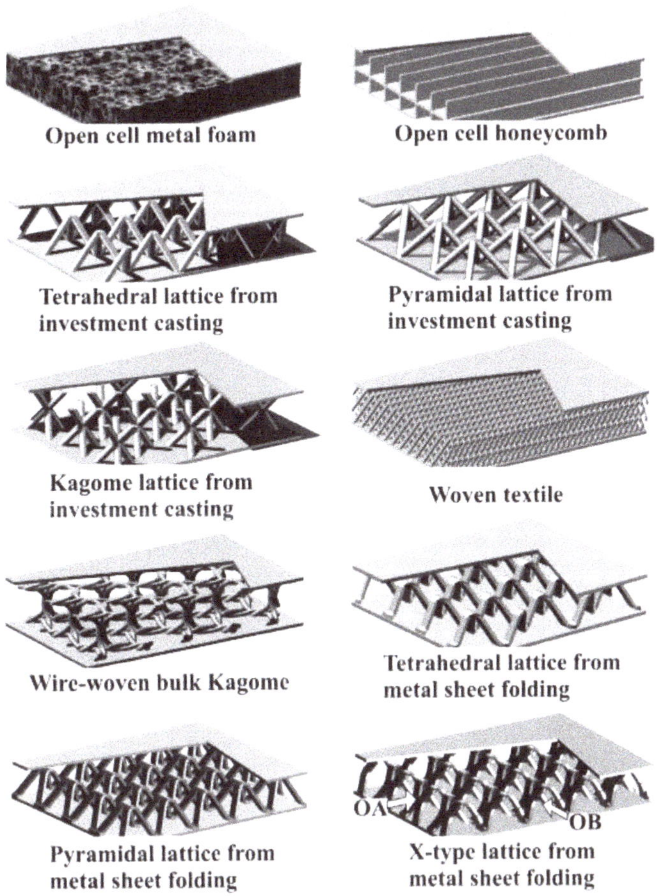

Figure 7. Several common lattice structures with different unit elements. Reprinted/adapted with permission from Ref. [22]. 2017, Elsevier Ltd.

In the past few years, several common porous/lattice structures have been investigated by researchers, such as octet truss lattice, face-centered cubic, body-centered cubic, micro-lattice, pyramid lattice, Kagome lattice, X-lattice, vertical lattices, tetrahedral lattices, slanted lattices, graded lattice, and composite lattice [19,21–23,67,69,98–108].

The effects of different lattice elements have been numerically investigated by researchers [21,101,106]. Shen et al. [21] studied the local and overall flow and heat transfer characteristics of the single-layered Kagome and wire-woven bulk Kagome lattice by a verified SST numerical model. They found the area-averaged Nusselt number on the ligaments of single-layered Kagome was lower than that of wire-woven bulk Kagome, which resulted from more high-momentum vortices and higher turbulent kinetic energy near the endwall as a result of the Kagome panel. Downing et al. [101] numerically analyzed the flow and heat transfer characteristics to optimize the manufacturing thermal process. The results provided a great reference for the effects of local geometry and topology on temperature field evolution in lattice structures during metal additive manufacturing. Yun et al. [106] numerically investigated the thermos-fluid-structural characteristics of the increase-type graded, V-type graded, and W-type graded lattice channels. The results provided strong evidence for heat transfer enhancement for graded porous/lattice structures.

Moreover, some researchers carried out experiments to analyze the effect of the porous/lattice elements [58,105,108]. The overall heat transfer performance of a vertical tube under a low flow rate and high heat flux with a body-centered cubic porous/lattice structure was investigated by Shi et al. [58] to suppress the heat transfer deterioration. Supercritical CO_2 is used in the experiments. Liang et al. [105] experimentally compared the thermo-fluid behaviors of a staggered pin fin array, a Kagome lattice array, and a body-centered cubic lattice array in a rectangular channel. The results indicated that the endwall-averaged Nusselt number of the body-centered cubic and Kagome lattices were higher than conventional pin fins but had similar heat transfer characteristics. Bai et al. [108] conducted a systematic experiment to obtain the heat transfer performance of five lattice core structures, i.e., vertical lattices, tetrahedral lattices, slanted lattices, pyramidal lattices, and Kagome lattices. Based on the experimental results, they finally proposed a novel lattice core structure named after the windward bend structure, with the lattices bent in the windward direction and arranged in a staggered pattern.

Some researchers combined porous/lattice structures with other heat transfer enhancement elements [67,69,102]. Li et al. [67] introduced dimples, protrusions, and pin fins into an X-lattice-cored sandwich panel downstream. The results indicated that the appropriate combination of the X-lattice and a certain element is beneficial for the overall heat transfer enhancement of the sandwich panel. Shen et al. [69] numerically compared the heat transfer capability of pin fins with that of Kagome porous/lattice structures on the trailing edge of a turbine blade. They found the Kagome porous/lattice structures can increase the overall Nusselt numbers but result in a similar pressure loss compared with the traditional cooling arrangements. Ma et al. [102] numerically investigated the forced convection heat transfer in a new sandwich panel mounted with a pyramidal lattice and plate fins. The thermal performances of different kinds of sandwich panels are compared. The results indicated that the combination of pin fins and pyramidal lattice produced a larger overall Nusselt number and better heat dissipation performance.

Overall, the octet truss lattice and the micro-lattice are suitable for heat exchangers or heat sinks because of the excellent structural characteristics and heat transporting ability [103]. Because of their higher bearing capability, the body-centered cubic (BCC) structure and face-centered cubic lattice (FCC) structure are widely used in aerospace fields for multifunctional applications such as load bearing, vibration isolation, and thermal protection. Literatures on different elements and related parameters of porous/lattice structures are listed in Table 2.

Table 2. Literature on different elements and related parameters of porous/lattice structures.

Ref.	Lattice Type and Geometry	Lattice Material	Coolant	Flow Parameter	Heat Transfer Parameter and Boundary Conditions
[21]	Single-layered Kagome and wire-woven bulk Kagome cores sandwich (d = 1.0/1.68 mm, l = 12.73 mm, h = 13 mm)	Copper-2% Beryllium (Be2Cu) alloy	Incompressible fluid with constant thermophysical properties	Re = 3995–8710	Nu = 50–200; $heat\ flux$ = 4000 W/m^2
[22]	X-type metallic lattice cored sandwich panel, tetrahedral lattice cored sandwich panel (w = 2.16 mm, l = 12 mm, h = 12 mm)	AISI 304 stainless steel	Air	Re = 2165-6043	Nu = 0–500; $heat\ flux$ = 4088-8529 W/m^2
[58]	Body-centered cubic lattice (h < 4.5 mm)	316L stainless steel	Supercritical CO_2	$mass\ flux$ = 130–419 g/(m^2·s)	Nu = 100–400; $heat\ flux$ = 41.94 kW/m^2
[67]	X-lattice-cored sandwich panels with pin fins, dimples or protrusions (l = 12 mm, w = 12 mm, h = 9.66 mm)	AISI 304 stainless steel	Air	Re = 3100–5700	Nu = 40–260; $heat\ flux$ = 9882.6 W/m^2
[69]	Kagome lattice structures, pin fins lattice structures (d = 3.52/3.28/2.43/5.8/4.25/2.5 mm, h = 3 mm, w = 30 mm)	ASTM type 310 stainless steel (0Cr25Ni20)	Air	Re = 5000–15,000	h = 0–150 W/(m^2·K); $heat\ flux$ = 1500 W/m^2
[98]	A new lattice with non-uniform wall roughness (p/D = 1.12)	Stainless steel	Air	Re = 38,754	Nu = 210–400; T_{in} = 47 °C
[99]	Composite lattice core sandwich structures (l = 10 mm, w = 10 mm, h = 1 mm)	CF-reinforced resin matrix composite (CFRP)	Incompressible fluid with constant thermophysical properties at 300 K	Re = 0–100,000	Nu = 10–1000; $heat\ flux$ = 10 kW/m^2
[100]	Multi-layered lattice structures (d = 2 mm, l = 17.3 mm)	Aluminium	Air	v = 5.4–11 m/s	h = 90–220 W/(m^2·K); T = 350 K
[102]	Plate fins or/and pyramidal lattice (l = 12 mm, h = 9.64 mm, w = 12 mm)	A kind of solid with thermal conductivity = 16.2 W/(m·K)	Air	Re = 3100–5700	Nu = 83.31–196.27; $heat\ flux$ = 9882.6 W/m^2
[105]	Staggered pin fin array, Kagome lattice array, body centered cubic lattice array (d = 7.4/4/3.4 mm, l = 20 mm, h = 15 mm)	Acrylic plastics	Compressed air	Re = 5000–20,000	Nu = 20–55; T = 293 K
[106]	Increase-type graded, V-type graded, W-type graded lattice (d = 0.3–0.57 mm)	17-4 PH steel	Water	v = 0.1–0.5 m/s	h = 5000–14,000 W/(m^2·K); $heat\ flux$ = 300 kW/m^2
[107]	Composite sandwich structure with lattice truss cores (d = 1/1.5/2.5/3 mm, h = 9/10/14/16 mm)	Carbon-fiber-reinforced resin matrix composites (T700/3234)	A specific fluid with T = 300 K	v = 5 m/s	h = 161.68–171.13 W/(m^2·K); $heat\ flux$ = 10 kW/m^2
[108]	Vertical lattices, slanted lattices, Kagome lattices, tetrahedral lattices and pyramidal lattices (d = 0.35 mm)	Metal foam	Air	v = 1–12 m/s	h = 0–660 W/(m^2·K); An equal pumping power
[109]	A new lattice with nonuniform wall roughness (e = 2.8 mm, p/e = 4–15)	Stainless steel	Air	Re = 2000–22,000	Nu = 10–105; $heat\ flux$ = 1390 W/m^2

3. Transpiration Cooling

3.1. Theory and Mechanism

Transpiration cooling is an effective active cooling method that causes the coolant to flow into the micro skeleton material of porous/lattice structures and consequently generate convective heat transfer [110]. At the same time, a thin film is formed covering the solid surface of the used piece when the coolant flow is ejected, which can isolate the protected surfaces from the high-temperature mainstream [111–113], as shown in Figure 8. Transpiration cooling can improve the cooling efficiency using a limited coolant and provide a large amount of contact surface for the solid surfaces and coolant flows [114]. Recently, some researchers have concentrated on the research of transpiration cooling in terms of cooling efficiency [37,115]. Compared with traditional film cooling, transpiration cooling has the advantages of more uniform coolant coverage, less coolant consumption, and a larger solid–fluid contact area.

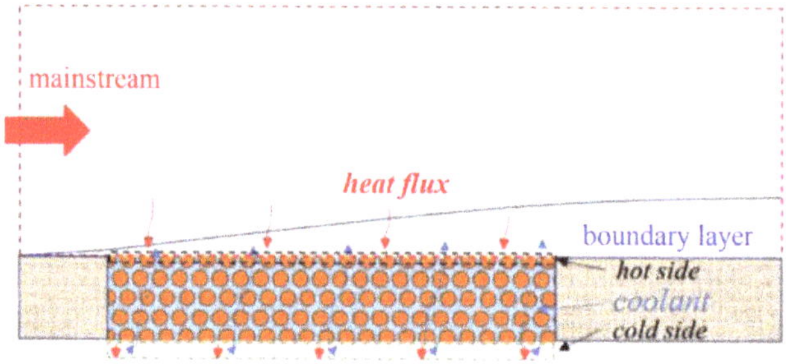

Figure 8. Mechanism description of transpiration cooling. Reprinted/adapted with permission from Ref. [113]. OPEN ACCESS.

3.2. Experimental Study

Experimental investigations on transpiration cooling can be divided into two groups. One group is based on porous/lattice plates to perform a fundamental study of transpiration cooling [10,35,45,115,116]. The other group is application study, including transpiration cooling in turbine blades [117–120], nose cones [33,121], combustion chambers [122], and leading regions [123].

Based on sintered metal porous/lattice plate, researchers have conducted a series of investigations on transpiration cooling [10,35,45,115,116]. These experiments were mainly conducted in a heated wind tunnel, and the coolant was supplied by an air compressor. Zheng et al. [35], using the selective laser metal sintering additive manufacturing technology, fabricated different kinds of lattice plates for transpiration cooling investigations. They experimentally analyzed the effects of geometry parameters on heat transfer performance, including five different types of porous media, three different coolant mass flow rates, and four different mainstream temperatures. Jiang et al. [115] proposed a combination of transpiration cooling and film cooling based on the leading edge. It was found that cooling efficiency increased by 25.7% for standard transpiration cooling and by 37.9% for combined cooling. Huang et al. [116] investigated the effects of the particle diameter of the sintered porous/lattice structure and coolant injection ratio on the transpiration cooling efficiency in a supersonic wind tunnel system in which the air was compressed to 0.55 MPa. The temperature distribution of the porous/lattice structure surface was captured by an IR camera.

Some researchers carried out experiments based on turbine blades to improve cooling efficiency [117–120]. Kim et al. [118] investigated micro cooling for s turbine blade by a DPSS laser and a high-speed CCD camera imaging system. They found transpiration cooling had a relatively stable boundary layer and verified it by simulation methods. Using steam as the coolant, Kumar et al. [119] performed an evaluation of the performance enhancement of a combined cycle based on transpiration cooling of gas turbine blades. The results indicated that steam was the superior coolant compared to air, and a gas/steam combined cycle could be an alternative adoption. Kim et al. [120] performed experiments on a single C3X blade with a multiple-hole array in a high-temperature subsonic wind tunnel. The surface temperature of the blade was measured, and the overall cooling effectiveness was analyzed. The results indicated that the adoption of transpiration cooling brought nearly one-quarter of cooling effectiveness improvement compared with internal cooling (shown in Figure 9).

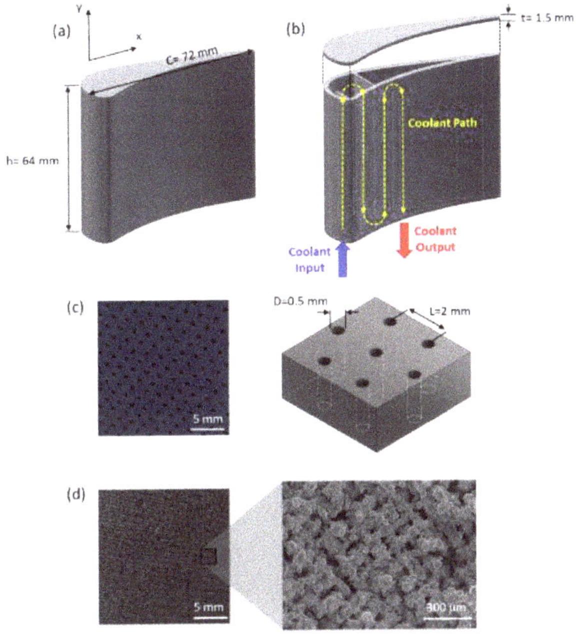

Figure 9. Illustration of transpiration cooling of a whole C3X blade. (**a**) dimensions of the exterior, (**b**) a typical internal cooling channel, (**c**) effusion cooling surface and (**d**) transpiration cooling surface. Reprinted/adapted with permission from Ref. [120]. 2020, Elsevier Ltd.

Transpiration cooling becomes necessary in some components, such as nose cones [33, 121], combustion chambers [122], and leading regions [123], because the extremely high heat flux is generated during high-speed flights. Wang et al. [33] investigated the transpiration cooling for a wedge-shaped nose cone with an unequal-thickness porous wall using liquid water as a coolant in a heated gas wind tunnel. The results indicated that the arrangement for the unequal-thickness walled nose cone can effectively solve the key issue of the cooling stagnation point. For transpiration cooling of the nose cone, Wu et al. [121] adopted two nose cone specimens, namely, "uniform porosity layout" and "gradient porosity layout", to carry out experiments. They found that using a "gradient porosity layout" could reduce the required coolant injection pressure (shown in Figure 10). The transpiration cooling efficiency in an annular slinger combustor was investigated by Zhang et al. [122]. The wall temperature and cooling effect of the combustor were observed to provide a reference of flow structures and heat transfer performance. Shen et al. [123] experimentally investigated the transpiration cooling of the leading edge in an arc-heated wind tunnel. An infrared thermal imaging system was used to capture surface temperature distributions using coolant water. They found interesting phenomena such as ice formation in a high enthalpy environment.

The effects of coolants are also evident, for example, the use of steam as a coolant is more effective than air [37]. Many researchers concentrated on the effects of different coolants to search for optimal cooling performance [124,125]. Qian et al. [125] used hydrogel as a coolant and experimentally compared the cooling characteristics using hydrogel and water as coolants, respectively. The results indicated that hydrogel can provide a considerable cooling effect and was superior for prolonging service time. In addition, He et al. [126] added propylene glycol to water to investigate the cooling effect. The results indicated that adding propylene glycol to water was an effective way to ease the unfavorable matters existing in transpiration cooling with liquid water, such as the stability of temperature distribution and the uniformity of coolant coverage. Literatures on different applications used in transpiration cooling are listed in Table 3.

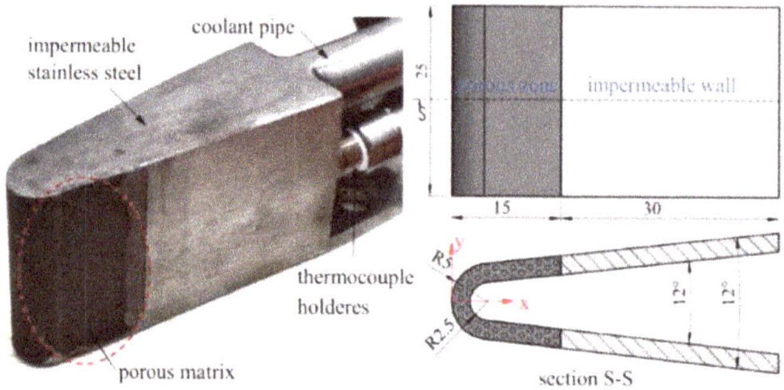

Figure 10. Schematic of transpiration cooling in nose cone regions. Reprinted/adapted with permission from Ref. [121]. 2019, Elsevier Inc.

Table 3. Literature on different applications used in transpiration cooling.

Ref.	Applications	Plate Material	Coolant	Mainstream
[10]	Sintered porous flat plates	Stainless steel or bronze	Air	Dry air supplied by a blower (30 m/s, 100 °C)
[46]	Sintered woven wire mesh structures	Stainless steel	Cooling air (20.5–22.3 °C)	Air (130–150 kg/h, 100–130 °C)
[115]	Leading edge	Stainless steel (1Cr18Ni9Ti)	Room temperature dry air	Air (Ma2.8, 398.15 K)
[116]	Sintered porous plates	Bronze	Liquid water (293.2 K)	Hot air (65 m/s, 800 K)
[117]	C3X blades	Stainless steel (SUS304)	Cooling air	Dry air (3197 m^3/h, heated by three electrical heaters)
[121]	Nose cone	316L alloy powder	Cooling air and liquid water (20 °C)	Air (20 m/s, 100 °C)
[122]	Annular Slinger Combustor	GH3044	Fresh air	Air (7.3 kg/s and 742 K)
[125]	A porous flat plate specimen	Nickel-based powders (Ni 73.46%, Cr 17.3%, Fe 7.5%, Si 1.7%, and C 0.04%)	Solid hydrogel	Air (523 K and 693 K)
[126]	A sintered porous plate and a solid frame	Nickel-based super alloy powders (73.46%Ni, 17.3%Cr, 7.5%Fe, 1.7%Si and 0.04%C)	Propylene glycol aqueous	Air (Re12000, 523 K)

3.3. Numerical Study

Many numerical simulations of transpiration cooling have also been performed by researchers, which are mainly grouped into single-phase transpiration cooling and phase-change transpiration cooling [123]. For single-phase transpiration cooling, different kinds of turbulence models are adopted, such as the Spalart–Allmaras turbulence model [43,127], the k-ε turbulence model [128–133], and the k-ω SST turbulence model [134–137].

The Spalart–Allmaras turbulence model is specifically designed for aerospace applications involving bounded flow on walls, which has shown excellent results for boundary layers subjected to adverse pressure gradients [43,127]. He et al. [43] proposed two optimization strategies and developed five locally enhanced transpiration cooling designs after analyzing the interaction performances in a three-dimensional hypersonic inlet using the Spalart–Allmaras model. Huang et al. [127] numerically investigated transpiration cooling for a sintered porous strut with different strut structures, material properties, and coolant conditions using the Spalart–Allmaras model. The results indicated that increasing the coolant blowing ratio could produce an increase in the retention time of a mixture of steam and injected coolant.

Apart from the one-equation turbulence model, two-equation models, such as the k-ε turbulence model [128–133] and the k-ω SST turbulence model [134–137], are also used in simulations of transpiration cooling. Jiang et al. [129] compared the turbulent velocity and thermal characteristics of the main flow in a rectangular channel without and with transpiration cooling using the k-ε SST model. The results indicated that transpiration cooling could produce a thicker boundary layer and less wall skin friction. Kilic [131] studied the influence of the Reynolds number of hot gas, the inlet temperature of the air, and the mass flow rate of water on the temperature distribution of a local wall and the cooling effectiveness of a porous flat plate using the k-ε RNG model. It was found that with the increased Reynolds number, the surface temperature increased and the cooling effectiveness of the porous plate and the efficiency of the system decreased. Bellettre et al. [133] examined the liquid transpiration cooling effect of a porous plate using the k-ε RNG model and confirmed the evaporation rates calculated by semi-empirical relations. The related results can be applied in the regenerative cooling of rocket combustors and other internal cooling of high-temperature components.

Yang et al. [134] investigated the phenomenon of plugging pores for transpiration cooling using the k-ω SST model. They found that, although the overall plugging ratio was a dominant parameter for the cooling effectiveness, this single parameter was not enough to improve the cooling effectiveness for all locations but did provide a quantitative understanding of the random plugging on the specific transpiration cooling configuration. Ding et al. [135] carried out investigations on heat transfer performances of a porous matrix using a combination of transpiration and film cooling using the k-ω SST model. They systematically investigated the downstream film cooling effect derived by upstream transpiration cooling using gaseous Nitrogen as the coolant. Shen et al. [136] used the k-ω SST model to analyze a combinational opposing jet and platelet transpiration cooling nose-tip. Compared with only using the opposing jet, they found that using the combinational opposing jet and platelet transpiration cooling could result in a reduction of peak heat flux at more than 6.6% with less coolant consumption.

Some researchers conducted simulations based on the coolant phase-change process, which fully exploits the latent heat of vaporization occurring in the phase change of the coolant. Two kinds of phase-change models are used, i.e., the semi-mixing model [138–141] and the Local Thermal non-Equilibrium Two-Phase Mixture Model [51,142].

In 2013, He et al. [138] studied the influences of the coolant injection rate and external heat flux on temperature distributions using a numerical method validated by experiments. Based on the theoretical results, they [139] investigated thermal protection under extremely high heat flux and estimated an approach to analyzing the maximum heat flux afforded and minimal coolant required for the desired conditions of transpiration cooling in 2014. Recently, they [140] suggested using semi-mixing to investigate the reliability and stability of transpiration cooling in changeable and complicated cases. Dong et al. [141] used the semi-mixing model to analyze transpiration cooling with the coolant phase change and found it can overcome the difficulties of tracking the phase-change interface and avoid calculation mistakes, which was validated by experiments.

The Local Thermal non-Equilibrium Two-Phase Mixture Model is also adopted by many researchers to describe the physical process of phase change in liquid coolants [51,142]. Yang et al. [51] investigated the mechanism of transpiration cooling and studied the effects of the coolant mass flow rate and various coolants. They found low-density and high-specific-heat coolants could generate a thicker film and produce a better cooling effect. Shi et al. [142] studied the influence of thermal conductivity, sphere diameter, and porosity on the temperature and saturation distributions within the matrix and numerically discussed the effect of hot surface pressure.

Recently, multi-discipline research methods have been proposed to improve transpiration cooling, such as biology [143,144]. Huang et al. [143] designed self-pumping transpiration cooling based on tree-like micro-channels. The results indicated the heated surface was efficiently cooled by the system, and upon increasing the heat flux, the coolant mass flow rate self-adaptively increased. Furthermore, Huang et al. [144] used a biomimetic non-smooth surface, which was inspired by earthworms, to compare the cooling efficiency of porous plates with three types of shapes. The results indicated that the non-smooth surface inspired by earthworms has important significance in heat transfer enhancement within porous/lattice materials. Literatures on different numerical models used in transpiration cooling are listed in Table 4.

Table 4. Literature on different numerical models used in transpiration cooling.

Ref.	Applications	Plate Material	Coolant	Numerical Model
[41]	Rocket thrust chambers	Composite carbon/carbon materials	Air, argon and helium	Darcy–Brinkman–Forchheimer model, local thermal equilibrium model
[43]	Three-dimensional hypersonic inlet	High temperature alloy Inconel-600	Purified water	Spalart–Allmaras model
[51]	Combustion chamber	Stainless steel	Methane	Local Thermal Non-Equilibrium Model
[127]	Combustor	Stainless steel	Methane	Spalart–Allmaras model
[129]	A rectangular channel	Sintered bronze	Air	k-ε Standard model, two-layer k-ε model
[130]	Nose zone	Stainless steel	Air or other gases (N_2, He, Ar, CO_2)	k-ε Standard model, k-ε RNG model, k-ω SST model
[131]	A porous plate	Porous plate with a porosity of 0.4	Water	k-ε RNG model
[132]	A porous plate	Bakelite	Compressed dry air	k-ε Standard model
[134]	Hot sectioncomponents	A porous metal plate with multiple rows of straight holes	Air	k-ω SST model
[135]	Leading edge and entire structure of hypersonic vehicles	single porous matrix and binary porous matrixes	Gaseous nitrogen	k-ω SST model
[136]	Nose-tip	Bonding thin metal platelet with a limited number of enlarged transpiration orifices	Air	k-ω SST model
[137]	A series of porous metal plates	Porous metal plate with large numbers of straight cylindrical holes	Air	k-ω SST model
[139]	An infinite porous plate	Porous plate with specific thickness, porosity, sphere diameter, and thermal conductivity	Water and vapor	Local thermal equilibrium model
[141]	A porous plate	Porous plate with specific thermal conductivity and permeability	Liquid water	Semi-mixing model

Table 4. *Cont.*

Ref.	Applications	Plate Material	Coolant	Numerical Model
[142]	A porous matrix	Porous matrix with specific thickness	Pure water	Two-phase mixture model with the local thermal nonequilibrium model
[145]	A porous plate	Porous plate with constant physical properties	Hot flow with constant thermal physical properties and density	Local thermal equilibrium model
[146]	A porous material	A cooled porous ceramic matrix composite material	Air	k-ω SST turbulence model
[147]	A turbulence channel	Metal plate composed of Fe, Cr, Ni, Si, Mn, and C	Cold air	Darcy–Forchheimer model, local non-equilibrium thermal mode
[148]	A heated copper plate	Copper	Slurry and nanofluid	k-ε Standard model and k-ω SST model

4. Developments and Challenges

4.1. Heat Transfer Enhancement by Porous/Lattice Structures

(1) For the applications of porous/lattice structures in scramjet/rocket engines, the investigations are still limited. The selection of the coolant is also limited by a relatively low heat flux. Most researchers select water or kerosene as coolants and the applied heat flux ranges from 0.5 to 1.5 MW/m^2, which is lower than the actual thermal environment.

(2) For the applications of porous/lattice structures on the trailing edge of a blade, the related investigations are almost all numerical calculations. The blade wall temperature is close to 335 K with a coolant temperature set at room temperature, as shown in Table 1. These parameters are far from the real thermal environment of the trailing edge of turbine blades. The effect of thermal-physical properties changing with temperature is ignored. For investigations of a whole blade with a latticework design, these studies were carried out at a temperature difference of approximately 20 K, and high-temperature effects are also ignored. In addition, reliable experimental data are also needed in this field.

(3) For heat exchangers and sinks, two lattice types, i.e., octet truss lattice and micro-lattice, are mainly used to obtain larger heat transfer areas. However, the current research on heat exchangers and sinks is mostly limited to the regular shape with low heat flux. Determining how to effectively dissipate heat in a high-temperature environment and with limited space, such as in aerospace applications, still needs further exploration.

(4) Although manufacturing technology has developed significantly in recent years, additive manufacturing technology has limited applications in porous/lattice metal materials. Determining how to control the porosity and uniformity of the porous/lattice structures is still a big problem for additive manufacturing technology. Therefore, the investigations of porous/lattice structures in regenerative cooling and cooling of a turbine blade are still at the theoretical level, and more investigations are needed before its applications in practice.

(5) Investigations on the heat transfer characteristics of porous/lattice structures mounting a cooling channel are mainly numerical calculations. The experimental research works are limited, especially for high-temperature and high-pressure-flow conditions, such as scramjet/rocket combustors and turbine blades.

(6) The porous/lattice structures are proposed to achieve heat transfer enhancement. For this kind of material, heat transfer enhancement, structure strength, and weight are contradictory in practice. Therefore, it is necessary to conduct a comprehensive evaluation of the performance of all aspects when designing the structure to meet the requirements of use under different working conditions.

(7) Highly efficient and accurate numerical models dealing with flow and heat transfer passing through porous/lattice structures are needed. Porous/lattice structure configuration optimization technology that considers multi-scale and multi-physical field effects needs to be developed.

4.2. Transpiration Cooling

As a new active cooling technology, transpiration cooling has developed rapidly in recent years and will face challenges in the future.

For experimental investigations:

(1) High-temperature experiments are hard to find in the literature, but they are necessary when transpiration cooling is applied in high-temperature components, such as combustors, the leading edge of a hypersonic vehicle, and turbine blades. The high-temperature field of porous media is difficult to capture because of the weakness of the high-temperature measurement technique. The flow fields of the porous structure measured by experiments are hardly found, which prevents the study of the physical mechanism of transpiration cooling from an experimental perspective.
(2) Transpiration cooling can be used for blades, leading edges, nose zones, combustion chambers, etc., but the existing experimental conditions find it difficult to meet the high-temperature environment and manufacture precision components, so many experiments are carried out based on a simplified flat plate structure made of stainless steel or bronze. In these experiments, the temperature of the stream mainly ranges from 373–800 K and the flow velocity mainly ranges from 20–65 m/s. Only in some investigations into the transpiration cooling of the leading edge of a hypersonic vehicle did the main steam total temperature reach 2310 K and the main steam velocity reach Ma = 4.2.
(3) For experimental studies, the coolants used for transpiration cooling are primarily air, water, solid hydrogel, and propylene glycol aqueous. The materials of porous/lattice structures are primarily plastic and metal. Apart from metal and plastic porous/lattice structures, there are also certain materials with a high melting point and good thermal conductivity used for transpiration cooling. In recent years, to meet the extremely high-temperature conditions, the porous materials used in the experimental research of transpiration cooling have been developed with composite materials such as C/C composite material in order to obtain better cooling efficiency.

For numerical investigations:

(1) For the numerical study of single-phase transpiration cooling, the porous/lattice structures region is often simplified as a fluid zone with resistance and source items. The simplified model cannot accurately describe the flow field and temperature field in the porous zone. When geometric details of the porous/lattice structures are considered in calculations, large computational efforts are needed to deal with the physical model. Therefore, a more accurate model of porous/lattice needs to be developed.
(2) The development of phase-change models of transpiration cooling is also limited, and the existing phase-change models require water as the coolant. With the extremely high thermal protection demand in aerospace applications, phase-change transpiration cooling has more potential to be used. The development of phase-change models should be coupled with a single-phase porous/lattice structure model to provide highly efficient and accurate predictions.

Author Contributions: Conceptualization, J.L.; writing—original draft preparation, M.X.; writing—review and editing, R.Z., X.Z., and W.X. All authors have read and agreed to the published version of the manuscript.

Funding: This work was supported by the Key Laboratory Stabilization Supporting Project (grant No. WDZC6142703202216) and the National Natural Science Foundation of China (Grant No. 12272133).

Conflicts of Interest: The authors declare no conflict of interest.

References

1. Yin, Y.; Mo, J.; Feng, J. Conductive fabric patch with controllable porous structure and elastic properties for tissue engineering applications. *J. Mater. Sci.* **2020**, *55*, 17120–17133. [CrossRef]
2. Vikulin, A.V.; Yaroslavtsev, N.L.; Zemlyanaya, V.A. Investigation into transpiration cooling of blades in high-temperature gas turbines. *Therm. Eng.* **2019**, *66*, 397–401. [CrossRef]
3. Huang, Z.; Zhu, Y.; Jiang, P.; Xiong, Y. Investigation of a porous transpiration cooled strut injector. *J. Propuls. Power* **2015**, *31*, 278–285. [CrossRef]
4. Pettes-Duler, M.; Roboam, X.; Sareni, B. Integrated optimal design for hybrid electric powertrain of future aircrafts. *Energies* **2022**, *15*, 6719. [CrossRef]
5. Han, Q.; Sun, C.; Tao, Y.; Li, Z.; Zhang, Y.; Chen, Y. Thermal protection of a hypersonic vehicle by modulating stagnation-point heat flux. *Aerosp. Sci. Technol.* **2020**, *98*, 105673. [CrossRef]
6. Shi, S.; Wang, Y.; Yan, L.; Sun, P.; Li, M.; Tang, S. Coupled ablation and thermal behavior of an all-composite structurally integrated thermal protection system: Fabrication and modeling. *Compos. Struct.* **2020**, *251*, 112623. [CrossRef]
7. Zhang, S.; Li, X.; Zuo, J.; Qin, J.; Cheng, K.; Feng, Y.; Bao, W. Research progress on active thermal protection for hypersonic vehicles. *Prog. Aerosp. Sci.* **2020**, *119*, 100646. [CrossRef]
8. Song, K.D.; Choi, S.H.; Scotti, S.J. Transpiration cooling experiment for scramjet engine combustion chamber by high heat fluxes. *J. Propuls. Power* **2006**, *22*, 96–102. [CrossRef]
9. Li, J.; Wang, X.; Liang, D.; Xu, N.; Zhu, B.; Li, W.; Yao, P.; Jiang, Y.; Min, X..; Huang, Z.; et al. A tandem radiative/evaporative cooler for weather-insensitive and high-performance daytime passive cooling. *Sci. Adv.* **2022**, *8*, eabq0411. [CrossRef] [PubMed]
10. Liu, Y.; Jiang, P.; Xiong, Y.; Wang, Y. Experimental and numerical investigation of transpiration cooling for sintered porous flat plates. *Appl. Therm. Eng.* **2013**, *50*, 997–1007. [CrossRef]
11. Liu, C.; Zhang, J.; Jia, D.; Li, P. Experimental and numerical investigation of the transition progress of strut-induced wakes in the supersonic flows. *Aerosp. Sci. Technol.* **2022**, *120*, 107256. [CrossRef]
12. Yu, J.; Song, Q.; Ma, X.; Zhao, Y.; Cao, J.; Chen, X. Study of heat transfer of composite lattice structure for active cooling used in the scramjet combustor. *Mater. Res. Innov.* **2015**, *19*, S5-843–S5-849. [CrossRef]
13. Baiju, A.P.; Jayan, N.; Nageswaran, G.; Suresh, M.S.; Narayanan, V. A technology for improving regenerative cooling in advanced cryogenic rocket engines for space transportation. *Adv. Astronaut. Sci. Technol.* **2021**, *4*, 11–18. [CrossRef]
14. Hussain, S.; Ghopa, W.A.W.; Singh, S.S.K.; Azman, A.H.; Abdullah, S. Experimental and numerical vibration analysis of octet-truss-lattice-based gas turbine blades. *Metals* **2022**, *12*, 340. [CrossRef]
15. Kaur, I.; Aider, Y.; Nithyanandam, K.; Singh, P. Thermal-hydraulic performance of additively manufactured lattices for gas turbine blade trailing edge cooling. *Appl. Therm. Eng.* **2022**, *211*, 118461. [CrossRef]
16. Grossmann, B.V.; Biermann, H.; Mughrabi, H. Measurement of service-induced internal elastic strains in a single-crystal nickel-based turbine blade with convergent-beam electron diffraction. *Philos. Mag. A* **2000**, *80*, 1743–1757. [CrossRef]
17. Lepoivre, A.; Edelin, D.; Baudin, N.; Soto, J.; Villière, M.; Redais, B.; Delaunay, D.; Sobotka, V. Rapid heat control for thermoplastic injection tooling using lattice structures as heat exchanger. *Key Eng. Mater.* **2022**, *926*, 1850–1863. [CrossRef]
18. Zhao, B.; Zhang, J.; Lian, W. Numerical modeling of heat exchanger filled with octahedral lattice frame porous material. *Aerospace* **2022**, *9*, 238. [CrossRef]
19. Qian, B.; Fan, H.; Liu, G.; Zhang, J.; Li, P. Microchannel liquid-cooled heat exchanger based on a nonuniform lattice: Study on structure calculation, formation process, and boiling heat transfer performance. *Materials* **2021**, *14*, 7248. [CrossRef]
20. Chen, H.; Mo, H.; Wan, Z.; Huang, S.; Wang, X.; Zhu, H. Thermal performance of a boiling and condensation enhanced heat transfer tube—Stepped lattice finned tube. *Appl. Therm. Eng.* **2020**, *173*, 115227. [CrossRef]
21. Shen, B.; Yan, H.; Xue, H.; Xie, G. The effects of geometrical topology on fluid flow and thermal performance in Kagome cored sandwich panels. *Appl. Therm. Eng.* **2018**, *142*, 79–88. [CrossRef]
22. Yan, H.; Yang, X.; Lu, T.; Xie, G. Convective heat transfer in a lightweight multifunctional sandwich panel with X-type metallic lattice core. *Appl. Therm. Eng.* **2017**, *127*, 1293–1304. [CrossRef]
23. Gao, L.; Sun, S.; Zhao, Y.; Sun, Y. Thermostructural multiobjective optimization of a composite sandwich panel with lattice truss cores. *Numer. Heat Transf. Part B Fundam.* **2016**, *70*, 233–250. [CrossRef]
24. Deng, H.; Zhao, J.; Wang, C. Bionic design method of a non-uniform lattice structure for a landing footpad. *Aerospace* **2022**, *9*, 220. [CrossRef]
25. Samipour, S.A.; Batrakov, V.V. Manufacture of aerospace lattice structures by radial braiding. *Russ. Eng. Res.* **2022**, *42*, 818–822. [CrossRef]
26. Li, Y.; Shen, B.; Yan, H.; Boetcher, S.K.S.; Xie, G. Heat transfer enhancement of rotating wedge-shaped channels with pin fins and Kagome lattices. *Numer. Heat Transf. Part A Appl.* **2020**, *77*, 1014–1033. [CrossRef]

27. Dixit, T.; Nithiarasu, P.; Kumar, S. Numerical evaluation of additively manufactured lattice architectures for heat sink applications. *Int. J. Therm. Sci.* **2021**, *159*, 106607. [CrossRef]
28. Skamniotis, C.; Courtis, M.; Cocks, A.C.F. Multiscale analysis of thermomechanical stresses in double wall transpiration cooling systems for gas turbine blades. *Int. J. Mech. Sci.* **2021**, *207*, 106657. [CrossRef]
29. Skamniotis, C.; Cocks, A.C.F. 2D and 3D thermoelastic phenomena in double wall transpiration cooling systems for gas turbine blades and hypersonic flight. *Aerosp. Sci. Technol.* **2021**, *113*, 106610. [CrossRef]
30. Alhatim, O. Simulation of a conjugate problem of friction and heat transfer in transpiration cooling of gas turbine blades. *High Temp. Mater. Process. Int. Q. High-Technol. Plasma Process.* **2015**, *19*, 169–178. [CrossRef]
31. Zhu, Y.; Jiang, P.; Sun, J.; Xiong, Y. Injector head transpiration cooling coupled with combustion in H2/O2 subscale thrust chamber. *J. Thermophys. Heat Transf.* **2013**, *27*, 42–51. [CrossRef]
32. Luo, S.; Miao, Z.; Liu, J.; Song, J.; Xi, W.; Liu, C. Effects of coolants of double layer transpiration cooling system in the leading edge of a hypersonic vehicle. *Front. Energy Res.* **2021**, *530*, 756820. [CrossRef]
33. Wang, J.; Zhao, L.; Wang, X.; Ma, J.; Lin, J. An experimental investigation on transpiration cooling of wedge shaped nose cone with liquid coolant. *Int. J. Heat Mass Transf.* **2014**, *75*, 442–449. [CrossRef]
34. Singh, O.; Prasad, B.N. Comparative performance analysis of cogeneration gas turbine cycle for different blade cooling means. *Int. J. Therm. Sci.* **2009**, *48*, 1432–1440.
35. Zheng, M.; Gan, H.; Narayan, P.S.; Li, Y.; Chyu, M.K. Experimental Investigation on Additively Manufactured Transpiration and Film Cooling Structures. *J. Turbomach.* **2019**, *141*, 031009.
36. Frank, G.; Pfitzner, M. Investigation of the heat transfer coefficient in a transpiration film cooling with chemical reactions. *Int. J. Heat Mass Transf.* **2017**, *113*, 755–763. [CrossRef]
37. Kumar, S.; Singh, O. Thermodynamic performance evaluation of gas turbine cycle with transpiration cooling of blades using air vis-à-vis steam. *Proc. Inst. Mech. Eng. Part A J. Power Energy* **2010**, *224*, 1039–1047. [CrossRef]
38. Liu, J.; Xu, M.; Xi, W.; Song, J.; Luo, S.; Sunden, B.A. Numerical investigations of endwall film cooling design of a turbine vane using four-holes pattern. *Int. J. Numer. Methods Heat Fluid Flow* **2021**, *32*, 2177–2197. [CrossRef]
39. Wambersie, A.; Wong, H.; Ireland, P.; Mayo, I. Experiments of transpiration cooling inspired panel cooling on a turbine blade yielding film effectiveness levels over 95%. *Int. J. Turbomach. Propuls. Power* **2021**, *6*, 16. [CrossRef]
40. Arai, M.; Suidzu, T. Porous ceramic coating for transpiration cooling of gas turbine blade. *J. Therm. Spray Technol.* **2013**, *22*, 690–698. [CrossRef]
41. Langener, T.; von Wolfersdorf, J.; Selzer, M.; Hald, H. Experimental investigations of transpiration cooling applied to C/C material. *Int. J. Therm. Sci.* **2012**, *54*, 70–81. [CrossRef]
42. Xiong, Y.; Zhu, Y.; Jiang, P. Numerical simulation of transpiration cooling for sintered metal porous strut of the scramjet combustion chamber. *Heat Transf. Eng.* **2014**, *35*, 721–729. [CrossRef]
43. He, F.; Wu, N.; Ran, F.; Wang, J. Numerical investigation on the transpiration cooling of three-dimensional hypersonic inlet. *Aerosp. Sci. Technol.* **2020**, *106*, 106152. [CrossRef]
44. Ding, R.; Wang, J.; He, F.; Wang, M.; Luan, Y.; Dong, G.; Tang, L. Numerical investigation on a double layer combined cooling structure for aerodynamic heat control of hypersonic vehicle leading edge. *Appl. Therm. Eng.* **2020**, *169*, 114949. [CrossRef]
45. Zhang, H.J.; Zou, Z.P. Numerical investigation of laminar-plate transpiration cooling by the preconditioned density-based algorithm. *Numer. Heat Transf. Part A Appl.* **2012**, *62*, 761–779. [CrossRef]
46. Xu, G.; Liu, Y.; Luo, X.; Ma, J.; Li, H. Experimental investigation of transpiration cooling for sintered woven wire mesh structures. *Int. J. Heat Mass Transf.* **2015**, *91*, 898–907. [CrossRef]
47. Liu, J.; Hussain, S.; Wang, L.; Xie, G.; Sundén, B.; Abrahamsson, H.; Arroyo, C. The effect of the pocket on the heat transfer of endwall with bluff body in the rear part of gas turbine. In Proceedings of the ASME Turbo Expo 2017: Turbomachinery Technical Conference and Exposition, Charlotte, NC, USA, 26–30 June 2017; Volume 50879, p. V05AT11A001.
48. Wu, H.; Ding, M.; Su, Y. The study of cavity flow and transpiration cooling in supersonic combustion. *Appl. Mech. Mater.* **2013**, *2644*, 370–374. [CrossRef]
49. Zhang, B.; Wang, H.; Jin, X.; Fan, X.; Huang, H.; Yao, J. Novel C/SiC porous ceramic with controllable properties served as transpiration cooling material. *Ceram. Int.* **2022**, *48*, 21616–21621. [CrossRef]
50. Wu, Y.; Zou, Z.; Fu, C.; Zhang, W. Numerical study of porous blunt nosecone transpiration cooling under supersonic incoming flow. *J. Porous Media* **2017**, *20*, 607–618. [CrossRef]
51. Yang, H.; Liu, X.; Bian, Y.; Wang, G. Numerical investigation on the mechanism of transpiration cooling for porous struts based on local thermal non-equilibrium model. *Energies* **2022**, *15*, 2091. [CrossRef]
52. Liu, T.; Su, H.; Chen, Z.; He, F.; Wang, J. Numerical investigation on the transient transport and heat transfer characteristics of transpiration cooling with liquid phase change during coolant adjustment. *Appl. Therm. Eng.* **2022**, *209*, 118277. [CrossRef]
53. Zhang, Z.; Sun, X.; Wang, X.; Xie, G. Flow structure and heat transfer of transpiration cooling by using a LBM: The effects of wall blowing and spatially nonuniform injection. *Int. Commun. Heat Mass Transf.* **2021**, *127*, 105491. [CrossRef]
54. Campbell, N.S.; Hanquist, K.; Morin, A.; Meyers, J.; Boyd, I. Evaluation of computational models for electron transpiration cooling. *Aerospace* **2021**, *8*, 243. [CrossRef]
55. Bandivadekar, D.; Minisci, E. Modelling and simulation of transpiration cooling systems for atmospheric re-entry. *Aerospace* **2020**, *7*, 29. [CrossRef]

56. Cheng, Z.; Li, X.; Xu, R.; Jiang, P. Investigations on porous media customized by triply periodic minimal surface: Heat transfer correlations and strength performance. *Int. Commun. Heat Mass Transf.* **2021**, *129*, 105713. [CrossRef]
57. da Silva, C.B.; MÉTais, O. On the influence of coherent structures upon interscale interactions in turbulent plane jets. *J. Fluid Mech.* **2002**, *473*, 103–145. [CrossRef]
58. Shi, X.; Yang, Z.; Chen, W.; Chyu, M.K. Investigation of the effect of lattice structure on the fluid flow and heat transfer of supercritical CO2 in tubes. *Appl. Therm. Eng.* **2022**, *207*, 118132. [CrossRef]
59. Zhao, J.; Chen, S. Non-darcy effects on nonparallel thermal instability of horizontal natural convection flow. *J. Thermophys. Heat Transf.* **2003**, *17*, 150–158. [CrossRef]
60. Yang, Z.; Luo, X.; Chen, W.; Chyu, M.K. Mitigation effects of body-centered cubic lattices on the heat transfer deterioration of supercritical CO_2. *Appl. Therm. Eng.* **2021**, *183*, 116085. [CrossRef]
61. Wang, J.; Ge, J.; Fan, Y.; Fu, Y.; Liu, X. Flow behavior and heat transfer in a rectangular channel with miniature riblets. *Int. Commun. Heat Mass Transf.* **2022**, *135*, 106049. [CrossRef]
62. Yi, L.; Yang, S.; Pan, M. Experimental investigation and parameter analysis of micro-jet impingement heat sink for improved heat transfer performance. *Chem. Eng. Process. Process Intensif.* **2022**, *174*, 108867. [CrossRef]
63. Song, J.; Yuan, Y.; Liu, J.; Luo, S.; Sunden, B. Heat transfer enhancement of regenerative cooling channel with pyramid lattice sandwich structures. *Heat Transf. Eng.* **2022**. [CrossRef]
64. Yun, S.; Kwon, J.; Lee, D.; Shin, H.H.; Kim, Y. Heat transfer and stress characteristics of additive manufactured FCCZ lattice channel using thermal fluid-structure interaction model. *Int. J. Heat Mass Transf.* **2020**, *149*, 119187. [CrossRef]
65. Furman, E.M.; Anghaie, S. Thermal hydraulic design analysis of ternary carbide fueled square-lattice honeycomb nuclear rocket engine. *AIP Conf. Proc.* **2008**, *458*, 1501.
66. Widargo, R.; Anghaie, S. Nuclear design analysis of square-lattice honeycomb space nuclear rocket engine. *AIP Conf. Proc.* **2008**, *458*, 1507–1512.
67. Li, Y.; Xie, G.; Boetcher, S.K.S.; Yan, H. Heat transfer enhancement of X-lattice-cored sandwich panels by introducing pin fins, dimples or protrusions. *Int. J. Heat Mass Transf.* **2019**, *141*, 627–642. [CrossRef]
68. Xu, L.; Shen, Q.; Ruan, Q.; Xi, L.; Gao, J.; Li, Y. Optimization design of lattice structures in internal cooling channel of turbine blade. *Appl. Sci.* **2021**, *11*, 5838. [CrossRef]
69. Shen, B.; Li, Y.; Yan, H.; Boetcher, S.K.S.; Xie, G. Heat transfer enhancement of wedge-shaped channels by replacing pin fins with Kagome lattice structures. *Int. J. Heat Mass Transf.* **2019**, *141*, 88–101. [CrossRef]
70. Moss, A.; Macquart, T.; Panesar, A.; Forrest, M.; Greaves, P.; Pirrera, A. Structural design of wind turbine blades with an additively manufactured graded lattice core using topology optimisation. *J. Phys. Conf. Ser.* **2022**, *2265*, 032004. [CrossRef]
71. Feng, X.B.; Xu, H.J. Modeling the propane combustion process within a micro-catalytic porous combustor by using the lattice Boltzmann method. *J. Therm. Anal. Calorim.* **2020**, *139*, 2659–2677. [CrossRef]
72. Du, W.; Luo, L.; Wang, S.; Liu, J.; Sunden, B. Numerical investigation of flow field and heat transfer characteristics in a latticework duct with jet cooling structures. *Int. J. Therm. Sci.* **2020**, *158*, 106553. [CrossRef]
73. Du, W.; Luo, L.; Wang, S.; Liu, J.; Sunden, B. Heat transfer and flow structure in a detached latticework duct. *Appl. Therm. Eng.* **2019**, *155*, 24–39. [CrossRef]
74. Zhang, Y.; Gu, W.; Han, J. Heat transfer and friction in rectangular channels with ribbed or ribbed-grooved walls. *J. Heat Transf.* **1994**, *116*, 58–65. [CrossRef]
75. Saha, K.; Acharya, S.; Nakamata, C. Heat transfer enhancement and thermal performance of lattice structures for internal cooling of airfoil trailing edges. *J. Therm. Sci. Eng. Appl.* **2013**, *5*, 011001. [CrossRef]
76. Wang, G. A review of research progress in heat exchanger tube rupture accident of heavy liquid metal cooled reactors. *Ann. Nucl. Energy* **2017**, *109*, 1–8. [CrossRef]
77. Mohamed, A.A.; Younis, O. Performance of drop shaped pin fin heat exchanger with four different fin dimensions. *J. Mech. Eng. Sci.* **2020**, *14*, 6934–6951. [CrossRef]
78. Zhao, L.; Ryan, S.M.; Ortega, J.K.; Ha, S.; Sharp, K.W.; Guest, J.K.; Hemker, K.J.; Weihs, T.P. Experimental investigation of 3D woven Cu lattices for heat exchanger applications. *Int. J. Heat Mass Transf.* **2016**, *96*, 296–311. [CrossRef]
79. Lai, X.; Wang, C.; Peng, D.; Yang, H.; Wei, Z. Analysis of heat transfer characteristics of a heat exchanger based on a lattice filling. *Coatings* **2021**, *11*, 1089. [CrossRef]
80. Ho, J.Y.; Leong, K.C.; Wong, T.N. Additively-manufactured metallic porous lattice heat exchangers for air-side heat transfer enhancement. *Int. J. Heat Mass Transf.* **2020**, *150*, 119262. [CrossRef]
81. Chaudhari, A.; Ekade, P.; Krishnan, S. Experimental investigation of heat transfer and fluid flow in octet-truss lattice geometry. *Int. J. Therm. Sci.* **2019**, *143*, 64–75. [CrossRef]
82. Ekade, P.; Krishnan, S. Fluid flow and heat transfer characteristics of octet truss lattice geometry. *Int. J. Therm. Sci.* **2019**, *137*, 253–261. [CrossRef]
83. Kaur, I.; Singh, P. Numerical investigation on conjugate heat transfer in octet-shape-based single unit cell thick metal foam. *Int. Commun. Heat Mass Transf.* **2021**, *121*, 105090. [CrossRef]
84. Akçay, F.A.; Wu, D.; Bai, Y. Comprehensive studies on strength of 3d-printed aluminum micro lattice structures. *Procedia Struct. Integr.* **2020**, *28*, 1399–1406. [CrossRef]

85. Maloney, K.J.; Fink, K.D.; Schaedler, T.A.; Kolodziejska, J.A.; Jacobsen, A.J.; Roper, C.S. Multifunctional heat exchangers derived from three-dimensional micro-lattice structures. *Int. J. Heat Mass Transf.* **2012**, *55*, 2486–2493. [CrossRef]
86. Wang, X.; Chen, M.; Tate, D.; Rahimi, H.; Zhang, S. Numerical investigation on hydraulic and thermal characteristics of micro latticed pin fin in the heat sink. *Int. J. Heat Mass Transf.* **2020**, *149*, 119157. [CrossRef]
87. Gopalan, K.S.; Eswaran, V. Numerical investigation of thermal performance of PCM based heat sink using structured porous media as thermal conductivity enhancers. *Int. J. Therm. Sci.* **2016**, *104*, 266–280. [CrossRef]
88. Shamvedi, D.; McCarthy, O.J.; O'Donoghue, E.; Danilenkoff, C.; O'Leary, P.; Raghavendra, R. 3D Metal printed heat sinks with longitudinally varying lattice structure sizes using direct metal laser sintering. *Virtual Phys. Prototyp.* **2018**, *13*, 301–310. [CrossRef]
89. Modrek, M.; Viswanath, A.; Khan, K.A.; Ali, M.I.H.; Al-Rub, R.K.A. An optimization case study to design additively manufacturable porous heat sinks based on triply periodic minimal surface (TPMS) lattices. *Case Stud. Therm. Eng.* **2022**, *36*, 102161. [CrossRef]
90. Wu, L.; Dai, N.; Wang, H. Evaluation of rods deformation of metal lattice structure in additive manufacturing based on skeleton extraction technology. *Math. Biosci. Eng. MBE* **2021**, *18*, 7525–7538. [CrossRef]
91. Niu, J.; Choo, H.L.; Sun, W. Finite element analysis and experimental study of plastic lattice structures manufactured by selective laser sintering. *Proc. Inst. Mech. Eng. Part L J. Mater. Des. Appl.* **2017**, *231*, 171–178. [CrossRef]
92. Tian, J.; Kim, T.; Lu, T.J.; Hodson, H.P.; Queheillalt, D.T.; Sypeck, D.J.; Wadley, H.N.G. The effects of topology upon fluid-flow and heat-transfer within cellular copper structures. *Int. J. Heat Mass Transf.* **2004**, *47*, 3171–3186. [CrossRef]
93. Cheraghipoor, M.; Mohebbi-Kalhori, D.; Noroozifar, M.; Maghsoodlou, M.T. Production of greener energy in microbial fuel cell with ceramic separator fabricated using native soils: Effect of lattice and porous SiO_2. *Fuel* **2021**, *284*, 118938. [CrossRef]
94. Ho, J.Y.; Leong, K.C.; Wong, T.N. Experimental and numerical investigation of forced convection heat transfer in porous lattice structures produced by selective laser melting. *Int. J. Therm. Sci.* **2019**, *137*, 276–287. [CrossRef]
95. Jin, X.; Li, Y.; Yan, H.; Xie, G. Comparative study of flow structures and heat transfer enhancement in a metallic lattice fabricated by metal sheet folding: Effects of punching location shift. *Int. J. Heat Mass Transf.* **2019**, *134*, 209–225. [CrossRef]
96. Wei, K.; He, R.; Cheng, X.; Pei, Y.; Zhang, R.; Fang, D. Fabrication and heat transfer characteristics of C/SiC pyramidal core lattice sandwich panel. *Appl. Therm. Eng.* **2015**, *81*, 10–17. [CrossRef]
97. Jebauer, S.; Czerwinska, J. Three-dimensional vortex structures on heated micro-lattice in a gas. *Int. J. Heat Mass Transf.* **2014**, *70*, 827–834. [CrossRef]
98. Yan, B. A new lattice with enhanced heat transfer characteristics. *Nucl. Eng. Des.* **2013**, *265*, 254–261. [CrossRef]
99. Gao, L.; Sun, Y. Fluid flow and heat transfer characteristics of composite lattice core sandwich structures. *J. Thermophys. Heat Transf.* **2014**, *28*, 258–269. [CrossRef]
100. Ernot, J.; Verdin, P.G.; Ahmad, H.; Indge, P. Analytical and numerical predictions of the thermal performance of multi-layered lattice structures. *Int. J. Heat Mass Transf.* **2019**, *145*, 118752. [CrossRef]
101. Downing, D.; Leary, M.; McMillan, M.; Alghamdi, A.; Brandt, M. Heat transfer in lattice structures during metal additive manufacturing: Numerical exploration of temperature field evolution. *Rapid Prototyp. J.* **2020**, *26*, 911–928. [CrossRef]
102. Ma, Y.; Yan, H.; Xie, G. Flow and thermal performance of sandwich panels with plate fins or/and pyramidal lattice. *Appl. Therm. Eng.* **2020**, *164*, 114468. [CrossRef]
103. Wang, X.; Wei, K.; Wang, K.; Yang, X.; Qu, Z.; Fang, D. Effective thermal conductivity and heat transfer characteristics for a series of lightweight lattice core sandwich panels. *Appl. Therm. Eng.* **2020**, *173*, 115205. [CrossRef]
104. Zhang, J.; Li, P.; Qian, B.; Li, B.; Qiu, Z.; Xuan, F. RETRACTED ARTICLE: Selective laser melting of G-surface lattice: Forming process and boiling heat transfer characteristics. *J. Nanoparticle Res.* **2020**, *22*, 1–10. [CrossRef]
105. Liang, D.; Chen, W.; Ju, Y.; Chyu, M.K. Comparing endwall heat transfer among staggered pin fin, Kagome and body centered cubic arrays. *Appl. Therm. Eng.* **2021**, *185*, 116306. [CrossRef]
106. Yun, S.; Lee, D.; Jang, D.S.; Lee, M.; Kim, Y. Numerical analysis on thermo-fluid–structural performance of graded lattice channels produced by metal additive manufacturing. *Appl. Therm. Eng.* **2021**, *193*, 117024. [CrossRef]
107. Gao, L.; Sun, Y. Thermal control of composite sandwich structure with lattice truss cores. *J. Thermophys. Heat Transf.* **2015**, *29*, 47–54. [CrossRef]
108. Bai, X.; Zheng, Z.; Nakayama, A. Heat transfer performance analysis on lattice core sandwich panel structures. *Int. J. Heat Mass Transf.* **2019**, *143*, 118525. [CrossRef]
109. Xiao, H.; Chen, W.; Yan, B. Heat transfer and friction in nonuniform wall roughness lattice. *Prog. Nucl. Energy* **2015**, *83*, 43–51. [CrossRef]
110. Liu, X.; Bian, Y.; Zhou, W.; Wu, C.; Jia, Z. Investigation on heat transfer during transpiration cooling with hydrocarbon fuel coolant. *Int. J. Heat Mass Transf.* **2023**, *200*, 123507. [CrossRef]
111. Glass, D.E.; Dilley, A.D.; Kelly, H.N. Numerical analysis of convection/transpiration cooling. *Proc. 9th Int. Space Planes Hypersonic Syst. Technol. Conf.* **2001**, *38*, 15–20. [CrossRef]
112. Gritskevich, M.S.; Garbaruk, A.V.; Schütze, J.; Menter, F.R. Development of DDES and IDDES Formulations for the k-ω Shear Stress Transport Model. *Flow Turbul. Combust.* **2012**, *88*, 431–449. [CrossRef]
113. Qian, K.; Liu, T.; He, F.; Wang, M.; Tang, L.; Zhou, J. Numerical investigation on radiation effect in transpiration cooling. *J. Phys. Conf. Ser.* **2021**, *2097*, 012011. [CrossRef]

114. Peng, Y.; Xu, G.; Luo, X.; He, J.; Liu, D. Experimental investigation on the transpiration cooling characteristics of sintered wire mesh in plain weave. *Micromachines* **2022**, *13*, 450. [CrossRef]
115. Jiang, P.; Huang, G.; Zhu, Y.; Liao, Z.; Huang, Z. Experimental investigation of combined transpiration and film cooling for sintered metal porous struts. *Int. J. Heat Mass Transf.* **2017**, *108*, 232–243. [CrossRef]
116. Huang, G.; Zhu, Y.; Liao, Z.; Ouyang, X.; Jiang, P. Experimental investigation of transpiration cooling with phase change for sintered porous plates. *Int. J. Heat Mass Transf.* **2017**, *114*, 1201–1213. [CrossRef]
117. Kim, M.; Shin, D.H.; Lee, B.J.; Lee, J. Experimental and numerical investigation of micro-scale effusion and transpiration air cooling on cascaded turbine blades. *Case Stud. Therm. Eng.* **2022**, *32*, 101892. [CrossRef]
118. Kim, M.; Shin, D.H.; Lee, B.J.; Lee, J. Flow characterization of microscale effusion and transpiration air cooling on single blade. *Case Stud. Therm. Eng.* **2022**, *31*, 101863. [CrossRef]
119. Kumar, S.; Singh, O. Enhancement of combined cycle performance using transpiration cooling of gas turbine blades with steam. *J. Mech. Sci. Technol.* **2014**, *28*, 2429–2437. [CrossRef]
120. Kim, M.; Shin, D.H.; Kim, J.S.; Lee, B.J.; Lee, J. Experimental investigation of effusion and transpiration air cooling for single turbine blade. *Appl. Therm. Eng.* **2021**, *182*, 116156. [CrossRef]
121. Wu, N.; Wang, J.; He, F.; Dong, G.; Tang, L. An experimental investigation on transpiration cooling of a nose cone model with a gradient porosity layout. *Exp. Therm. Fluid Sci.* **2019**, *106*, 194–201. [CrossRef]
122. Zhang, T.; Feng, Q.; Cai, W.; He, W. Numerical analysis of transpiration cooling in annular slinger combustor. *IOP Conf. Ser. Earth Environ. Sci.* **2019**, *234*, 012058. [CrossRef]
123. Shen, L.; Wang, J.; Dong, W.; Pu, J.; Peng, J.; Qu, D.; Chen, L. An experimental investigation on transpiration cooling with phase change under supersonic condition. *Appl. Therm. Eng.* **2016**, *105*, 549–556. [CrossRef]
124. Xing, Y.; Weigand, B. Experimental investigation of impingement heat transfer on a flat and dimpled plate with different crossflow schemes. *Int. J. Heat Mass Transf.* **2010**, *53*, 3874–3886. [CrossRef]
125. Qian, K.; Wang, J.; He, F.; Wu, Y.; Zhou, Z. An experimental investigation on transpiration cooling performances using solid hydrogel as coolant. *Appl. Therm. Eng.* **2019**, *158*, 113753. [CrossRef]
126. He, F.; Wu, N.; Zhou, Z.; Wang, J. Experimental investigation on transpiration cooling using propylene glycol aqueous solution. *Int. J. Therm. Sci.* **2021**, *164*, 106890. [CrossRef]
127. Huang, Z.; Zhu, Y.; Xiong, Y.; Jiang, P. Investigation of transpiration cooling for sintered metal porous struts in supersonic flow. *Appl. Therm. Eng.* **2014**, *70*, 240–249. [CrossRef]
128. Chamoli, S. ANN and RSM approach for modeling and optimization of designing parameters for a V down perforated baffle roughened rectangular channel. *Alex. Eng. J.* **2015**, *54*, 429–446. [CrossRef]
129. Jiang, P.; Yu, L.; Sun, J.; Wang, J. Experimental and numerical investigation of convection heat transfer in transpiration cooling. *Appl. Therm. Eng.* **2004**, *24*, 1271–1289. [CrossRef]
130. Liu, Y.; Jiang, P.; Jin, S.; Sun, J. Transpiration cooling of a nose cone by various foreign gases. *Int. J. Heat Mass Transf.* **2010**, *53*, 5364–5372. [CrossRef]
131. Kilic, M. A numerical analysis of transpiration cooling as an air cooling mechanism. *Heat Mass Transf.* **2018**, *54*, 3647–3662. [CrossRef]
132. Jiang, P.; Liao, Z.; Huang, Z.; Xiong, Y.; Zhu, Y. Influence of shock waves on supersonic transpiration cooling. *Int. J. Heat Mass Transf.* **2019**, *129*, 965–974. [CrossRef]
133. Bellettre, J.; Bataille, F.; Lallemand, A.; Andoh, H.Y. Studies of the transpiration cooling through a sintered stainless steel plate. *Exp. Heat Transf.* **2005**, *18*, 33–44. [CrossRef]
134. Yang, L.; Chen, W.; Chyu, M.K. A convolution modeling method for pore plugging impact on transpiration cooling configurations perforated by straight holes. *Int. J. Heat Mass Transf.* **2018**, *126*, 1057–1066. [CrossRef]
135. Ding, R.; Wang, J.; He, F.; Dong, G.; Tang, L. Numerical investigation on the performances of porous matrix with transpiration and film cooling. *Appl. Therm. Eng.* **2019**, *146*, 422–431. [CrossRef]
136. Shen, B.; Yin, L.; Liu, H.; Liu, W. Thermal protection characteristics for a combinational opposing jet and platelet transpiration cooling nose-tip. *Acta Astronaut.* **2019**, *155*, 143–152. [CrossRef]
137. Yang, L.; Min, Z.; Yue, T.; Rao, Y.; Chyu, M.K. High resolution cooling effectiveness reconstruction of transpiration cooling using convolution modeling method. *Int. J. Heat Mass Transf.* **2019**, *133*, 1134–1144. [CrossRef]
138. He, F.; Wang, J.; Xu, L.; Wang, X. Modeling and simulation of transpiration cooling with phase change. *Appl. Therm. Eng.* **2013**, *58*, 173–180. [CrossRef]
139. He, F.; Wang, J. Numerical investigation on critical heat flux and coolant volume required for transpiration cooling with phase change. *Energy Convers. Manag.* **2014**, *80*, 591–597. [CrossRef]
140. He, F.; Dong, W.; Wang, J.; Zhou, J.; Tang, L. Transient model and its application to investigate the injection mode and periodical operation of transpiration cooling with liquid coolant phase change. *Appl. Therm. Eng.* **2020**, *181*, 115956. [CrossRef]
141. Dong, W.; Wang, J. A new model and its application to investigate transpiration cooling with liquid coolant phase change. *Transp. Porous Media* **2018**, *122*, 575–593. [CrossRef]
142. Shi, J.; Wang, J. A numerical investigation of transpiration cooling with liquid coolant phase change. *Transp. Porous Media* **2011**, *87*, 703–716. [CrossRef]

143. Huang, G.; Zhu, Y.; Liao, Z.; Xu, R.; Jiang, P. Biomimetic self-pumping transpiration cooling for additive manufactured porous module with tree-like micro-channel. *Int. J. Heat Mass Transf.* **2019**, *131*, 403–410. [CrossRef]
144. Huang, G.; Zhu, Y.; Liao, Z.; Huang, Z.; Jiang, P. Transpiration cooling with bio. *Bioinspiration Biomim.* **2020**, *15*, 036016. [CrossRef]
145. Shi, J.; Wang, J. A numerical investigation on the laminar boundary flow layer with transpiration cooling. *Transp. Porous Media* **2008**, *78*, 37–46. [CrossRef]
146. Dahmen, W.; Müller, S.; Rom, M.; Schweikert, S.; Selzer, M.; von Wolfersdorf, J. Numerical boundary layer investigations of transpiration-cooled turbulent channel flow. *Int. J. Heat Mass Transf.* **2015**, *86*, 90–100. [CrossRef]
147. Xiao, X.; Zhao, G.; Zhou, W.; Martynenko, S. Large-eddy simulation of transpiration cooling in turbulent channel with porous wall. *Appl. Therm. Eng.* **2018**, *145*, 618–629. [CrossRef]
148. Rehman, M.M.U.; Qu, Z.G.; Fu, R.P.; Xu, H.T. Numerical study on free-surface jet impingement cooling with nanoencapsulated phase-change material slurry and nanofluid. *Int. J. Heat Mass Transf.* **2017**, *109*, 312–325. [CrossRef]

aerospace

MDPI

Article

Atomization Characteristics of Gelled Fuels Containing Different Concentrations of Metal Particles

Penghui Li [1,2], Dong Yang [3], Zixuan Fang [1], Qingfei Fu [1,*] and Lijun Yang [1]

[1] School of Astronautics, Beihang University, Beijing 100191, China
[2] Shenyuan Honors College, Beihang University, Beijing 100191, China
[3] Xi'an Aerospace Composites Research Institute, Xi'an 710025, China
* Correspondence: fuqingfei@buaa.edu.cn

Abstract: Gelled fuels have promising applications in the aerospace field. Higher density and calorific value can be achieved with the addition of energetic metal particles to gelled fuels, which can also effectively improve the combustion efficiency of the fuel and thus enhance the engine performance. However, the addition of metal particles can also make the rheological properties of gelled fuels more complex, which introduces difficulties regarding their atomization and combustion. In order to investigate the effect of the concentration of metal particles on the rheological and atomization characteristics of gelled fuels, the gelled fuel was prepared with three metal particle concentrations of 0%, 15%, and 30%. In this paper, the rheological properties of the gelled fuel were tested by a rotational rheometer, and the atomization properties (spray cone angle, Sauter mean diameter (SMD), and droplet size distribution) of the gelled fuel were measured experimentally. In this paper, three nozzle structures were designed, including a DC nozzle, a swirl nozzle, and a self-excited oscillation nozzle. The effects of different nozzle structures and metal particle concentrations on the atomization of gelled fuels are compared and discussed.

Keywords: gelled fuel; rheological properties; atomization characteristics; Sauter mean diameter

Citation: Li, P.; Yang, D.; Fang, Z.; Fu, Q.; Yang, L. Atomization Characteristics of Gelled Fuels Containing Different Concentrations of Metal Particles. *Aerospace* **2023**, *10*, 221. https://doi.org/10.3390/aerospace10030221

Academic Editor: Kevin Lyons

Received: 7 February 2023
Revised: 22 February 2023
Accepted: 23 February 2023
Published: 25 February 2023

1. Introduction

Gelled fuels can behave as solids at rest and can also flow under external shear like conventional liquid fuels, which makes them promising with regard to potential aerospace applications [1–4]. However, gelled fuels are typically non-Newtonian fluids, and both their viscosity and elasticity significantly affect atomization performance. A study by Padwal and Mishra [5] showed that even if the apparent viscosities of two gelled fuels are equal, the difference in their elasticity results in a large difference between their respective Sauter mean diameters (SMD, D_{32}). The atomization problem remains one of the key issues for the application of gelled fuels.

The atomization mechanism indicates that the increase in fuel viscosity increases the viscous dissipation during the atomization process and inhibits destabilization of the jet surface, which, in turn, hinders the fragmentation and atomization of the fuel. The shear thinning property means that the viscosity of gelled fuel decreases with the increase in external shear stress. The atomization effect can be improved by increasing the shear stress on the fuel inside the nozzle to reduce its viscosity. The experimental gel engine developed by Rahimi and Natan [6] used a nozzle structure with conical channels. They initially calculated the velocity and viscosity distribution of the gelled fuel in the nozzle, and the results showed that the average apparent viscosity of the fuel decreased significantly at the nozzle exit, and that increasing the channel convergence angle could achieve better spray quality. Madlener et al. [7] experimentally investigated the effect of the convergence angle on the fuel flow and spray behavior. However, the improvement of atomization by increasing the convergence angle was also limited. The results of Rahimi and Natan's experiment [8] showed that, at the same fuel mass flow rate, the mean diameter of droplets produced by gelled fuel atomization at cone angles of 30° and 2° appeared to be the same.

Yang et al. [9] investigated the atomization characteristics of gel simulating fluids using swirl nozzles. They found that it is more difficult for gelled fuel to form a liquid film compared to a Newtonian fluid. Under the same pressure conditions, the gel simulating fluid was ejected as a nearly cylindrical jet, while the aqueous solution was able to form a stable conical liquid film. The breakup length of the gel simulant gradually decreased as the injection pressure gradually increased, which is consistent with the law of Newtonian fluids. However, for gelled fuels, even at higher injection pressure, there are still a large number of unbroken liquid filaments in the spray field, and fine droplets cannot be obtained.

The addition of energetic solid particles, such as aluminum, boron, and magnesium, to gelled fuels is an important means of improving the energy characteristics of the fuel [10]. The volumetric calorific value of these energetic solid particles is significantly higher than that of conventional liquid hydrocarbon fuels, and thus gelled fuels with additional energetic solid particles have a higher density and volumetric calorific value [11]. However, the introduction of energetic solid particles also makes the rheological properties of the fuel more complex, which makes the atomization of gelled fuels more difficult than that of conventional fuels. Kim et al. [12] investigated the effect of metal particle concentration and average particle size on the atomization performance of slurry fuels using a swirl nozzle. The research results showed that, with the increase in metal particle concentration, the thickness of the film gradually increased, the instability of the liquid film was suppressed, and the position of the liquid filament and droplet generation in the spray field gradually moved downstream. Furthermore, the breakup length of the liquid film gradually decreases with the increase in the average diameter of the particles, which is due to the change of the breakup mechanism of the liquid film caused by the increase in the particle diameter [12]. Additionally, the authors point out that the spray cone angle of the gelled fuel is determined by the fuel viscosity, and the increase in both particle concentration and average diameter increases the fuel viscosity, which, in turn, affects the spray cone angle of the fuel. Kampen et al. [13] investigated the effect of aluminum particle concentration on the atomization characteristics of gelled fuels in detail. Their results showed that gelled fuels with high particle concentration (40%) could be atomized by impinging nozzles, and the form of fuel atomization was related to the concentration of aluminum particles and Reynolds number. Baek et al. [14] measured the atomized droplet size of water and C934 Carbopol gels with/without nanoparticles using an image processing method. It was noted that the addition of nanoparticles made it possible to reduce the strength of the gelled fuel and make its breakup length much smaller than that of the pure gel.

Due to the complex characteristics of the gelled fuel containing metal particles, fuel atomization by conventional atomization means is not satisfactory [15–18]. Jia et al. [19] proposed a self-excited oscillation nozzle for nanoparticle-containing slurry fuels and conducted a detailed experimental study on the effects of different self-excited oscillation nozzle structures. The results show that the self-excited oscillation nozzle has a smaller discharge coefficient, better spray quality, and a more uniform droplet size distribution compared to the conventional DC nozzle. Additionally, the authors point out that the self-excited oscillation nozzle is more stable and reliable than the conventional DC nozzle, due to the strong internal turbulence. Li et al. [20] proposed an improved single-phase atomization nozzle based on the rheological properties of gelled fuels. It was shown that the new nozzle can effectively reduce the SMD of gelled fuel containing solid particles by introducing a cone-like structure into the nozzle.

The aim of this paper is to investigate the effect of the concentration of metal particles on the rheological and atomization properties of gelled fuels. This paper is structured as follows: Section 2 introduces the preparation of gelled fuels and describes the effect of metal particle concentration on their rheological properties. Section 3 describes the experimental system and method, in which three nozzle structures are designed, including a DC nozzle, a swirl nozzle, and a self-excited oscillation nozzle. Section 4 compares the atomization characteristics of different nozzles and discusses the effect of metal particle concentration on fuel atomization characteristics. Section 5 summarizes the research results of this paper.

2. Gelled Fuel Preparation and Their Rheological Properties

For this paper, the effect of different aluminum nanoparticle concentrations on the rheological properties of gelled fuel was investigated, using JP-10 as the base liquid fuel. The metal particles were selected as Al nanoparticles. The preparation process of Al/JP-10 gelled fuel mainly involves: the preparation of aluminum nanoparticles, the preparation of a small molecule gelling agent and the preparation of gelled fuel. In the experiments, a small molecule gelling agent and nanoparticles were added into the JP-10 solution separately, and the gelled fuel was obtained with mechanical stirring and standing processes. The preparation process is detailed in a study by Cao et al. [21]. Three concentrations of AL/JP-10 gelled fuel were selected for this paper: 0%, 15%, and 30%. The three solutions were named GF-0, GF-15, and GF-30, respectively, in accordance with their concentration of aluminum nanoparticles. The density of the three gelled fuels was 0.935 g/mL, 1.0582 g/mL, and 1.1758 g/mL, respectively.

The rheological curves of different gelled fuels, i.e., apparent viscosity versus shear rate, were measured for this paper using an Anton Paar rotary rheometer (MCR92). The measurement temperature was set at 25 °C. The shear rate was taken in the range of 0.1–1000 s^{-1}, and the sampling time of the experimental data was varied logarithmically. The rheometer is a tapered plate test system, which is suitable for measuring samples with high viscosity, and only very small solid particles. Gelled fuel has very complex rheological properties, including yield stress and thixotropy. Therefore, under different tests, its rheological curve may behave differently, but its overall law is unchanged.

Figure 1 shows the rheological curves of gelled fuels containing different concentrations of Al nanoparticles. The gelled fuel as a whole shows a shear thinning characteristic, i.e., the apparent viscosity of the gelled fuel gradually decreased with the increasing shear rate. This is because the external shear can disturb or even disrupt the mesh structure of the gel, which can release the liquid fuel trapped therein and thus reduce the apparent viscosity of the gelled fuel [21]. Furthermore, the addition of aluminum nanoparticles did not change the shear thinning characteristics of the gelled fuels. Generally, gelled fuels undergo different levels of shear action during the process of preparation and use, and it is generally believed that the typical shear rate of gelled fuel during atomization is larger than 1000 s^{-1} [1]. It can also be seen that gelled fuels retain their shear thinning properties at high shear rates, and their viscosity is still higher than that of conventional Newtonian fluids. The apparent viscosity of three gelled fuels at a high shear rate of 1000 s^{-1} is 8.376, 13.841, and 62.384 mPa s, which indicates that the viscosity of gelled fuel is increased by the addition of Al nanoparticles. To describe the rheological properties of gelled fuels, the power-law flow model was used to fit the experimental data. The power-law model is:

$$\eta = K \cdot \dot{\gamma}^{n-1} \tag{1}$$

where K is the consistency coefficient and n is the power-law index. Table 1 shows the consistency coefficient and power-law index of gelled fuels containing different concentrations of nanoparticles. In general, a higher consistency coefficient indicates a more viscous fuel, while a higher power law index indicates a nearly Newtonian behavior. When n is equal to 1, the fuel is a typical Newtonian liquid. It can be seen that both the consistency coefficient and the power-law index increased with the concentrations of nanoparticles, which means that the addition of nanoparticles makes the gelled fuels more viscous and less shear thinning. This is also in agreement with the results of Cao et al. [21] and Madlener et al. [22]. Additionally, some local irregularities exist in the flow curve, particularly for GF−15 at a shear rate between 40 and 50 s^{-1}. This is caused by relative displacement of particles, which may occur during shearing because of the difference in density between liquid fuel and solid particles. This results in the agglomeration of small particles or the disintegration of large particle clusters. Therefore, the flow curve becomes not smooth, creating some local irregularities.

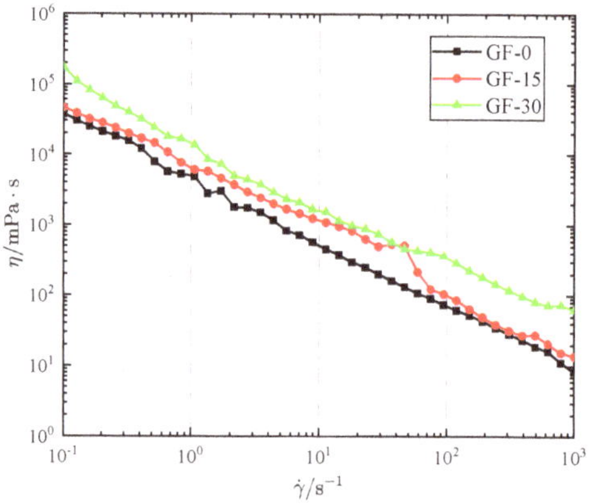

Figure 1. Rheological curves of gelled fuels containing different concentrations of Al nanoparticles.

Table 1. The consistency coefficient and power-law index of gelled fuels containing different concentrations of nanoparticles.

Index	GF-0	GF-15	GF-30
Consistency coefficient $K/\text{Pa s}^n$	4466.8	7413.1	13,489.6
Power-law index n	0.0994	0.1017	0.1566

3. Experimental Setup

Figure 2 shows the experimental system used in this paper. The system consists of a fuel supply unit, an injector, measuring devices, and piping. The fuel supply unit contains an electrical motor, fuel tank, and piston rod. In the experiments, the motor drove the piston in a linear motion to supply the gelled fuel in the storage tank to the atomization injector.

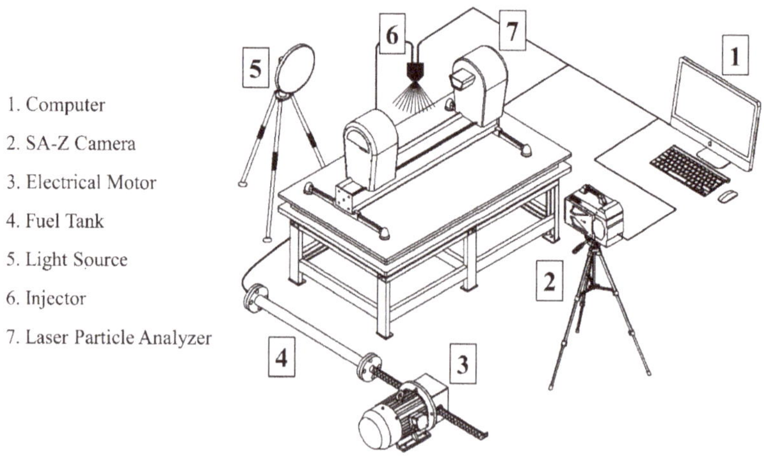

Figure 2. Experimental setup.

In order to investigate the atomization characteristics of different gelled fuels, three kinds of atomizers were designed for this paper: a DC nozzle, a swirl nozzle, and a self-excited oscillation nozzle. The DC nozzle and swirl nozzle are traditional atomizers, and they have their own advantages and disadvantages. The structure of the DC nozzle is a simple one, which is widely used in the combustion chamber of liquid rocket engines. The fuel behaves as a continuous liquid jet after ejecting from the nozzle outlet, which usually has a long penetration distance. However, the spray cone angle of the DC nozzle is typically small, and the atomization performance of the DC nozzle is generally poor. A swirl nozzle can have a better spray performance, with a larger spray cone angle, smaller droplet diameter, and more uniform droplet distribution. The disadvantage of the swirl nozzle is that the penetration depth is short, which brings the fuel mixing and combustion zone closer to the injection panel. The structure of the DC nozzle and the swirl nozzle used in this paper is shown in Figure 3. The outlet diameters of the DC nozzle and swirl nozzle are 0.42 mm and 0.46 mm, respectively. The swirl nozzle contains a cyclone, which has two 1 × 1 mm cyclonic flow grooves.

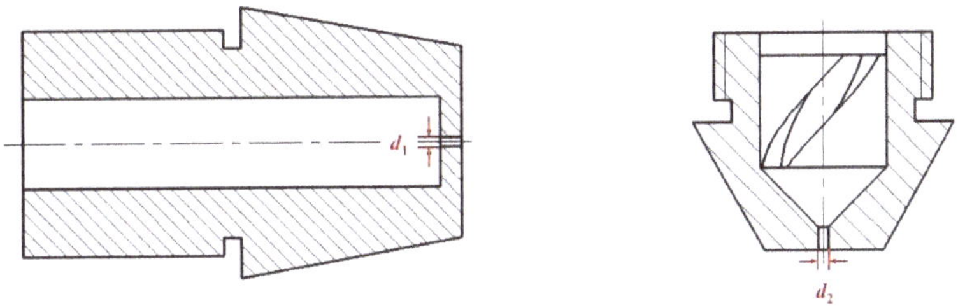

(a) DC Nozzle (b)Swirl Nozzle

Figure 3. Structure of the DC nozzle and swirl nozzle.

The self-excited oscillation nozzle is an improved single-phase atomization nozzle, which introduced a resonant chamber in the nozzle. The resonant chamber generates self-excited oscillations at a specific frequency to excite the destabilization of the jet, which could enhance the atomization of liquid fuel. Jia et al. [19] showed that self-excited oscillation nozzles are ideal atomizers for nanoparticle suspensions, while, in this paper, we tested the atomization performance of the self-excited nozzle when using the gelled fuels containing nanoparticles. This nozzle has four key parameters: oscillation chamber inlet diameter d_1, resonant chamber diameter D, resonant chamber length L, and nozzle outlet diameter d_2. The design process of the self-excitation nozzle mainly involved [19]:

(1) Calculating the unstable frequency interval of the cylindrical jet based on the physical parameters of the fuel and linear instability theory.
(2) Designing the resonant chamber size, and calculating the self-excited oscillation frequency of the resonant chamber according to the following equation [23]:

$$f = \frac{\omega}{2\pi} = \frac{\alpha d_0 \sqrt{1 + 0.64(d_2/d_1)^4}}{2\pi D \sqrt{L l_0}} \quad (2)$$

where α is the wave velocity, d_0 is the diameter of pipe before the nozzle, and l_0 is the length of pipe before the nozzle.
(3) Comparing whether the oscillation frequency of the resonant chamber is in the unstable frequency interval of the liquid jet. If not, redesigning the size of the resonant chamber.

Based on the study by Jia et al. [19], the key structure parameters of the self-excited oscillation nozzle are d_1 = 0.38 mm, D = 1.2 mm, L = 2 mm, and d_2 = 0.44 mm. The structure of the self-excited oscillation nozzle is shown in Figure 4.

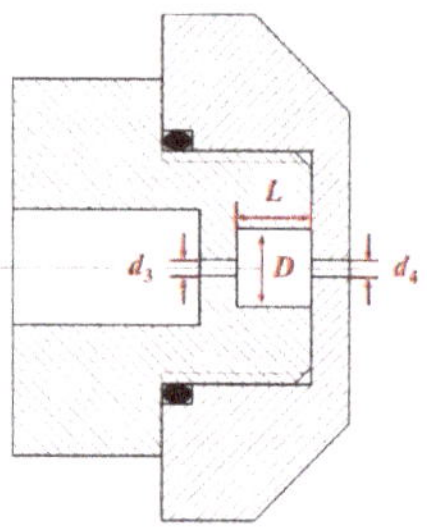

Figure 4. Structure of the self-excited oscillation nozzle.

In the experiments, the spray images of the gelled fuels were captured by a FASTCAM SA-Z high-speed camera from Photron, Japan, with a resolution of 1024 × 1024 pixels and a maximum frame rate of 200 kfps. The frame rate was set to 4000 fps in the experiments. SMD, spray field particle size distribution, and other atomization data were measured by a laser particle size analyzer from the Malvern company, with a particle size measurement range of 0.1 μm~1000 μm. The sampling rate of the particle size analyzer was 10 kHz. In the experiment, the flow rate of the gelled fuel was obtained indirectly through the displacement signal of the piston in the fuel tank, and the injection pressure of the fuel was obtained through a pressure sensor. More details of the experimental setup can be found in the study by Li et al. [20].

4. Results and Discussion

4.1. Atomization Characteristics of Different Gelled Fuels Using DC Nozzles

We tested the atomization performance of DC nozzles for different gelled fuels at the same volume flow rate. In the experiments, the volume flow rate of gelled fuels was controlled by changing the speed of the piston in the fuel tank, which is proportional to the electrical motor speed. The set volume flow rate is 7 mL/s, and the deviation of the actual flow rate from the set flow rate does not exceed 5%. Figure 5 shows spray images of different gelled fuels using DC nozzles at the same liquid volume flow rate. As can be seen in Figure 5, the atomization characteristics of the DC nozzle is similar for different types of gelled fuels, and a continuous cylindrical jet is formed when the gelled fuel is injected from the DC nozzle. Due to turbulence inside the liquid jet and the shearing effect at the gas-liquid interface, the droplets are stripped from the jet surface. However, the breakup length of the liquid jet produced by the DC nozzle is large, and large liquid clusters and filaments still exist downstream of the spray field. The large sized droplets are difficult to evaporate and burn quickly, which is harmful to engine performance. The spray angle of GF-0, GF-15, and GF-30 is 5.862°, 5.791°, and 5.574°, respectively. Table 2 shows the injection pressure and the mean droplet diameters for different gelled fuels. It is notable that the injection pressure of the fuel increases with the increase in solid particle content, while the SMD first decreases and then increases with the increase in solid particle content. This can be explained in that the apparent viscosity of GF-0 and GF-15 is approximately equal at a high shear rate, while the injection pressure of GF-15 is higher and the velocity of the liquid jet is large, so the jet breaks up more fully and therefore the SMD is smaller. The viscosity of GF-30 is much larger than that of GF-0 and GF-15, and has a greater effect on the liquid jet break-up, so the SMD of GF-30 is larger than that of the other two gelled fuels. In order to characterize the microscopic characteristics of the spray field, it is not sufficient to use the Sauter mean diameter D_{32} alone. The values of Dv50 and Dv90 are also listed in Table 2. Dv50 is the particle diameter corresponding to 50% of the volume distribution, Dv90 is the particle diameter corresponding to 90% of the volume distribution. As can be seen in the table, Dv50 is around 300 μm, which means that 50% of the droplets in the spray field are larger than 300 μm. Dv90 of the three gelled fuels are all above 600 μm, larger than the diameter of the DC nozzle exit 400 μm.

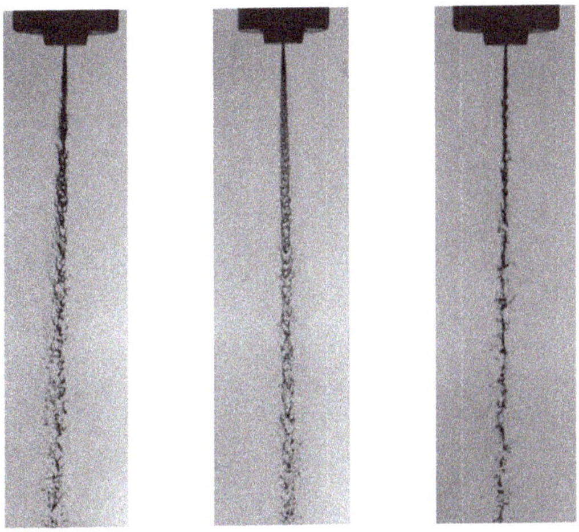

(a)GF-0 **(b) GF-15** **(c) GF-30**

Figure 5. Spray images of different gelled fuels using DC nozzles.

Table 2. The injection pressure and the mean droplet diameters of the DC nozzle for different gelled fuels.

Type	Volumetric Flow Rate/(mL/s)	Pressure/MPa	D_{32}/μm	Dv50/μm	Dv90/μm
GF-0	6.82	1.65 ± 0.005	137.42 ± 17.83	344.15	696.21
GF-15	6.66	1.94 ± 0.010	88.29 ± 9.53	239.28	632.99
GF-30	6.90	2.44 ± 0.005	167.71 ± 19.00	353.33	697.70

Figure 6 shows the droplet size distribution for different fuel types when using DC nozzles. The bar graph indicates the volume frequency of droplets whose size lie within a specific range. The solid line indicates the cumulative volume of droplets whose size is less than a specific value. As can be seen in Figure 6, the spray field of the DC nozzle has a higher percentage of large diameter droplets, and the largest volume frequency droplet has a diameter of approximately 500 μm. The droplet size distribution of GF-15 is a bimodal distribution, as predicted and seen in the spray images. In the spray field of the DC nozzle, there are both large liquid filaments and much smaller droplets which peeled off from the liquid jet. The number of these two sizes of droplets are similar to each other, so the droplet size distribution shows a bimodal distribution. Though the SMD are all around 100 μm for the three gelled fuels, the spray uniformity and fineness of the DC nozzle is not good, and a large number of un-atomized droplets still exist in the spray field. Overall, the atomization performances of DC nozzles for different gelled fuels are not satisfactory.

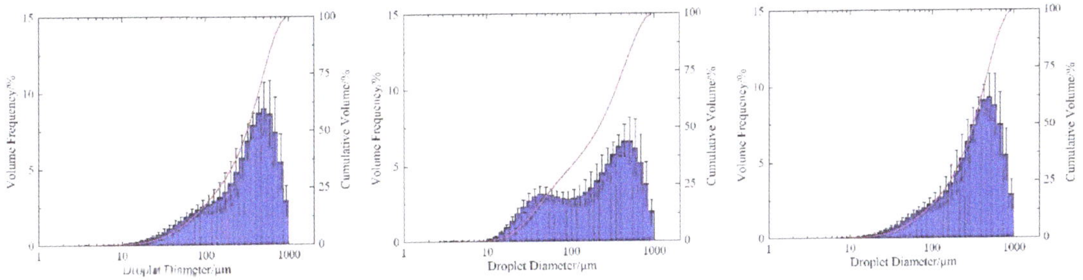

(a)GF-0 **(b) GF-15** **(c) GF-30**

Figure 6. Droplet size distribution of different gelled fuels using DC nozzles.

4.2. Atomization Characteristics of Different Gelled Fuels Using Swirl Nozzles

The atomization characteristics of different gelled fuels using swirl nozzles were also tested at the same volumetric flow rate of 7 mL/s. Figure 7 shows the spray images of swirl nozzles for three different gelled fuels at the same liquid volume flow rate. It is obvious that the atomization performance of the swirl nozzle is significantly better than that of the DC nozzle, which can produce a more uniform spray field for the gelled fuels containing different concentrations of nanoparticles. The centrifugal force causes the fuel to be ejected from the nozzle exit as a conical liquid film, which creates a larger spray cone angle than the DC nozzle. The conical liquid film is rapidly broken, forming a fine and uniform spray field. As can be seen in Figure 7, the increase in Al nanoparticle concentration makes the droplet diameter in the spray field larger, which can also be seen in the data from the laser particle size analyzer. The spray angles of GF-0, GF-15, and GF-30 are 42.2°, 34.504°, and 16.1°, respectively. It is more evident that the spray cone angle of the swirl nozzle gradually decreases with the increase in nanoparticles. This can be attributed to the increase in metal nanoparticles, which makes the fuel viscosity increase and the radial velocity of the nozzle decrease.

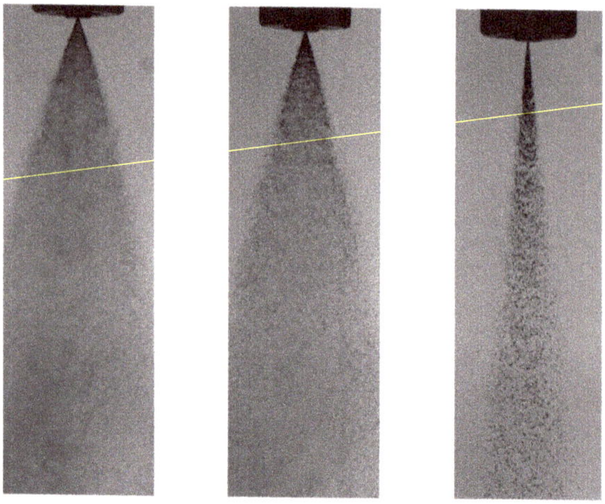

Figure 7. Spray images of different gelled fuels using swirl nozzles.

Table 3 shows the injection pressure and mean droplet diameters of the swirl nozzle for different gelled fuels. As can be seen in Table 3, D_{32}, Dv50, and Dv90 produced by the swirl nozzle for different gelled fuels are all smaller than that of the DC nozzle. As the Al nanoparticle concentrations increase, the mean droplet diameter gradually increases. Figure 8 shows the corresponding droplet size distribution of different gelled fuels. It can be seen in Figure 8 that the volume proportion of large droplets gradually increases with the increase in nanoparticle concentration, and the atomization performance gradually becomes worse, which is consistent with the spray images.

Table 3. The injection pressure and mean droplet diameters of the swirl nozzle for different gelled fuels.

Type	Volumetric Flow Rate/(mL/s)	Pressure/MPa	D_{32}/μm	Dv50/μm	Dv90/μm
GF-0	7.05	1.72 ± 0.003	36.42 ± 2.03	57.88	136.78
GF-15	6.94	1.39 ± 0.038	48.29 ± 8.18	71.60	169.57
GF-30	7.01	1.38 ± 0.010	146.53 ± 8.10	174.34	318.15

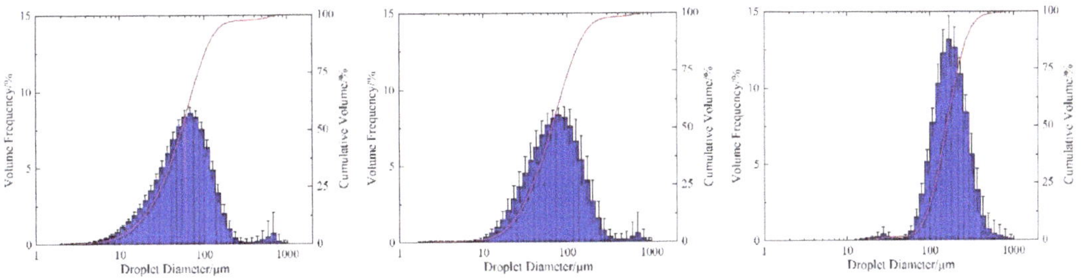

(a)GF-0 **(b)GF-15** **(c)GF-30**

Figure 8. Droplet size distribution of different gelled fuels using swirl nozzles.

Although the atomization performance of swirl nozzle is better, due to the unstable physical properties of the fuel, and the difference in density between the fuel and the energy-containing additive, the aggregation and adhesion of solid particles are significant. The fuel flow rate inside the swirl nozzle is low and some of the metal particles can obstruct the cyclone. However, the centrifugal effect of the swirl nozzle will create a separation effect between the solid particles and the fuel, which will eventually increase the deposition of metal particles on the walls of the thrust chamber.

4.3. Atomization Characteristics of Different Gelled Fuels Using Self-Excited Oscillation Nozzles

As discussed above, the use of DC nozzles alone cannot overcome the high viscosity characteristics of gelled fuels and achieve good atomization quality. To improve its atomization characteristics, it is necessary to modify the DC nozzle, such as the use of multiple DC nozzles hitting each other or changes to the nozzle structure. Figure 9 shows the spray images of different gelled fuels using self-excited oscillation nozzle at the same liquid volume flow rate. It can also be seen in Figure 9 that the atomization characteristics of the self-excited oscillation nozzle are similar for the three gelled fuels containing different concentrations of Al nanoparticles, which can all produce a uniform spray field. Near the nozzle exit of the self-excited oscillation nozzle, the fuel jet breaks up more violently, due to the introduction of the resonant chamber. As pointed by Jia et al. [19], the flow inside the nozzle is more disordered and the vortex scale inside the liquid is larger. At the same time, the spray angle of GF-0, GF-15, and GF-30 is 15.774°, 17.319°, and 14.432°, respectively. The spray cone angle is larger than that of the DC nozzle, but smaller than that of the swirl nozzle. However, as the nanoparticles increased, the spray cone angle of self-excited oscillation nozzle did not change significantly.

Table 4 shows the injection pressure and the mean droplet diameters of self-excited oscillation nozzle for different gelled fuels, while Figure 10 shows their corresponding droplet size distribution. By comparing Table 4 and Figure 10, it can be seen that the droplet diameter size and droplet size distribution produced by self-excited oscillation nozzle are similar for different types of fuels. Therefore, it can be concluded that the self-excited oscillation nozzle designed in this paper has good stability, and can be applied to fuels with different metal particle concentrations.

4.4. Comparison of Atomization Characteristics of Different Nozzles

The comparison of the spray images shows that the atomization performance of the swirl nozzle and the self-excited oscillation nozzle is significantly better than that of the DC nozzle. The introduction of a resonant chamber can improve the atomization quality of the DC nozzle, which makes the breakup of the liquid jet more obvious and more intense. At the same time, the spray cone angle of the self-excited oscillation nozzle is significantly larger than that of the DC nozzle, and even tends to be similar than that of the swirl nozzle. It should be also noted that the conventional DC nozzle and swirl nozzle cannot overcome the change of fuel viscosity caused by addition of metal particles. As the concentration of nanoparticles increases, the DC nozzle produces more liquid clusters and filaments in the spray field downstream, while the swirl nozzle obviously decreases the spray cone angle of gelled fuels. However, the atomization effect of the self-excited oscillation nozzle does not change significantly with increasing nanoparticle concentration, which indicates that the self-excited oscillation nozzle can offer a more stable atomization performance for gelled fuels.

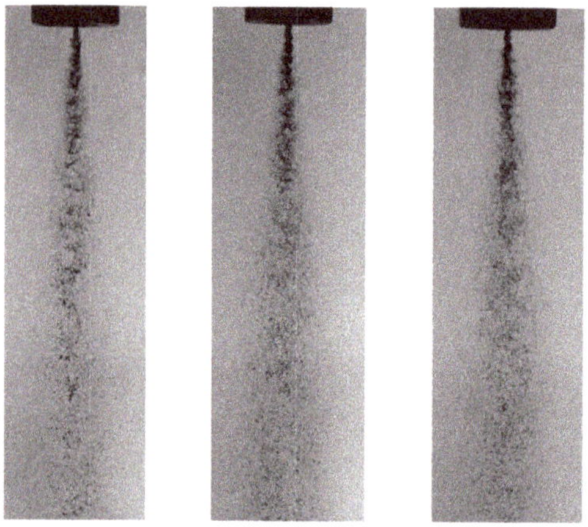

(a)GF-0 **(b)GF-15** **(c)GF-30**

Figure 9. Spray images of different gelled fuels using self-excited oscillation nozzles.

Table 4. The injection pressure and the mean droplet diameters of the self-excited oscillation nozzle for different gelled fuels.

Type	Volumetric Flow Rate/(mL/s)	Pressure/MPa	D_{32}/μm	Dv50/μm	Dv90/μm
GF-0	6.91	2.79 ± 0.019	97.52 ± 11.07	156.10	427.85
GF-15	6.81	3.29 ± 0.009	70.99 ± 4.51	109.68	282.78
GF-30	6.80	3.66 ± 0.013	93.02 ± 8.10	135.98	318.52

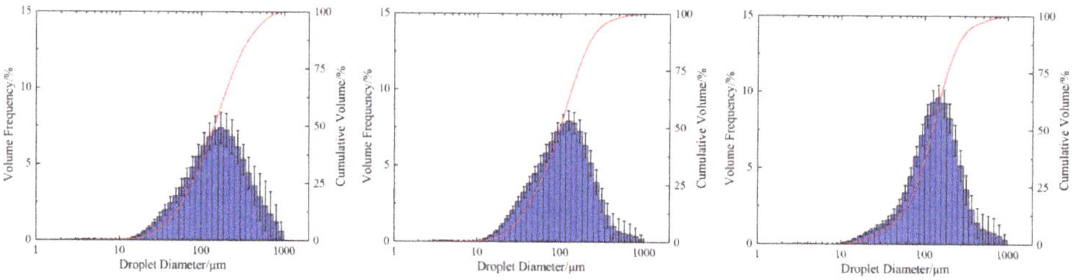

(a)GF-0 **(b)GF-15** **(c)GF-30**

Figure 10. Droplet size distribution of different gelled fuels using self-excited oscillation nozzles.

The comparison of the mean droplet diameter shows that the droplet of the swirl nozzle is significantly smaller than that of the DC nozzle and the self-excited oscillation nozzle. The Dv50 and droplet size distribution show that the swirl nozzle also produces a higher percentage of small droplets and a more uniform spray field. However, it should also be noted that when the concentration of metal nanoparticles increases from 15% to 30%, the SMD, Dv50, and Dv90 of the swirl nozzle also increases rapidly. The SMD data of the self-excited oscillation nozzle is close to that of the DC nozzle, but its Dv50 and Dv90 values are significantly smaller than that of the DC nozzle. It can also be seen from the droplet size distribution of the self-oscillation nozzle that the peak of the volume frequency shifts towards smaller values, which also means that the self-oscillation nozzle can effectively reduce the large size droplets and liquid filaments in the spray field, thus improving the atomization characteristics of gelled fuels.

It can be seen in Table 4 that the fuel injection pressure of the self-excited oscillation nozzle is higher when compared with the DC nozzle and swirl nozzle. This can be attributed to the turbulence of the flow in the resonant chamber, where the strong oscillation effect will have some feedback on the upstream pipeline, resulting in the phenomenon of flow oscillation, and make the self-excited oscillation nozzle discharge coefficient more unstable. At the same time, due to the small size of the resonant chamber, the aggregation and sedimentation of metal particles inside the chamber will also have an impact on the flow coefficient of the nozzle. Generally speaking, the discharge coefficient of self-excited oscillation nozzle is smaller than that of conventional nozzles, so the working pressure of self-excited oscillation nozzle is higher.

5. Conclusions

In this paper, the rheological and atomization characteristics of gelled fuels with different Al nanoparticle concentrations were investigated. Three single-phase injectors were used for comparison. The main findings of the study are as follows:

1. The addition of metal particles does not change the shear thinning characteristics of the gelled fuels, but makes their rheological curves more complex. The consistency coefficient and power law index of the gelled fuels gradually increase with the increase in nanoparticle concentration.
2. The atomization performance of the self-excited oscillation nozzle is better than that of the DC nozzle. The introduction of a resonant chamber can enhance the fragmentation of the liquid jet. At the same time, the spray cone angle of the self-excited oscillation nozzle is significantly larger than that of the DC nozzle, and even tends to be similar to that of the swirl nozzle.
3. The droplet diameter produced by the swirl nozzle is significantly smaller than that of the DC nozzle and the self-excited oscillation nozzle, producing a higher percentage of small droplets and a more uniform spray field. When the concentration of nanoparticles increases from 15% to 30%, the mean droplet diameter of the swirl nozzle will also increase rapidly.
4. The conventional DC nozzle and swirl nozzle cannot overcome the change of fuel viscosity caused by the addition of metal particles. As the nanoparticle concentration increases, the large size droplets and liquid filament increase in the spray field downstream of the DC nozzle, while the spray cone angle decreases significantly when using a swirl nozzle. However, the spray quality of self-excited oscillation nozzle is similar for different gelled fuels and does not change significantly.

Author Contributions: Conceptualization, D.Y. and L.Y.; Data curation, P.L. and Z.F.; Funding acquisition, Q.F. and L.Y.; Investigation, P.L. and Z.F.; Methodology, P.L., D.Y., Z.F. and Q.F.; Project administration, Q.F.; Resources, Q.F.; Software, P.L.; Supervision, Q.F. and L.Y.; Validation, Z.F. and Q.F.; Visualization, Z.F.; Writing—original draft, P.L. and Z.F.; Writing—review & editing, D.Y. and Q.F. All authors have read and agreed to the published version of the manuscript.

Funding: This research was supported by the National Natural Science Foundation of China (grant No. 11922201, 11872091, and U1837211) and the Academic Excellence Foundation of BUAA for Ph.D. Students.

Data Availability Statement: Not applicable.

Acknowledgments: We would like to express our thanks to the editors of Aerospace and the reviewers for their work in processing this article.

Conflicts of Interest: The authors declare no conflict of interest.

References

1. Padwal, M.B.; Natan, B.; Mishra, D.P. Gel propellants. *Prog. Energy Combust. Sci.* **2021**, *83*, 100885. [CrossRef]
2. Xue, K.; Cao, J.; Pan, L.; Zhang, X.; Zou, J.J. Review on design, preparation and performance characterization of gelled fuels for advanced propulsion. *Front. Chem. Sci. Eng.* **2022**, *16*, 819–837. [CrossRef]
3. Glushkov, D.; Paushkina, K.; Pleshko, A. Gel Fuels: Preparing, Rheology, Atomization, Combustion. *Energies* **2022**, *16*, 298. [CrossRef]
4. Jin, Y.; Xu, X.; Yang, Q.; Dou, S.; Wang, X.; Fu, Q.; Pan, L. Combustion Behavior of Hydrocarbon/Boron Gel-Fueled Scramjet. *AIAA J.* **2022**, *60*, 3834–3843. [CrossRef]
5. Padwal, M.B.; Mishra, D.P. Interactions among synthesis, rheology, and atomization of a gelled propellant. *Rheol. Acta* **2016**, *55*, 177–186. [CrossRef]
6. Rahimi, S.; Natan, B. Flow of gel fuels in tapered injectors. *J. Propuls. Power* **2000**, *16*, 458–464. [CrossRef]
7. Madlener, K.; Moser, H.A.; Ciezki, H.K. Influence of the Injection Inlet Angel on the Flow and Spray Behaviorof Shear Thinning Fluids in Impinging Jet Injectors. In Proceedings of the 38th In-ternational Annual Conference of ICT, Karlsruhe, Germany, 26–29 June 2007.
8. Rahimi, S.; Natan, B. Atomization characteristics of gel fuels. In Proceedings of the 34th AIAA/ASME/SAE/ASEE Joint Propulsion Conference and Exhibit, Cleveland, OH, USA, 13–15 July 1998; Volume 3830.
9. Yang, L.; Fu, Q.; Qu, Y.; Wang, W.; Du, M.; Xu, B. Spray characteristics of gelled propellants in swirl injectors. *Fuel* **2012**, *97*, 253–261. [CrossRef]
10. Mehta, R.N.; Chakraborty, M.; Parikh, P.A. Nanofuels: Combustion, engine performance and emissions. *Fuel* **2014**, *120*, 91–97. [CrossRef]
11. Ojha, P.K.; Karmakar, S. Boron for liquid fuel Engines-A review on synthesis, dispersion stability in liquid fuel, and combustion aspects. *Prog. Aerosp. Sci.* **2018**, *100*, 18–45. [CrossRef]
12. Kim, H.; Ko, T.; Kim, S.; Yoon, W. Spray characteristics of aluminized-gel fuels sprayed using pressure-swirl atomizer. *J. Non-Newton. Fluid Mech.* **2017**, *249*, 36–47. [CrossRef]
13. Von Kampen, J.; Alberio, F.; Ciezki, H.K. Spray and combustion characteristics of aluminized gelled fuels with an impinging jet injector. *Aerosp. Sci. Technol.* **2007**, *11*, 77–83. [CrossRef]
14. Baek, G.; Kim, S.; Han, J.; Kim, C. Atomization characteristics of impinging jets of gel material containing nanoparticles. *J. Non-Newton. Fluid Mech.* **2011**, *166*, 1272–1285. [CrossRef]
15. Jejurkar, S.Y.; Yadav, G.; Mishra, D.P. Visualizations of sheet breakup of non-Newtonian gels loaded with nanoparticles. *Int. J. Multiph. Flow* **2018**, *100*, 57–76. [CrossRef]
16. Lee, J.G.; Fakhri, S.; Yetter, R. Atomization and spray characteristics of gelled-propellant simulants formed by two impinging jets. In Proceedings of the 45th AIAA/ASME/SAE/ASEE Joint Propulsion Conference & Exhibit, Denver, CO, USA, 2–5 August 2009; Volume 5241.
17. Desyatkov, A.; Adler, G.; Prokopov, O.; Natan, B. Atomization of gel fuels using impinging-jet atomizers. *Int. J. Energetic Mater. Chem. Propuls.* **2011**, *10*, 55–65. [CrossRef]
18. Fang, Z.; Fu, Q.; Xu, X.; Yang, L.; Jia, B.; Li, P. Spray characteristics of Al-nanoparticle-containing nanofluid fuel in a Y-jet injector. *Energetic Mater. Front.* **2021**, *2*, 249–257. [CrossRef]
19. Jia, B.; Fu, Q.; Xu, X.; Yang, L.; Zhang, D.; Wang, T.; Wang, Q. Spray characteristics of Al-nanoparticle-containing nanofluid fuel in a self-excited oscillation injector. *Fuel* **2021**, *290*, 120057. [CrossRef]
20. Li, P.; Yang, L.; Fu, Q.; Fang, Z. Spray characteristics of the nanoparticle-containing gel propellants by using an improved single-phase nozzle. *Fuel* **2022**, *315*, 122968. [CrossRef]
21. Cao, J.W.; Pan, L.; Zhang, X.W.; Zou, J. Physicochemical and rheological properties of Al/JP-10 gelled fuel. *Chin. J. Energetic Mater.* **2020**, *28*, 382–390.
22. Madlener, K.; Sinatra, C.; Ciezki, H. Investigation on the influence of particle size and concentration on the spray and combustion characteristics of aluminium particle containing gels with an impinging jet injector. In Proceedings of the 2nd European Conference for Aerospace Sciences, Brussels, Belgium, 27 July 2007.
23. Liao, Z.; Li, J.; Chen, D.S. Theory and experimental study of the self-excited oscillation pulsed jet nozzle. *Chin. J. Mech. Eng.* **2003**, *16*, 379–383. [CrossRef]

aerospace

MDPI

Article

Fundamental Study of Premixed Methane Air Combustion in Extreme Turbulent Conditions Using PIV and C-X CH PLIF

Md. Amzad Hossain [1,*], Md Nawshad Arslan Islam [1], Martin De La Torre [2], Arturo Acosta Zamora [3] and Ahsan Choudhuri [1]

[1] Aerospace Center, Department of Aerospace and Mechanical Engineering, The University of Texas at El Paso, El Paso, TX 79968, USA; mislam12@utep.edu (M.N.A.I.); ahsan@utep.edu (A.C.)
[2] Engineer, Blue Origin, Cape Canaveral, FL 32920, USA; made2@miners.utep.edu
[3] Engineer, Intel Inc., Portland, OR 97124, USA; aacosta29@miners.utep.edu
* Correspondence: mhossain12@utep.edu

Abstract: This paper presents the flow and flame characteristics of a highly turbulent reactive flow over a backward-facing step inside a windowed combustor. Flow and combustion experiments were performed at *Re* = 15,000 and *Re* = 30,000 using high-resolution 10 kHz PIV and 10 kHz PLIF diagnostic techniques. Grid turbulators (Grid) with two different hole diameters (HD of 1.5 mm and 3 mm) and blockage ratios (BR of 46%, 48%, 62%, and 63%) were considered for the turbulence study. Grids introduced different turbulent length scales (L_T) in the flow, causing the small eddies and turbulence intensity to increase downstream. The backward-facing step increased the turbulence level in the recirculation zone. This helped to anchor the flame in that zone. The small HD grids (Grids 1 and 3) produced continuous fluid structures (small-scale), whereas the larger HD grids (Grids 2 and 4) produced large-scale fluid structures. Consequently, the velocity fluctuation was lower (~25.6 m/s) under small HD grids and higher (~27.7 m/s) under large HD grids. The flame study was performed at Φ = 0.8, 1.0, and 1.2 using C-X CH PLIF. An Adaptive MATLAB-based flame imaging scheme has been developed for turbulent reacting flows. Grids 1 and 3 induced more wrinkles in the flame due to higher thermal instabilities, pressure fluctuation, and diffusion under those grids. The flamelet breakdown and burnout events were higher under Grids 2 and 4 due to higher thermal diffusivity and a slower diffusion rate. It was observed that the flame wrinkling and flame stretching are higher at *Re* = 30,000 compared to *Re* = 15,000. The Borghi–Peters diagram showed that the flames were within the thin reaction zone except for Grid 1 at *Re* = 15,000, where flames fell in the corrugated zone. It was observed from PIV and PLIF analyses that *Re* and L_T mostly controlled the flame and flow characteristics.

Keywords: premixed combustion; closed flame system; extreme turbulence; PIV; C-X CH PLIF

Citation: Hossain, M.A.; Islam, M.N.A.; De La Torre, M.; Zamora, A.A.; Choudhuri, A. Fundamental Study of Premixed Methane Air Combustion in Extreme Turbulent Conditions Using PIV and C-X CH PLIF. *Aerospace* **2023**, *10*, 620. https://doi.org/10.3390/aerospace10070620

Academic Editor: Jian Liu

Received: 27 May 2023
Revised: 29 June 2023
Accepted: 6 July 2023
Published: 8 July 2023

1. Introduction

The design and development of next-generation high-speed engine combustors such as scramjets, ramjets, and diesel engines require an in-depth understanding of high-intensity turbulent combustion [1,2]. The study of turbulent premixed combustion is always difficult, especially when it comes to stabilizing the combustion and detecting the flames, or finding whether is it the instability or the turbulence that controls the flame. Instabilities such as the Darrieus–Landau instability, flame wrinkling, flame straining pattern (tangential and normal), stretching rate and duration, turbulence and vorticity levels, and heat release-induced pressure gradients play an important role in turbulent flame behavior [3–6].

To better understand turbulent flame behavior, researchers developed highly accurate non-intrusive flame measurement techniques with high repetition rates in the last two decades. Among those techniques, particle image velocimetry (PIV) and planar laser-induced fluorescence (PLIF) have been widely used [3–6]. In the following discussion, the authors aim to present the advancements made with PIV and PLIF techniques, the limitations of this research work, and the rationale behind their selection of C-X CH PLIF for the flame study.

G. Troiani [7] used the PIV system to study the wrinkling of flame fronts caused by the Darrieus–Landau (DL) instability and its interaction with turbulence across various Reynolds numbers (*Re*) in a Bunsen setup. The instability was induced by adjusting the mixture ratio (MR) in laminar conditions and by varying the Bunsen nozzle diameter (DN) in turbulent conditions. They compared stable and unstable Bunsen flames by analyzing the vorticity generation, total flame strain, and the straining pattern of a single DL cusp. The results indicated that unstable flames with wrinkling displayed statistical characteristics typically seen in stable flames under high turbulence levels, suggesting a unified regime with some remaining distinctions. This study demonstrates that there is a correlation between flame-front instability with flame wrinkling, flame straining, vorticity level, etc. Lapenna et al. [8] have examined the impact of DL instability on turbulent propane–air premixed flames. The theoretical analysis identifies two regimes: one dominated by instability and the other by turbulence. In the turbulence-dominated regime (also known as the unified regime), the presence of turbulence hinders the formation of large-scale cusp-like structures in the flame front caused by the DL instability, resulting in reduced effects on turbulent flame propagation and front curvature statistics. As a result, the characteristics of stable and unstable flame configurations become indistinguishable across laminar to turbulent conditions. Similarly, Chaudhuri et al. [9] utilized a high-speed PIV system to examine how flames interact with turbulence in expanding premixed flames using a CH_4–air mixture. The study presents information on mean flow velocity, fluctuating flow velocity, stretch rate contributions, and the movement of flame edge points. Their findings indicate non-Gaussian tails in normal straining and near-Gaussian behavior in tangential straining. Lipatnikov et al. [10] demonstrated that the pressure gradients from the heat release in surrounding flamelets can generate unburned mixture fingers that penetrate deeply into the combustion products, eventually increasing the flame surface area, flame velocity, and mean flame brush thickness. This mechanism is distinct from the DL instability and made the flame fronts move towards the leading edge of the flame brush. Chaudhuri [11] utilized the DNS to examine the finite lifetime of flame particles residing on iso-temperature surfaces of statistically planar H2–air flames interacting with near-isotropic turbulence. The particles encounter increasing tangential straining rate (Kt) and negative curvature (κ) and the flame surface mostly shifted because of the tangential strain rate. Zhou et al. [12] investigated the flow fields of premixed and stratified CH_4–air flames under turbulent conditions using laser Doppler anemometry (LDA) and PIV. Their study revealed significant changes in flow behavior with varying levels of swirl and stratification, incorporating axial, radial, and tangential velocities to derive key turbulence parameters. Fan et al. [13] investigated the impact of water droplets on strained methane–air laminar flames using PIV and measured the flame speed and motion of the liquid phase simultaneously. They found that the addition of water droplets significantly reduces the flame speed and is more effective in flame suppression. These studies clearly show that turbulent flow characterization and flame detection mechanisms are very complex and demand more fundamental studies.

PLIF system has been extensively used for turbulent reacting flow combustion studies. PLIF involves the use of laser-induced fluorescence to visualize and quantify various properties, such as temperature, species concentration, and flow velocity. The PLIF system typically consists of a laser source, optics for shaping the laser beam, a fluorescent dye for tagging the fluid, and a high-resolution imaging system. When the laser beam interacts with the tagged fluid, the fluorescent molecules emit light, which is captured by the imaging system and converted into meaningful data [14]. PLIF targets the species present in the flame, such as hydroxyl (OH), CH_2O, and methylidyne (CH). Among these radicals, OH PLIF is widely used in the lab [15]. The concentration gradient of OH radicals in flames is high [16]. Additionally, OH radicals are present near the flame fronts and combustion products [17]. OH also sustains for a long time in the flame. Because of these characteristics, OH radicals are suitable for tracing down the flame fronts in a highly turbulent environment. For the same reasons, OH radicals have also been used in high-repetition-rate PLIF diagnostics [18,19]. While these advantages seem prevailing, there are a few questions about the use of OH as a flame front tracer [20]. For instance, OH is slowly removed away from the flame by three-body collision reactions [21–23] making it difficult to trace the primary fuel reaction zone. Additionally, OH is a tracer of the radical-recombination zone, not necessarily a marker of flame front tracing [24].

The alternative way of tracing flames is to use methylidyne (CH) PLIF. Generally, CH is a suitable marker for flame front tracing. CH radicals are formed during the initial fuel decomposition. Additionally, CH is removed away by rapid two-body collision reactions. Thus, the CH radicals are present during the main course of combustion. There are three bands of CH radicals as reported by Carter's group [25,26]: A-X ($A^2\Sigma^+$-$X^2\Pi$ (0,0)), B-X (BΣ—XΠ), and C-X ($C^2\Sigma^+$-$X^2\Pi$ (0,0)). Carter's group stated that the C-X CH band is better than the other two bands. The reason for this is that the C-X band is effective in detecting the flame during the main phase of combustion. It provides a clear visualization of the flame's central region, outer boundaries, and the way it extends in turbulent reacting flows. The C-X CH radical could be easily excited using the frequency-doubler dye laser commonly used for OH detection [22,27,28]. Therefore, a laser power system comprised of a dye laser and a pump laser (main power laser) can be used for C-X CH excitation. The excitation and emission spectrum of C-X CH is normally held at $\geq$314 nm [9].

Acosta-Zamora [28] studied the highly turbulent reacting flow over a backward-facing step. This study examined the flow and flame interaction at low to moderate Re conditions using both PIV and OH PLIF. Although the researcher was able to excite and detect the OH signal, the study experienced several issues in flame imaging. For example, the flame profiles were diffuse, and the flame core was not thin. This is because there was a presence of some foreign quantity, which was named the "ghost mixture quantity". This ghost quantity gets infused into the flame core and makes the core diffuse rather than a thin, flamelet-like structure. It was very difficult to differentiate the flame reaction zone from the overall flame profile. In addition, at higher turbulence conditions, OH PLIF does not give sharp flame-front structures. It was also hard to investigate how the turbulent eddies infused into the flame core. To better detect the flame profile at turbulent premixed combustion, the study recommended using CH PLIF. Acosta et al. [28] conducted a preliminary study but were not able to implement CH PLIF because of the system's limitation and lack of time to do the iterations. Similar recommendations have been reported in [22,27,29]. The authors in this study made changes to the laser diagnostics system described by Acosta-Zamora at The University of Texas at El Paso (UTEP) Aerospace Center. They then implemented CH PLIF for the turbulent premixed combustion study.

Based on the literature review [22,25,27,29], it is observed that CH PLIF has the potential to accurately detect flame profiles in turbulent conditions. Furthermore, most researchers, including Carter's group, use PLIF for open flame systems. Based on a statistical analysis of 100 random papers [30], it is seen that 95% of the papers talk about open flame systems (Bunsen burners, flat plate burners, flat swirl burners, etc.) [18,25,26], and only 5% of papers talk about closed flame systems [3,28]. This clearly shows the importance of researching closed flame systems. Hossain [30] also stated that out of these 100 papers, 68% considered OH PLIF, 22% considered CH PLIF and 10% addressed other types of PLIF systems. Out of the 22 papers on CH PLIF, 5 papers used C-X CH PLIF, 11 papers used A-X CH PLIF, and 6 papers used B-X CH PLIF. This indicates that more research should be carried out using C-X CH PLIF. In addition, experimental resources on flame–flow interaction at higher-order turbulence (higher *Re*) remain sparse [4,29], and [31]. More study needs to be carried out under low to moderate *Re* or Mach conditions.

Motivated by the above-mentioned issues and limitations, the present effort aimed to demonstrate experimental methodology and post-processing techniques for imaging CH in a high-intensity turbulent flow field within a windowed combustor. A high repetition rate PLIF (10 kHz) and PIV (10 kHZ) were used to study the flame and flow characteristics. Grid turbulators (Grids) or perforated plates were used to investigate the effect of turbulent length scale (L_T) on flame growth. Different Reynold numbers (*Re* = 15,000 and 30,000) and equivalence ratios (Φ = 0.8, 1.0, and 1.2) were considered to study the effect of mixture inlet velocities and fuel loading on flame propagation. Two Matlab-based programs were developed for flame post-processing: one for flame profile detection and another for flame edge detection. The ultimate goal of this research is to see the compatibility of the present laser system for high-speed combustion.

2. Experimental Setup and Methodology

2.1. Experimental Setup

A 304 stainless steel (SS) combustor was used to study high-speed combustion. The combustor has optical access through quartz windows located on the top and both lateral sides. These windows were designed with a factor of safety (FOS) of 3.5. The maximum chamber operating pressure is 6 bar (90 psi), and the maximum temperature is 500 °K (440 °F). The maximum thermal firing input of this combustor is 250 kW. Methane and air are premixed in a mixing chamber before entering the combustor. Air enters the mixing chamber axially, while methane enters radially in a cross-mixing arrangement. The circular flow path inside the mixing chamber is converted to a square at the mixing chamber outlet via a converter flange attached to the combustor inlet. Once inside the combustor, the mixture moves over the backward-facing step. This step offers a sudden expansion in the flow path area. Consequently, a high number of small eddies form in the recirculation zones. The eddies help to mix the burned gases with the fresh mixtures, stabilizing the flame in this area. The methane–air mixture ignition point is located at the recirculation zone. A pilot flame subsystem operating with a hydrogen–air flame is used to ignite the flammable mixture inside the combustor. After ignition, the exhaust combustion products are routed through the exhaust cooling tower and finally to the laboratory exhaust. Figure 1 shows an exploded view of the small-scale windowed combustor with the backward-facing step.

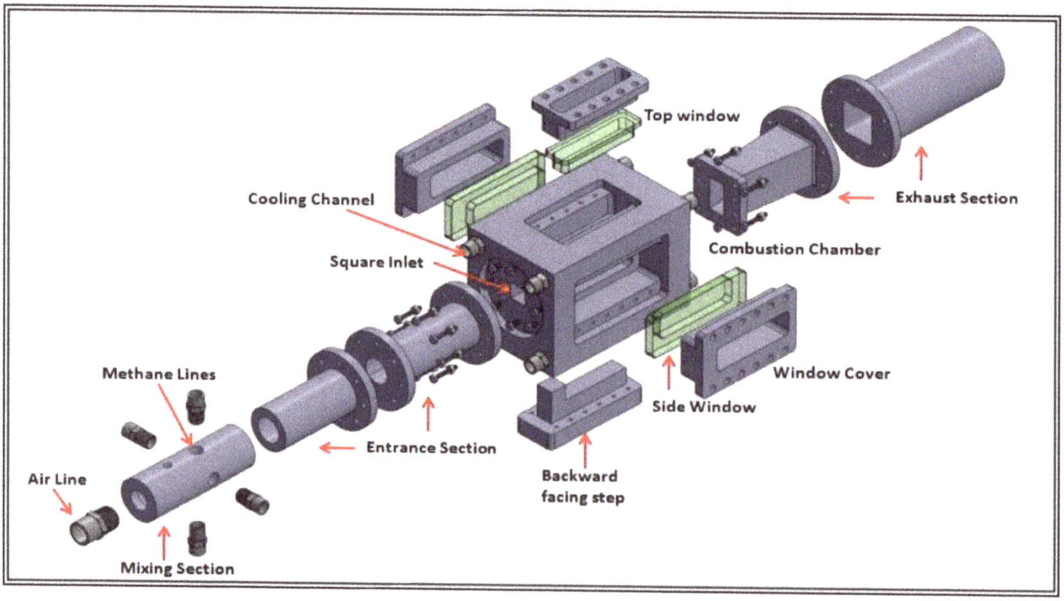

Figure 1. The exploded 3D view of the combustor system and its components. Adapted with permission from Ref. [32].

At the combustor inlet, perforated SS plates called grids (or grid turbulators) induce isotropic homogeneous turbulence inside the combustor. Grids also produce different turbulent length scale levels in the flow. Figure 2 shows the grid and the pilot flame subsystem.

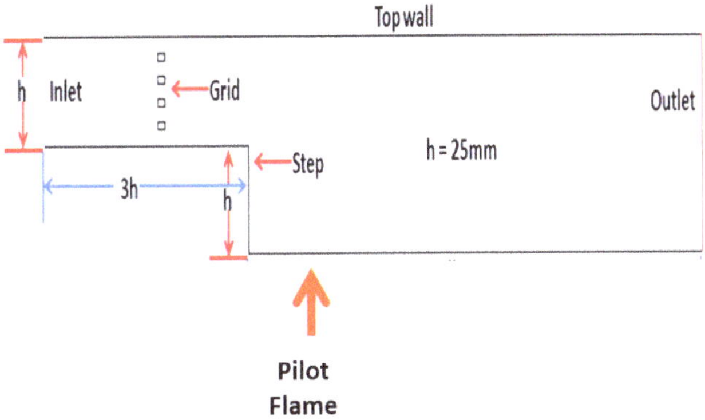

Figure 2. The 2D geometry of the combustor and the position of the grid and pilot flame igniter. Adapted with permission from Ref. [28].

2.2. PLIF Methodology

The planar laser-induced fluorescence (PLIF) system consists of three main components: a Q-switched DPSS laser (neodymium-doped yttrium aluminum garnet (Nd-YAG) Edgewave IS series), a frequency-doubled dye laser (Radiant Dyes NarrowScan HighRep Dye Laser), and an intensified high frame rate CCD camera (Speed sense 9070 with Hamamatsu C10880-01C). The pump laser produces a 532 nm beam through the tunable dye laser. The Nd-YAG pump laser has a pulse energy of 5.6 mJ/pulse (28 W) at 5 kHz and 2 mJ/pulse (20 W) at 10 kHz. The dye laser uses different dyes, oscillators, and a BBO crystal assembly to tune the laser to the desired wavelength. DCM in ethanol with a concentration of 0.55 g/L is used for CH radical excitation. The dye laser produces energy of approximately 0.25 mJ/pulse at 5 kHz. The fluorescence is detected using an intensifier equipped with a Cerco 100 mm f/2.8 lens coupled with a 3 kHz high-speed camera. The camera has a resolution of 640 $\times$ 480 pixels at 10 kHz, providing a resolution of 105.48 um/pixel. A UG-5 colored glass filter was placed on the camera lens to block visible light but allow UV fluorescence to pass through. The intensifier gate was set to 15 ns. A wavelength scan identified the 314.415 nm wavelength to be optimum for tracing flame front with C-X CH PLIF. At this wavelength, the excitation and emission coefficients of CH radicals are high. Figure 3 shows the complete schematic of the PLIF. For a complete overview of the high-speed combustion test facility and the laser diagnostic systems, readers are referred to the authors' previous articles [28,32–35].

Several challenges were encountered during the experimentation and post-processing stages of this study. The greatest challenge was to focus the laser sheet onto the interrogation area with optimum accuracy. It was also hard to get the right excitation wavelength for the CH C-X band. In this present work, the authors have excited CH using wavelengths from 314 nm to 320 nm. The CH signals are collected at 314.415 nm as it gives better flame profiles. The authors are aware that there is a chance of having resonance in the CH PLIF system; similar issues are reported in [18,26]. In addition to the above-mentioned challenges, it was particularly difficult to detect and extract the CH signal as it tends to overlap with OH. The OH signal was reduced by optimizing cameras and intensifier parameters. The optimization of both camera and intensifier parameters depends on the flow turbulence. Dantec Dynamics studio was used to acquire the data.

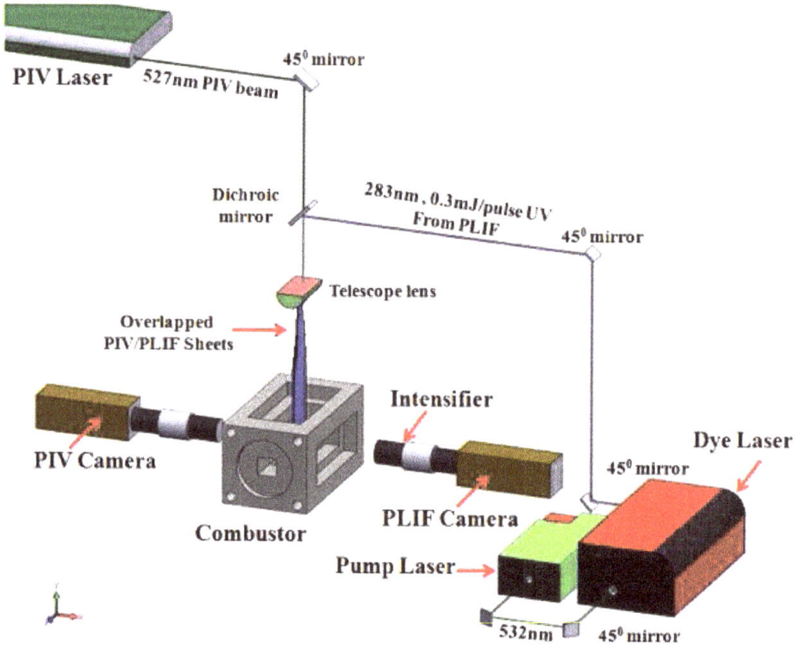

Figure 3. The integration of the PLIF and PIV systems used in this study. Adapted with permission from Ref. [28].

The acquired data were post-processed using in-house MATLAB image processing codes. The raw images contained unwanted laser reflections, background noise, and soot. In addition, the post-processing of the raw images can also contribute to scaling errors. A MATLAB image processing code was developed to address these challenges. First, the MATLAB code was adjusted to the flame image pixel matrix to mitigate unwanted laser reflections and background noises. This adjustment protects the target area from external influence. Next, a separate MATLAB code was implemented to improve the visibility of the flame profile. The base images (without flame) were subtracted from the raw images (with flame) to improve visibility. After that, the images were converted to grayscale and webinar scale. Colormap was then implemented on both grayscale and webinar-scale images. The flame profile generated by the grayscale colormap shows higher resolution and almost no background noises. For this reason, the grayscale-induced HSV colormap is reported in this paper. Additionally, the flame profiles were magnified to observe the flame edge events. A magnification factor of two to four was used in this research. Although it makes the flame core slightly dense, this does not obstruct or influence the flame structures.

The flame edges provide a better representation of flame evolution. For this, a separate MATLAB code was established. This code flipped the HSV colormap contours. Then, the flipped images were converted to grayscale images. After that, binarization and edge detection commands were implemented in the images. For simplicity, flame edges detected by the Canny command are presented in this research. Finally, the MATLAB complement command gives a final flame edge profile. The flow chart of MATLAB image processing codes is shown in Figure 4.

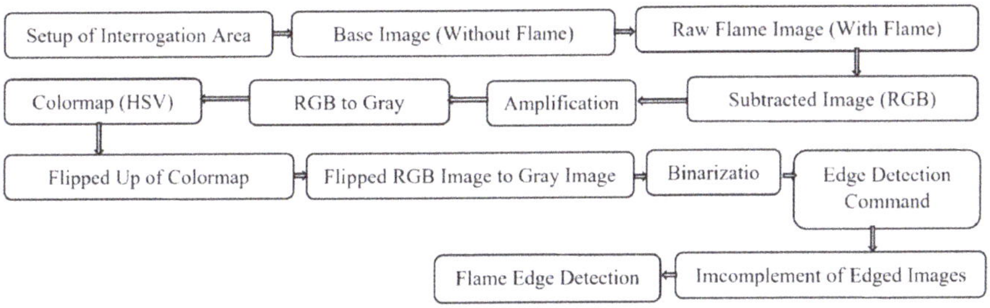

Figure 4. The flow chart used to post-process images to generate the HSV colormap flame profiles and flame edges.

2.3. PIV Methodology

A 15–1000 dual-power neodymium-doped yttrium lithium fluoride (Nd-YLF) laser with a pulse energy of 15 mJ was used for flow characterization. A series of optical mirrors guided the laser from the laser head to the combustor interrogation area. A compressed air-driven fine alumina particle (1 micron) was used as a seeding particle for flow visualization. The seeding particles were introduced at the mixing chamber to mix with the air–fuel mixture and then introduced to the combustor. Nd-YLF is a double-pulse laser. The camera records two instances when the particles travel a fixed distance. The software cross-correlates to the interrogation area and generates the velocity fields. Statistical analysis and adaptive PIV tools generate other flow characteristics such as turbulent intensity and vorticity. A complete overview of the PIV laser system, synchronization parameters, and PIV post-processing can be found elsewhere [33–35].

The main challenge in the PIV system was to redirect the laser beam to the laser optics sheet by 45°. The target was to create a laser plane area of 1 mm × 52 mm × 25 mm. Another challenge was to send the tracer particles to the combustor without flow interruption. The pressure on the seeding flow has to be higher than the bulk flow pressure for seeding particle introduction. The camera must also be calibrated accurately to obtain the flow field events.

Dynamic Studio was used to process the acquired images from PIV experiments. Dynamic Studio implements the image balance map to all raw images to address non-uniformities in the images. This generated light-balanced images with high contrast for easy distinguishment of the location of the particles. Dynamic Studio also utilizes a masking tool to minimize background noise in the processed images. A wall windowing function was used to resolve the wall bias. This function provides a symmetrical uniform distribution of the seeding particle within the interrogation area. An adaptive PIV function was used to adapt the interrogation area to the local seeding densities and flow gradients. The adaptive PIV was also used for frequency filtering and universal outliners detection.

2.4. Operating Conditions

The operating conditions were calculated based on the Reynolds numbers (*Re*), equivalence ratios (Φ), and mixture flow rates. The following two equations are used to calculate the Reynolds number and mixture viscosity:

$$Re = \frac{V_{mixture} \times L}{\nu_{mixture}} \quad (1)$$

$$\log\nu_{mixture} = x_{air}\log\nu_{air} + x_{CH_4}\log\nu_{CH_4} \quad (2)$$

where $V_{mixture}$ is bulk mixture velocity, x is the mole fractions, $\nu_{mixture}$ is the bulk mixture viscosity, and L is a characteristics dimension of the combustor inlet. The operating conditions are presented in Tables 1 and 2. The pump laser, camera, and intensifier unit are synchronized so that the right flow and flame sequence was captured. The synchronization parameters are presented in Table 3.

Table 1. Overview of the flow and associated mixture flow rates.

Re	Bulk Mixture Flow Rate	
	LPM	**CFM**
15,000	353.7	12
30,000	707.31	25

Table 2. The test operating conditions.

Equivalence Ratios (Φ)	*Re* = 15,000		*Re* = 30,000	
	$Q_{methane}$ (LPM)	Q_{air} (CFM)	$Q_{methane}$ (LPM)	Q_{air} (CFM)
0.8	27.4	11.5	54.8	23.0
1.0	33.6	11.3	67.2	22.6
1.2	39.6	11.1	79.2	22.2

Table 3. Synchronization parameters of the pump laser, camera, and intensifier unit.

Controlling Parameters	Magnitudes with Units
Pump Laser Capacity and Pulse Energy	532 nm and 2 mJ/pulse
Dye Laser Maximum Output and Pulse Energy	355 nm and 0.2 mJ/pulse
Camera Exposure Time	70 μs
Intensifier Parameters: Gain, Gate Delay Time, Gate Pulse Width	700–750, 10 ns, and 15 ns–100 ns
Excitation Wavelength for CH-CX Band	314.415 nm

PIV analysis generates the velocity vector contours, turbulent intensity, and vorticity contours. It also allows observing the flow characteristics at the interrogation line. The equations used to measure the flow characteristics at the interrogation line are presented in Table 4. For detailed PIV analysis, refer to articles previously published by the authors of this paper, [28,32–35].

Table 4. List of equations used for flow characteristics measurement.

Turbulence Properties	Formulas	Equations Numbers
RMS of u velocity	$u_{RMS} = \sqrt{\frac{(u-\bar{u})^2}{N}}$	(3)
RMS of v velocity	$v_{RMS} = \sqrt{\frac{(v-\bar{v})^2}{N}}$	(4)
Velocity Fluctuation	$u' = \sqrt{\frac{u^2{}_{RMS}+2v^2{}_{RMS}}{3}}$	(5)
Turbulent intensity	$I = \frac{u'}{U}$	(6)
Vorticity	$\overline{\omega} = \text{rot}(\overline{U}) = \text{curl}(\overline{U}) = \nabla \times \overline{U} = \left(\frac{\partial w}{\partial y} - \frac{\partial v}{\partial z}\right)\bar{i} + \left(\frac{\partial u}{\partial z} - \frac{\partial w}{\partial x}\right)\bar{j} + \left(\frac{\partial v}{\partial x} - \frac{\partial u}{\partial y}\right)\bar{k}$	(7)
	$\omega_z = \left(\frac{\partial v}{\partial x} - \frac{\partial u}{\partial y}\right)$	(8)
Karlovitz Number	$Ka = \frac{\text{Chemical Time Scale}}{\text{Kolmogorov Time Scale}} = \left(\frac{L_T}{\delta_L}\right)^{-\frac{1}{2}}\left(\frac{u'}{S_L}\right)^{\frac{3}{2}}$	(9)
Damkohler Number	$Da = \frac{\text{Reaction Rate}}{\text{Diffusion Rate}} = \frac{\text{Chemical Time Scale}}{\text{Turbulent Time Scale}} = \frac{L_T S_L}{u'\delta_L}$	(10)
Turbulent Reynolds Number	$Re_T = \frac{u' L_T}{S_L \delta_L}$	(11)

3. Results and Discussion

This section reports the fundamental study of high-speed combustion under different grids (see Table 5 below) at various mixture inlet velocities (10 m/s (*Re* = 15,000) and 20 m/s (*Re* = 30,000)) and equivalence ratio of Φ = 0.8, 1.0, and 1.2. The PIV results at *Re* = 15,000 and *Re* = 30,000 are used to understand the flow dynamics better. The effect of turbulent length scale (L_T) on flame–flow interaction and flame regime boundaries and scale factors on reaction zone are reported in this section.

Table 5. Characteristics of grids used to induce turbulence inside the combustor.

Grid #	Blockage Ratio (BR)	Hole Diameter (HD)	Mixture Velocity, m/s
1	48%	1.5 mm	10, 20
2	46%	3.0 mm	
3	63%	1.5 mm	
4	62%	3.0 mm	

3.1. Flow Field Characteristics Analysis

In this part, the authors aim to present the observed flow field characteristics obtained through experiments conducted with four grids at *Re* = 15,000 and 30,000. The velocity vector field was plotted to observe the flow events due to the backward-facing step. Figure 5 shows a dark background that was used to create a more impactful flow area and improve image contrast, resulting in clearer visualization of the flow. The grid at the combustor inlet induces homogenous turbulence in the flow because of interactions between wakes and jets behind the grid. This usually needs a significant distance downstream for the flow to become well-mixed and uniform. The authors optimized the combustor design by testing for the distance downstream that produced the best mixture. A detailed understanding of these inquiries can be found in [2,28,32]. The flow accelerated once it passed the backward-facing step, as seen in Figure 5. This happens because of the thermal dilation effects that the flow encountered downstream. The velocity contours showed the presence of a mixing or shear layer between the recirculation zone and the main flow. This layer was composed of vortices that were shed downstream from the edge of the backward-facing step. This low-speed recirculation zone helped to anchor the flame. Grids 3 and 4 had higher blockage ratios (BR), as expected flows showed fewer interacting vortices in the wake and jets of the grids, especially in the recirculation zone. At *Re* = 15,000 and 30,000, the occurrence of vortex shedding downstream of the sharp corner of the step was observed consistently, regardless of the presence of grids or perforated plates. The flow experienced more fluctuations at *Re* = 30,000 compared to *Re* = 15,000 due to the higher turbulence present at *Re* = 30,000.

In this article, the authors extracted the local velocities, vorticities, and other turbulent characteristics from the interrogation line and points distributed across the flow domain, as illustrated in Figure 5. (For a detailed understanding of the PIV data extraction, acquisition, and post-processing methods, the readers are referred to [32].) The velocities (U) and velocity fluctuations (u') were then calculated using Equations (3)–(5) and are presented in Figure 6. The maximum velocities of 15 m/s and 25 m/s are measured at *Re* = 15,000 and *Re* = 30,000, respectively. This indicates an increase of 5 m/s in flow velocity at both *Re* conditions. The maximum velocity fluctuations of 0.5 m/s and 0.25 m/s were observed for the u and v components, respectively. As expected, the perforated plate installed at the inlet of the combustor helped increase the flow velocity and flow fluctuation.

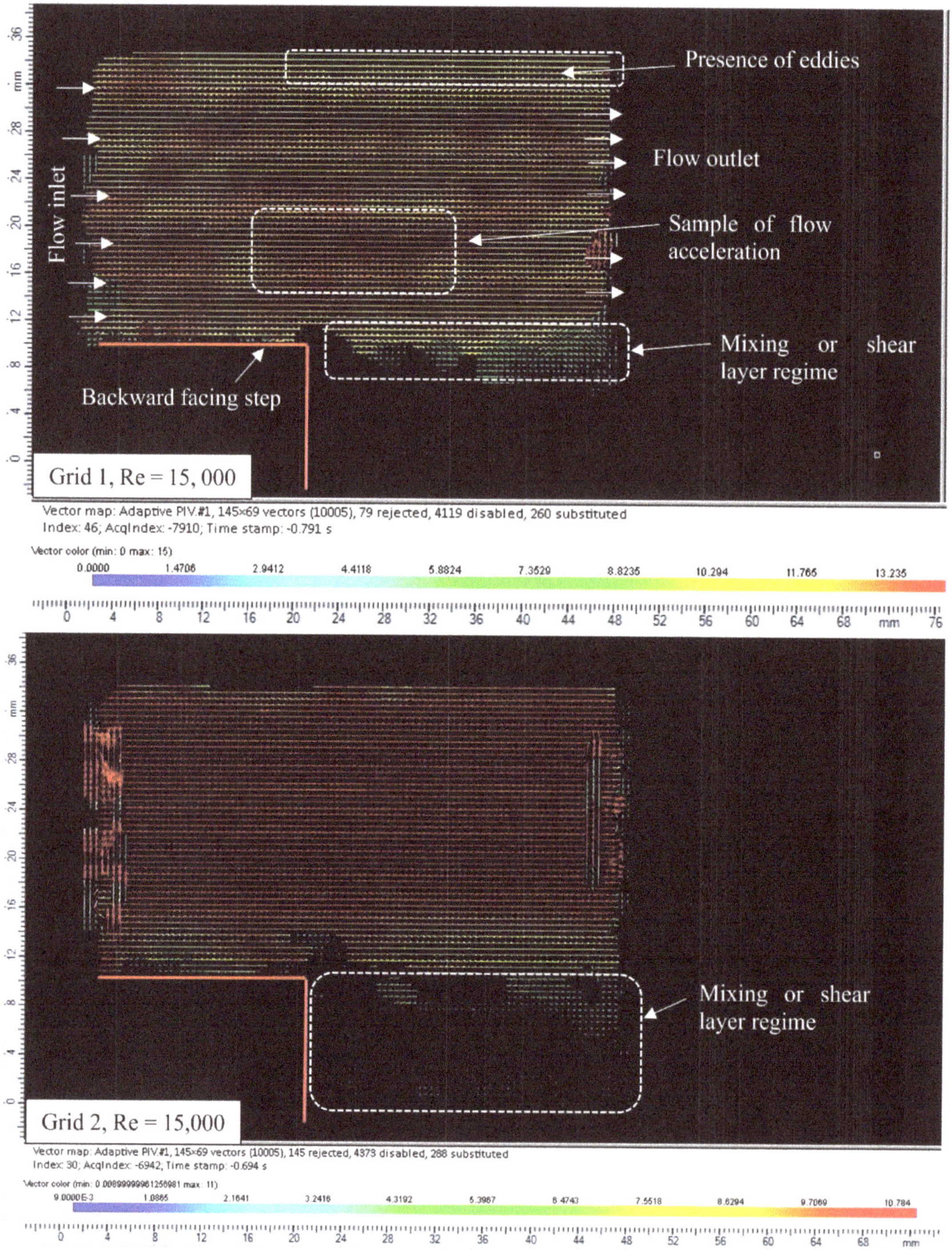

Figure 5. *Cont.*

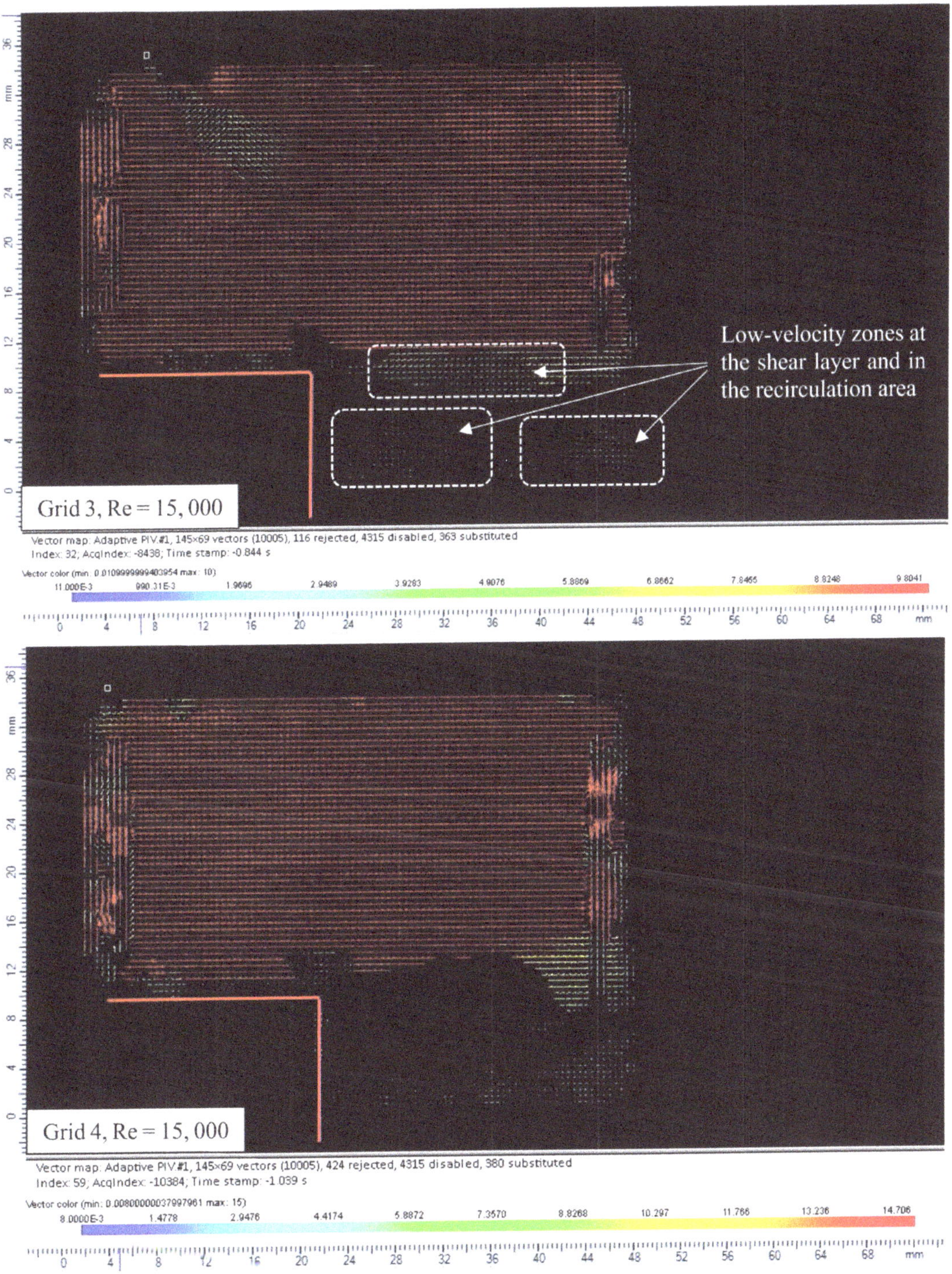

Figure 5. *Cont.*

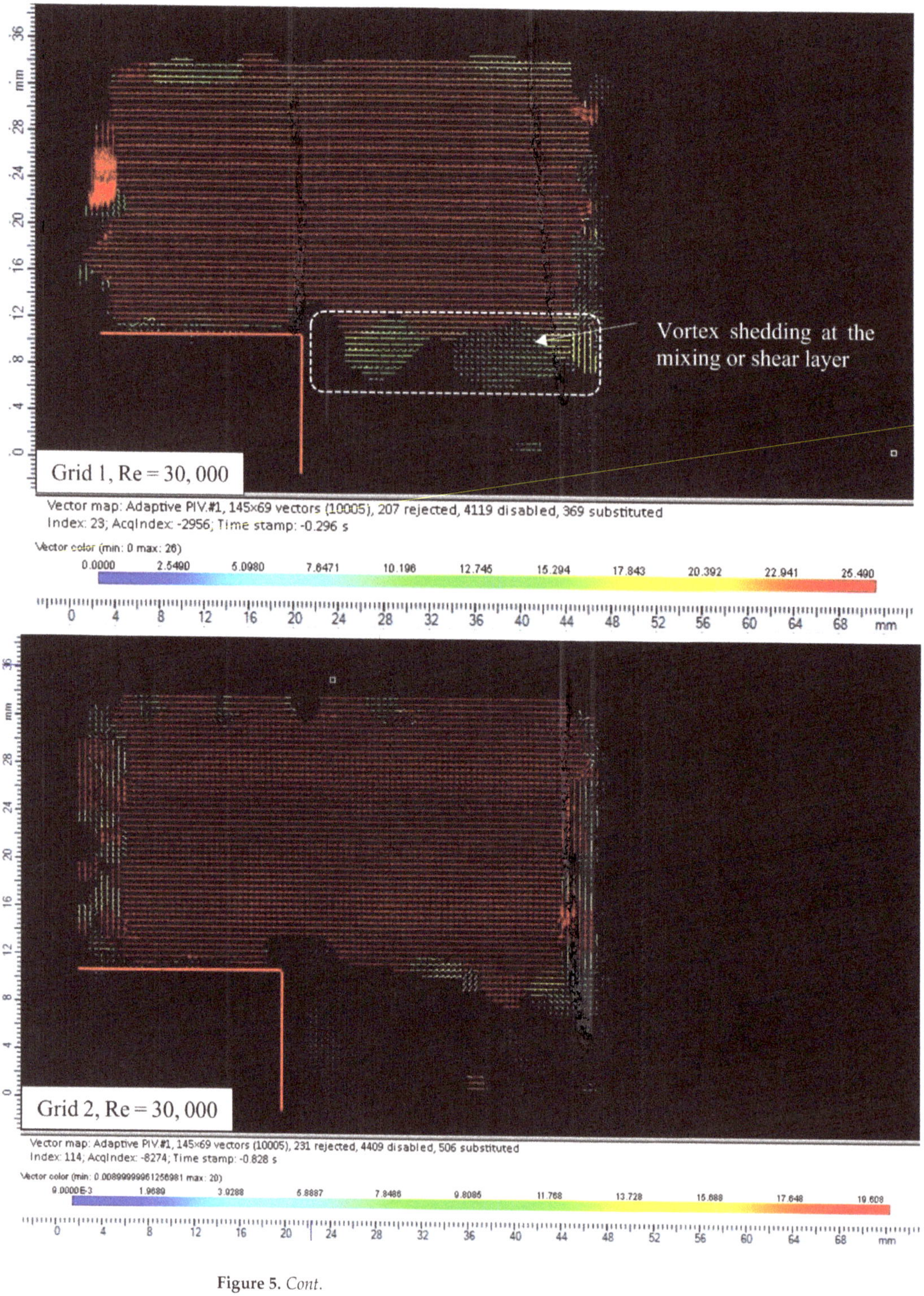

Figure 5. *Cont.*

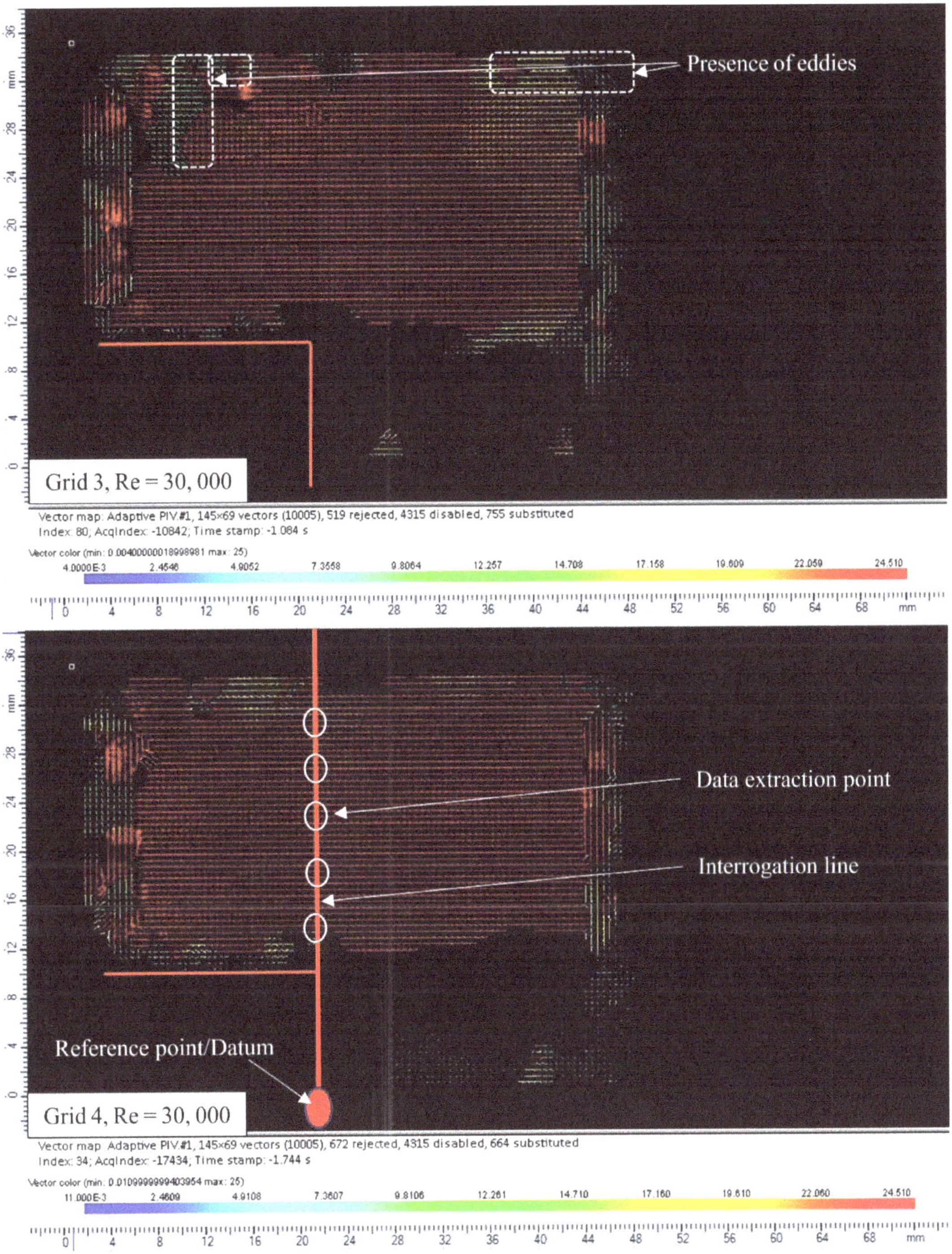

Figure 5. The velocity vector field maps at *Re* = 15,000 (**first four contours**) and *Re* = 30,000 (**last four contours**). Color print required.

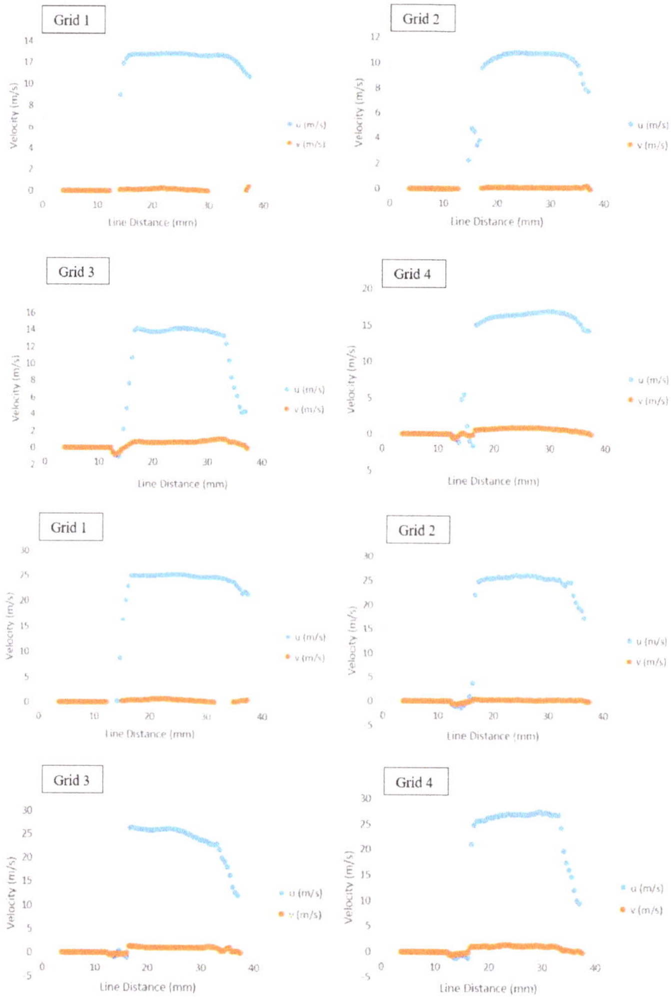

Figure 6. Average velocity acquired at the interrogation line for *Re* = 15,000 (**first two rows**) and *Re* = 30,000 (**last two rows**). Color print required.

Turbulent intensity contour maps are instrumental in improving comprehension of the structure of combusting flows by visually depicting the turbulent intensity distribution throughout the combustion system. These maps are used to identify and analyze significant flow structure features, including flame stabilization mechanisms, flame front propagation, and flame extinction. By investigating variations in turbulent intensity, one can gain a deeper understanding of the intricate interactions between turbulence and combustion, ultimately facilitating enhanced efficiency, reduced emissions, and safer operation through improved design and optimization of combustion systems. With that in mind, the authors generated the turbulent intensity contour map in Figure 7. These maps show how the recirculation zone behaves at different *Re* and unveil how turbulence is distributed in the fluid domain. Figure 7 also helps to locate the areas of lower turbulence (reduced mixing or recirculation) and higher turbulence (higher mixing or recirculation). Figure 7 shows that when the blockage ratio is increased, the turbulent intensity increases too. Such outcomes indicate the increased amount of induced eddies in the recirculation zone at the foot of the step. In addition, when the hole diameter of the grid is increased, the turbulent intensity also changes; in both cases, increasing the hole diameter decreases the intensity. This is important because the pilot flame used to ignite the premixed methane–air mixture is located near the foot of the backward-facing step. Figure 7 shows that turbulence can influence flame anchoring or stability. It has been observed that the increase in turbulence has both pros and cons in the flame anchoring regime. While at a lower *Re* of 15,000, the higher turbulence has shown an improvement in flame anchoring, which is due to the better mixing of the pilot flame charge. However, when the *Re* is increased, a lower BR tends to help anchor the flame better compared to a higher BR. This is due to the infusion of too many eddies, which partly break up the pilot flame and causes instability. Consequently, it ultimately leads to flame blowout due to pilot flame extinction.

Grids 1 and 2 experienced less flow restriction (low BR) and thus have diffused turbulence at the recirculation zones. This aids in stabilizing the flame in the recirculation zone. On the other hand, Grids 3 and 4 exhibit concentrated turbulence. The higher flow restriction governs this behavior (high BR). Consequently, flame stabilization under Grids 3 and 4 would be challenging. The turbulence level (I) is high at the edge of the backward-facing step (refer to Equation (6), Table 4) and in the recirculation zone. Figure 7 also indicates how the flow fluctuation varies within the flow domain. The vortices in the shear layer (mixing layer) control the flow fluctuation at the lower part of the fluid domain. However, the overall flow fluctuation is controlled by grids and the Reynolds number. Grids 3 and 4 show more turbulent intensities compared to Grids 1 and 2 in the recirculation zone. This is again due to the higher blockage ratio of Grids 3 and 4.

The vorticities at the edge of the step are calculated using Equations (7) and (8). Figure 8 presents the vorticity profiles. The vorticity profiles show the areas of lower and higher vorticity fluctuations. The vorticity fluctuation is very high between the 10 mm to 20 mm (the first peak) and 30 mm to 40 mm (the second peak) line distance from the reference/datum point. This is due to the significant change in velocity gradient and angular momentum in those areas, which reveals that the areas close to the step and upper part of the flow domain are susceptible to higher vorticity fluctuation. The turbulence level was found to be higher at the shear layer and is expected to have a higher vorticity value. In addition, maximum vorticity of 4100 rad/s and 50,000 rad/s was observed at $Re = 15{,}000$ and $Re = 30{,}000$, respectively. As expected, the vorticity level was high under Grids 2 and 4 (large HD), whereas it was low under Grids 1 and 3 (low HD). The PIV-derived vorticity contour maps are shown in Figure 9. These maps show how the flow was detached from the backward-facing step. Once the flow was detached, it formed low-speed recirculation zones downstream. The vorticity contours were successful in resolving the vortex pairs. The high vorticity level at the shear layers indicates that the system is designed and tested with high standards. As expected, the vorticity maps also show that the vorticity in the recirculation zone is higher under Grids 2 and 4. This is again due to the velocity gradients that are steep along the flow path under Grids 2 and 4. The y-velocity component dominated the axial component, which increased the rotation of the velocity vector. The opposite was true under Grids 1 and 3. In the colormap scales, the positive and negative values of vorticity are seen indicating the presence of both counterclockwise and clockwise vortices in the flow. Furthermore, the vorticity scale was six times higher at $Re = 30{,}000$ compared to $Re = 15{,}000$. The vortex pairs look like a straight line at $Re = 30{,}000$, whereas it looks like a spotted dot at $Re = 15{,}000$. In general, the authors found that the vorticity orientation and magnitude in the contour maps were mostly controlled by velocity gradients and rotation or angular momentum within the fluid elements. The vorticities were induced through the interaction of different flow components such as shear layers, boundary layers, or swirling flows, which created regions with intense vorticity where fluid elements rotate or circulate. The presence, strength, and spatial distribution of vortices in the vorticity maps helps understand the flow dynamics under turbulent environments and guide optimization of the design of the combustor.

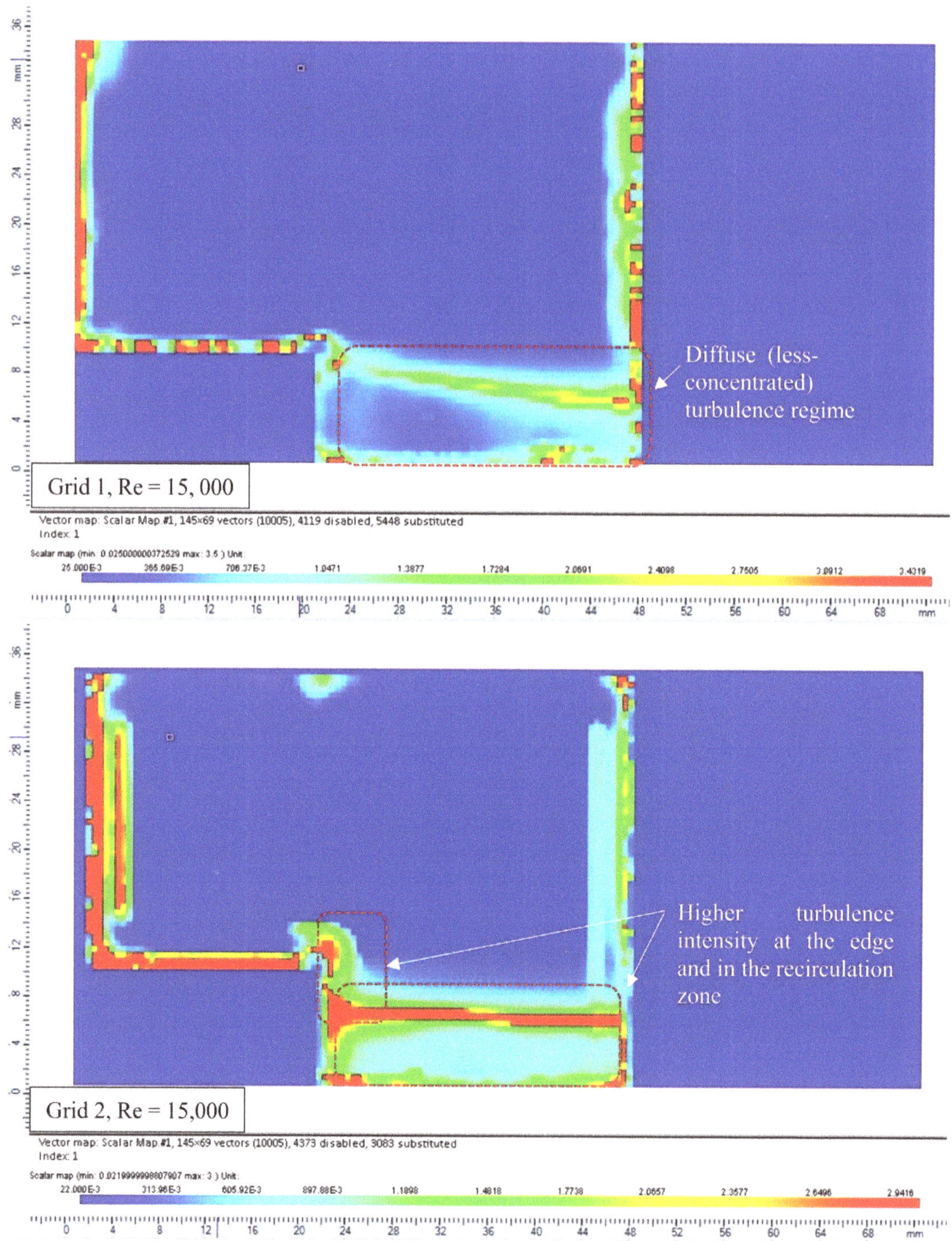

Figure 7. *Cont.*

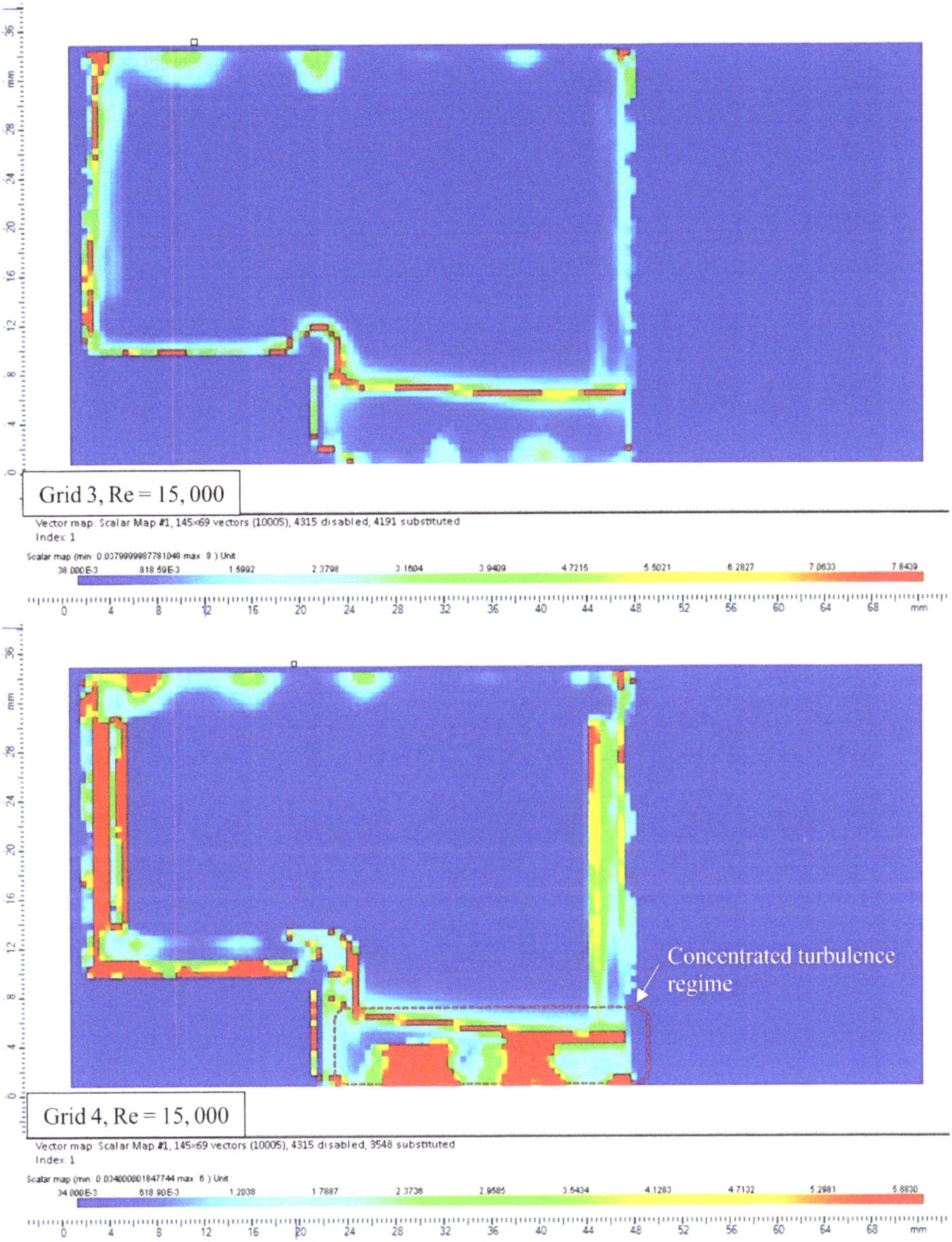

Figure 7. *Cont.*

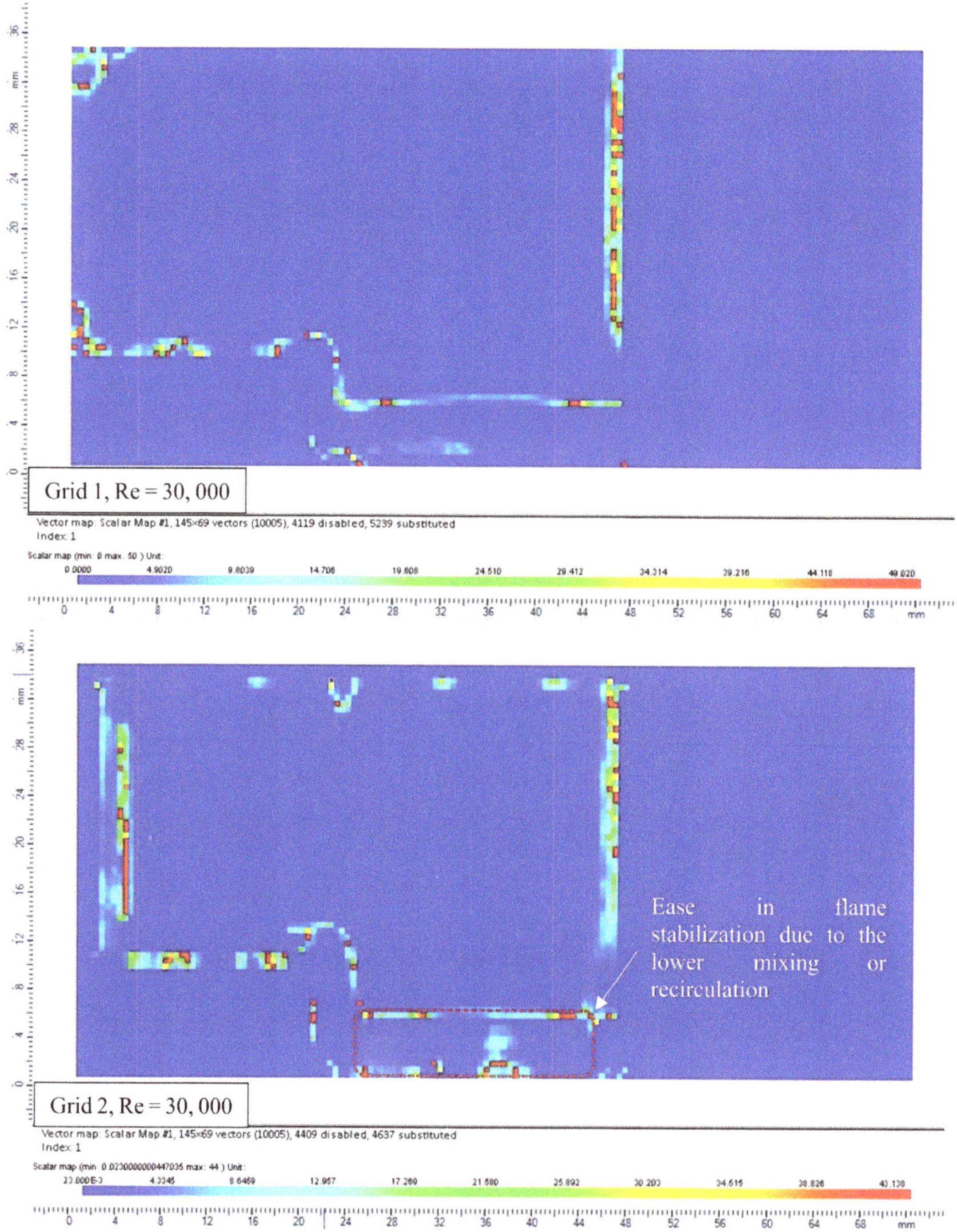

Figure 7. *Cont.*

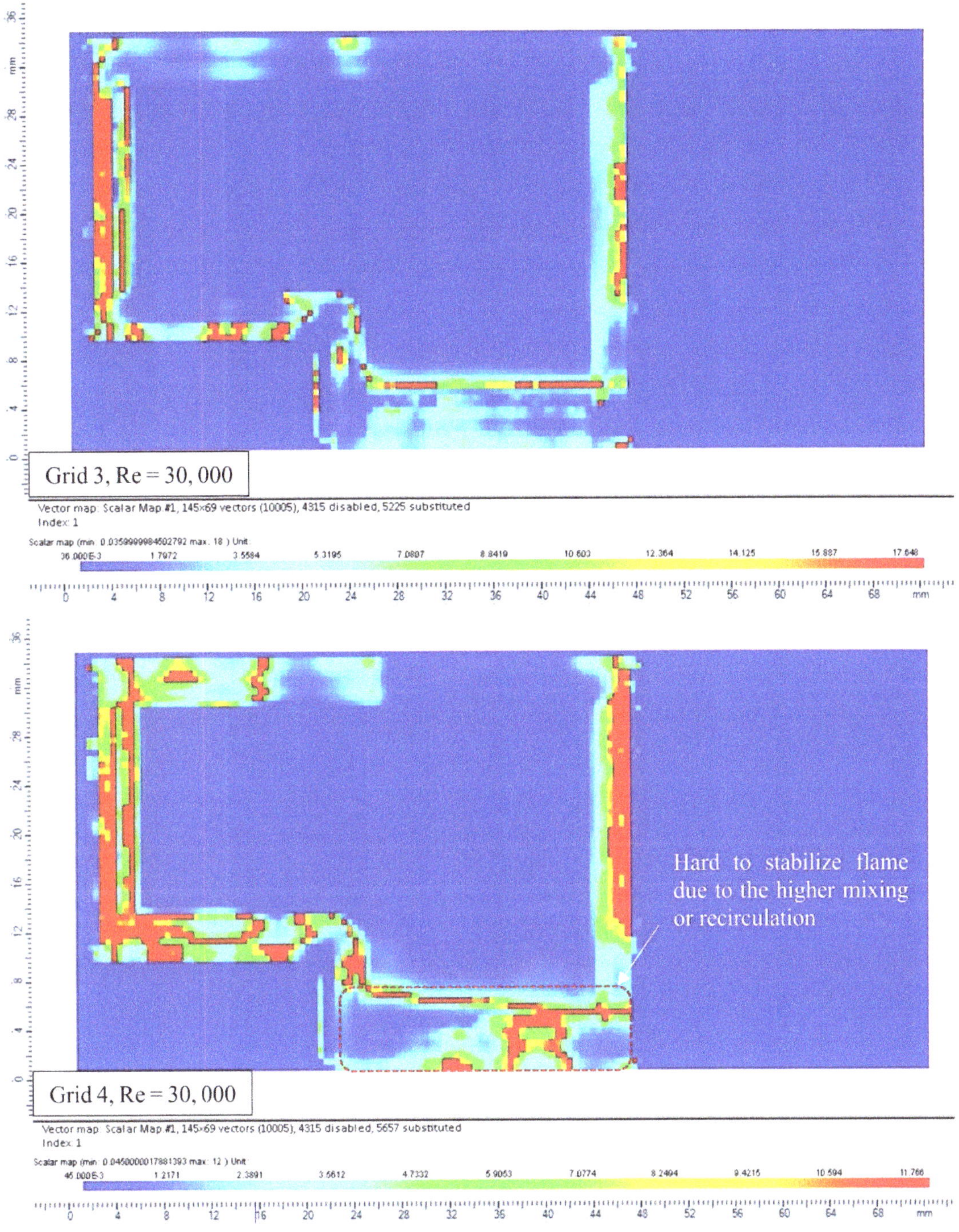

Figure 7. Turbulent intensity contour plot at *Re* = 15,000 (**first four contours**) and *Re* = 30,000 (**last four contours**). Color print required.

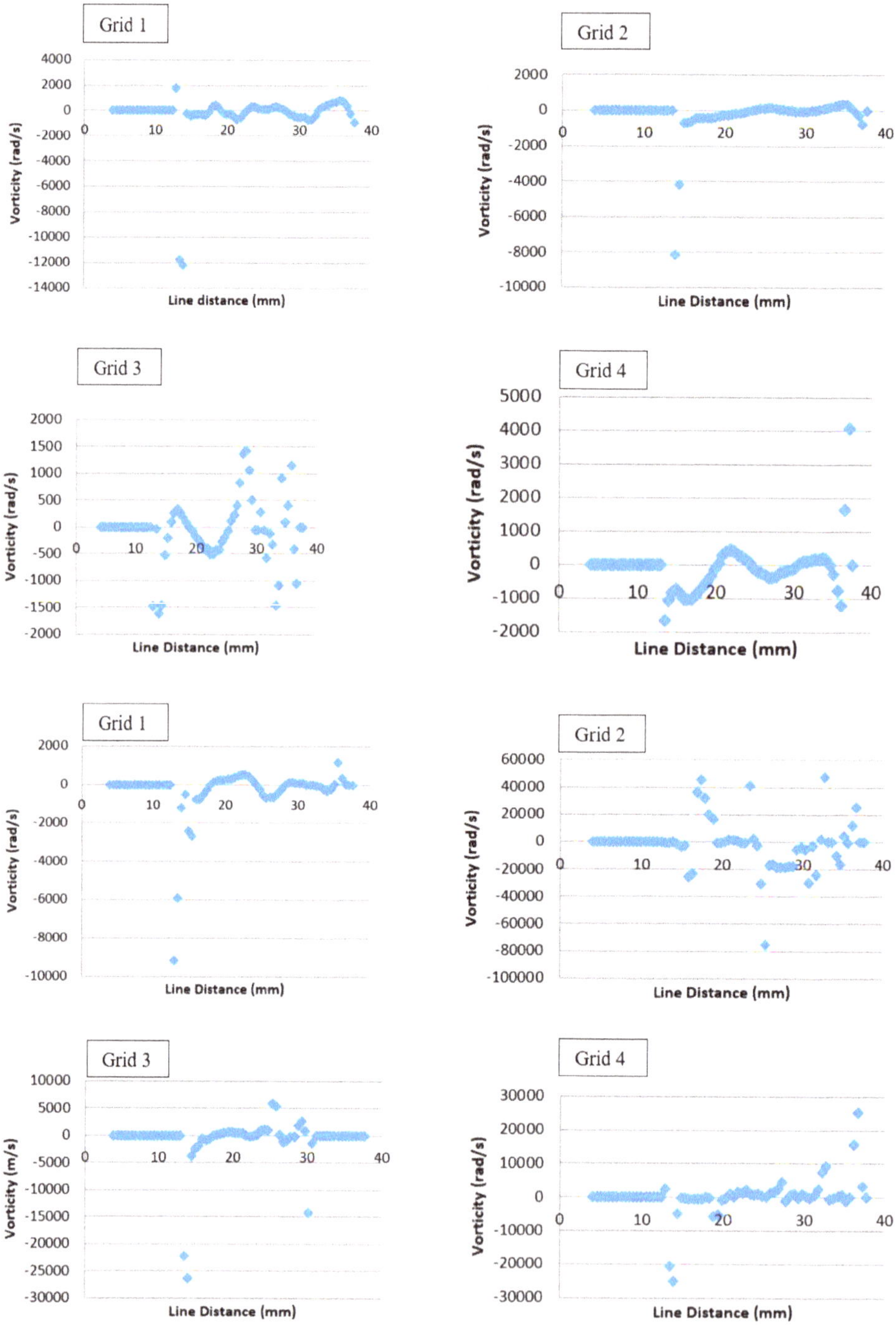

Figure 8. Vorticity levels at Re = 15,000 (**first two rows**) and Re = 30,000 (**last two rows**). Color print required.

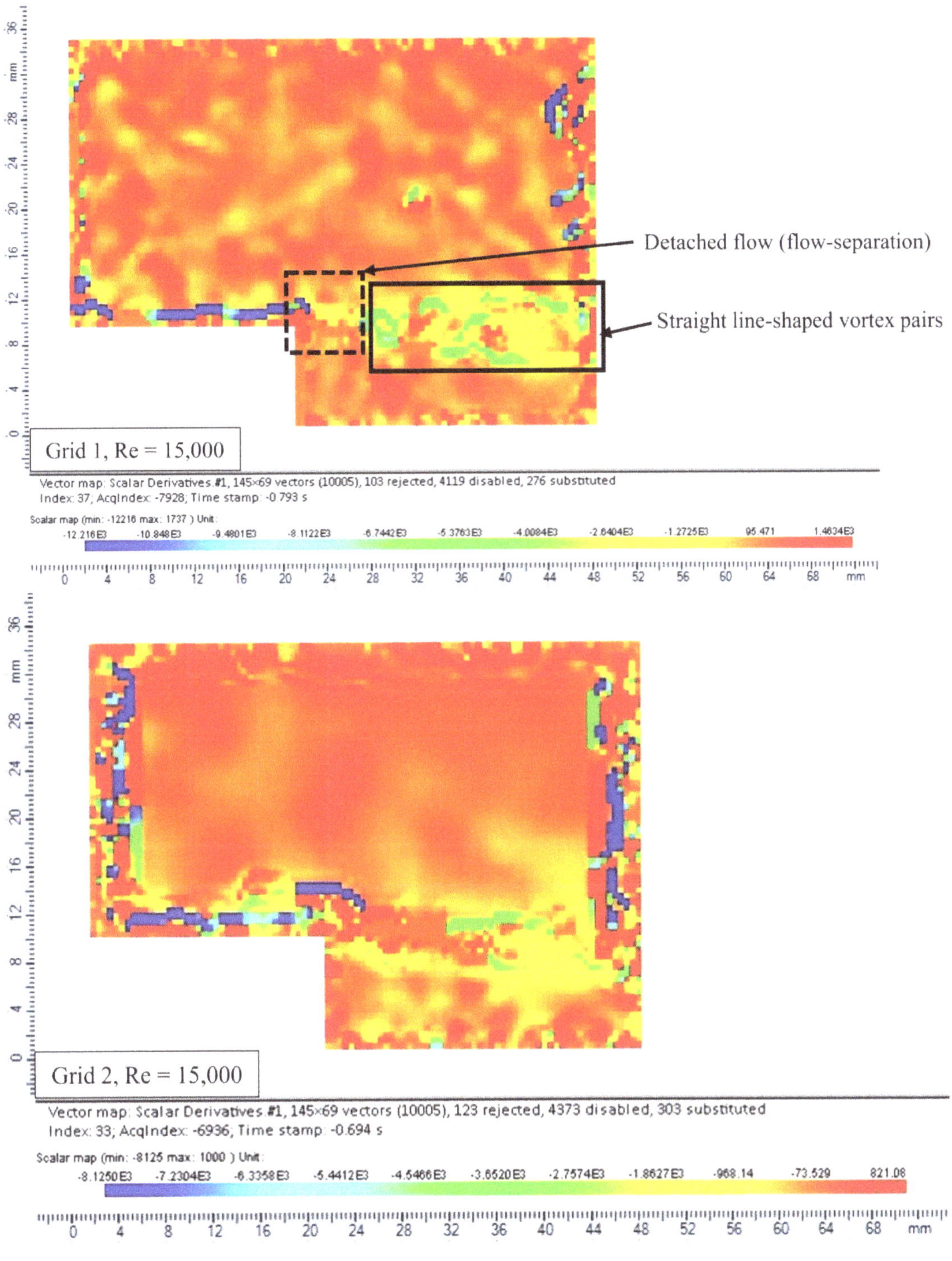

Figure 9. *Cont.*

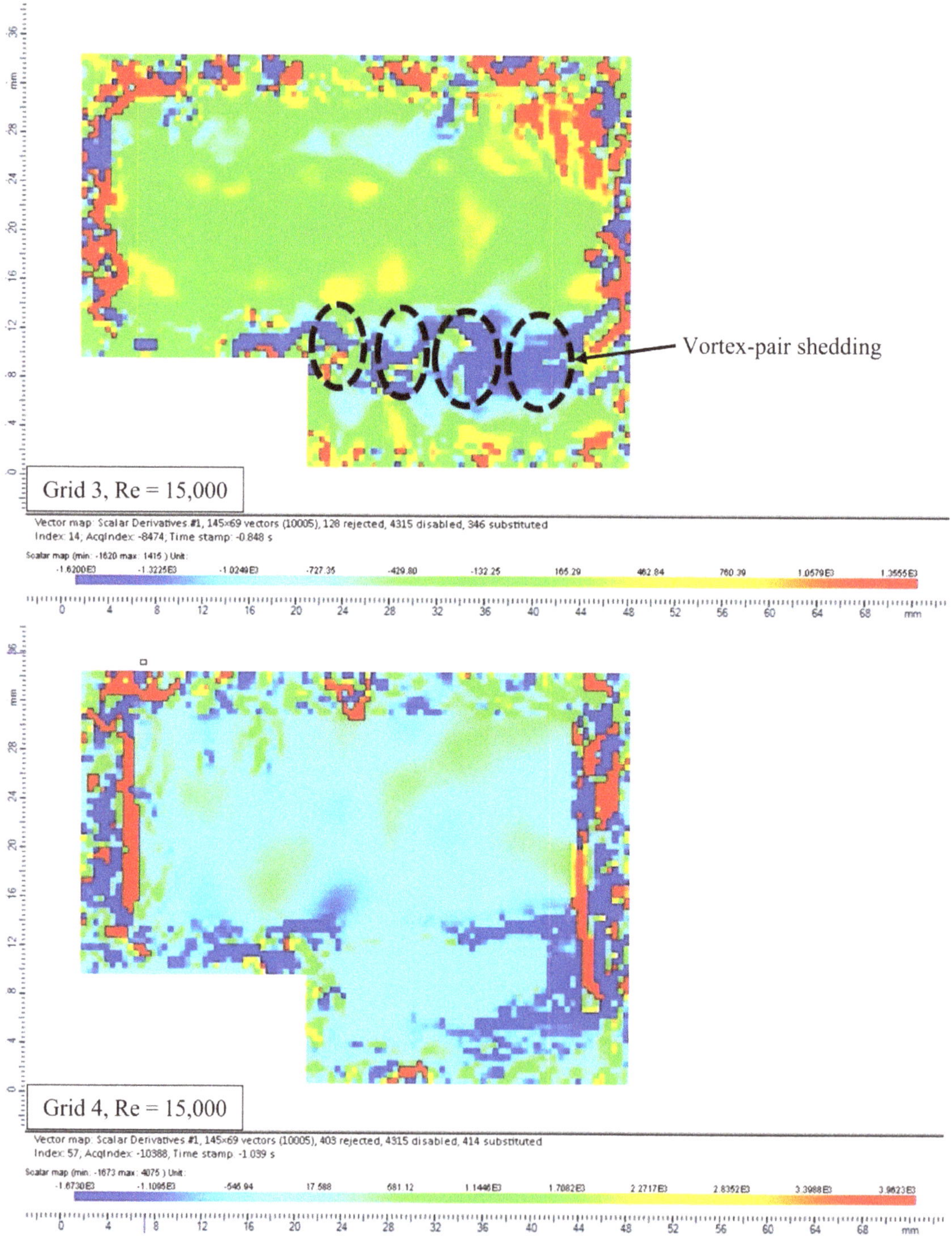

Figure 9. *Cont.*

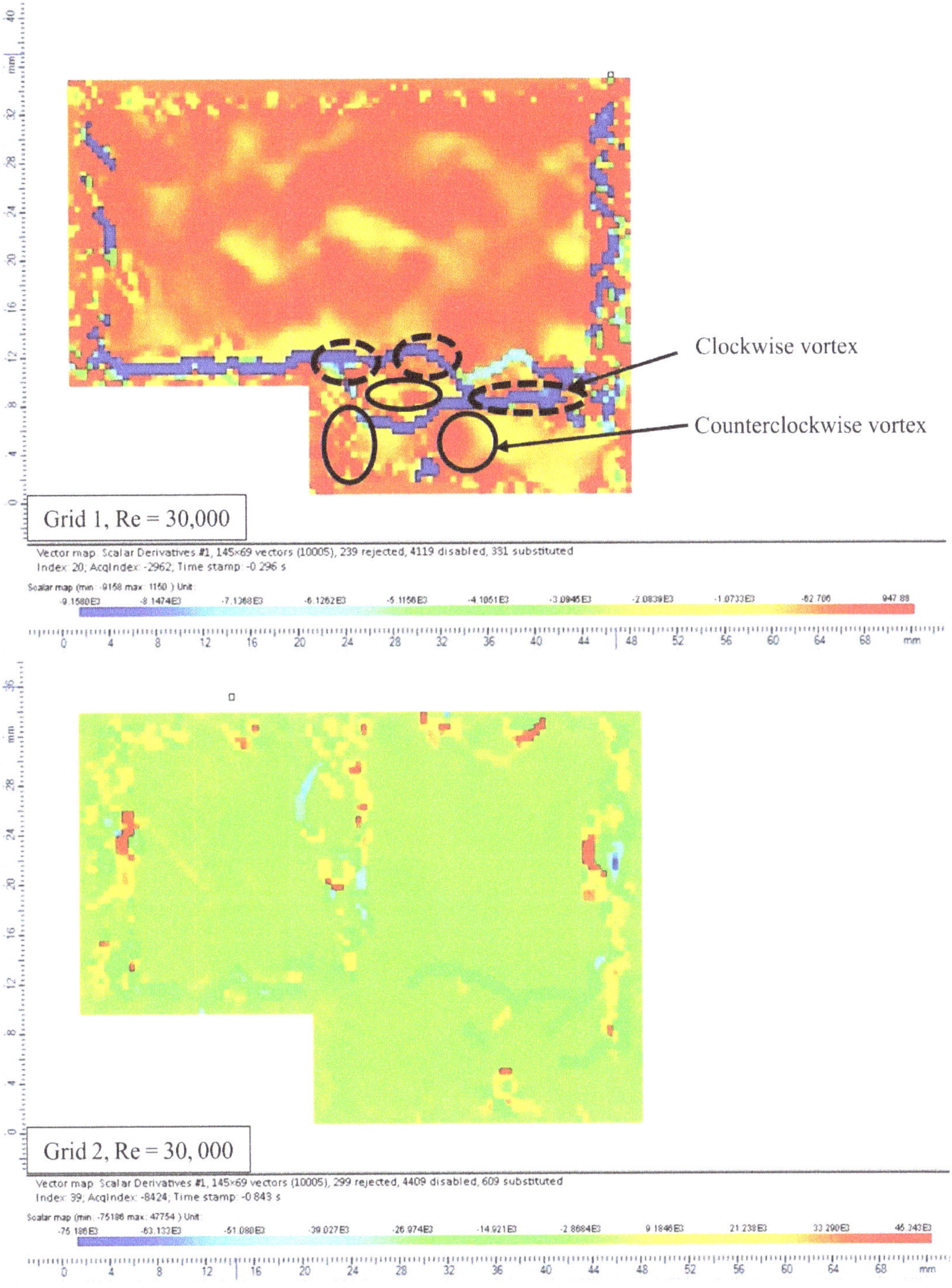

Figure 9. *Cont.*

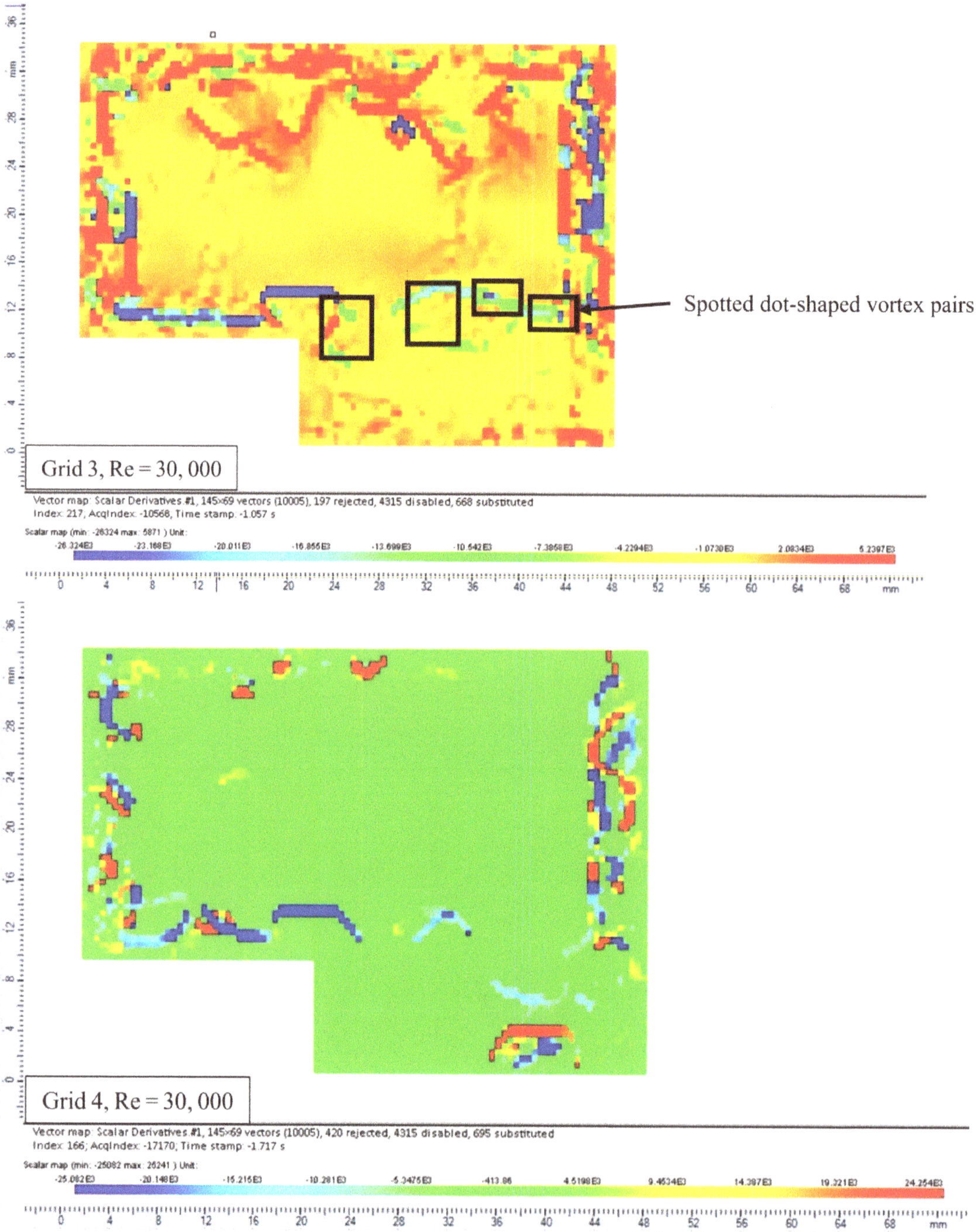

Figure 9. Vorticity contour plot at *Re* = 15,000 (**first four contours**) and *Re* = 30,000 (**last four contours**). Color print required.

The PIV results showed how the flow concentration was uniformly distributed inside the combustion chamber. The flow homogeneity was achieved using the different grids at the combustor inlet. The sharp velocity gradients were seen at the backward-facing step, which aids in accelerating flame growth. The bottom wall close to the step was densely populated by small eddies. These eddies controlled the flame anchoring mechanism inside the combustor. The PIV gave a detailed statistical explanation of the flow behavior, which Table 6 lists below. However, to understand the relation of flow behavior with flame dynamics, a PLIF study was carried out.

Table 6. The flow characteristics for all grids at Stoichiometric conditions.

Grid	*Re*	Φ	U_{bulk} (m/s)	U_{PIV} (m/s)	u′ (m/s)	Da	Ka	Re_T	u'/S_L	L_T	L_T/δ_L
1	15,000	1.0	10	13	1.1	21.3	0.8	147.5	3.0	0.0173	49.4
	30,000		20	25	3.5	18.8	7.0	458.4	9.3	0.0173	49.4
2	15,000		10	11	1.7	34.2	1.7	234.2	4.4	0.0186	33.2
	30,000		20	26	9.8	2.2	18.2	1364.7	25.7	0.0186	33.2
3	15,000		10	14	5.5	5.1	10.3	604.5	14.5	0.0146	41.8
	30,000		20	26	7.5	4.3	16.8	824.0	19.7	0.0146	41.8
4	15,000		10	17	2.5	13.9	3.4	300.5	6.6	0.0160	45.6
	30,000		20	27	11.0	1.8	24.6	1315.9	28.9	0.0160	45.6

3.2. Flame Characteristics Analysis

A flame study was conducted at *Re* = 15,000 (10 m/s) and *Re* = 30,000 (20 m/s) using the 10-kHz rate PLIF. The base image (without the flame) and the raw image (with the flame) were post-processed using the MATLAB image processing tools described previously in Figure 4. Figure 10i–iv below shows how a typical flame profile and flame edge were generated using those image processing tools.

The base image without the flame is shown in Figure 10i. This is when the igniter was off and there was no active combustion in the system. Figure 10ii is the raw image with flame when combustion occurred in the system. Figure 10iii shows the flame profiles that were generated using the MatLab-based image processing procedures listed in Figure 4. Flame profiles locate the reaction zones, show the thickness of the flame core, help to understand the flame expansion and stretching, etc. Figure 10iv shows the flame edges. The flame edges revealed the location, shape and size of flame wrinkling, and the presence of vortices, and showed how vortices infused the flame core and made the flame thin and stretch.

The flame progression is shown in Figure 10v. Images (a) to (e) show the progression of the flame growth from ignition initiation through blowout. Image (a) indicates the initial combustion stage; (b) shows the initiation of wrinkles; (c) reveals the formation of more wrinkles; (d) shows the breakdown of the flamelets; and (e) shows the burnout of the flamelets.

The presence of an unburned air–fuel mixture inside the burned gases and vice versa are observed in Figure 10(va). Eventually, combustion completely burns all the air–fuel mixture. The authors studied the recess length and made sure that the mixing, residence time and air–fuel ratio were adequate for the complete combustion of methane air. There were different levels of turbulence in the flame. Additionally, the flame profiles show irregularly shaped wrinkles (handgrip-like structures), see Figure 10(vb). The wrinkles continued to form as the flame progressed downstream as shown in Figure 10(vc). The flow fluctuation, thermal instabilities caused by buoyancy and expansion, and local flame viscosity controlled the flame wrinkling. The turbulent and velocity scales—for example, velocity fluctuation (u′) and turbulent length-scale (L_T)—generated high compressive forces in the flame, leading to accelerated flame–flame interactions. This interaction enhanced the reactant-product pocket formation rate [36], causing many finger-shaped wrinkles to form in the flame. It also causes to change in flame thickness, (δ_L) and flame expansion.

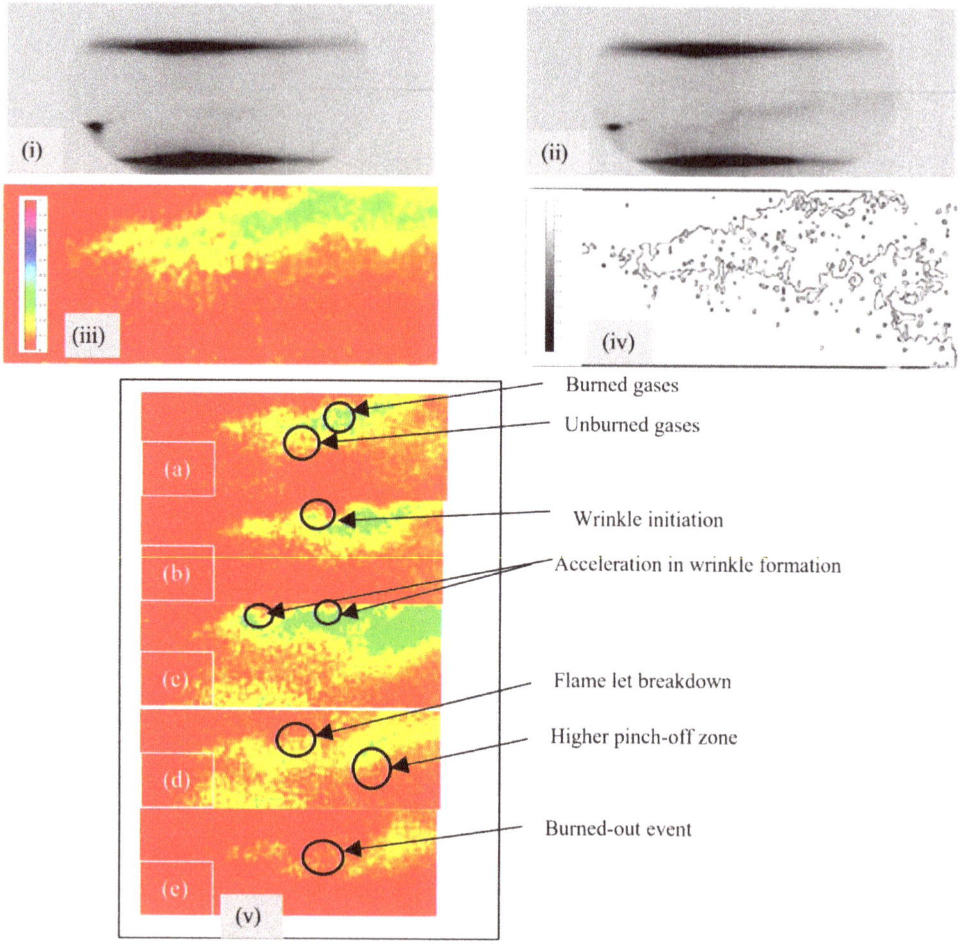

Figure 10. A typical flame profile and flame edge images generated using C-X CH PLIF at *Re* = 15,000 and Φ = 0.8 (**i**) base image (without flame); (**ii**) raw image (with flame); (**iii**) flame profile with HSV color map; (**iv**) flame edge; and (**v**) flame progression overview.

After the flame wrinkling, the next event is called the breakdown of the flamelet or pinch-off event Figure 10(vd). The large-scale fluid structures caused the reaction zone to shatter, leading to flamelet detachment from the edges. Local velocity fluctuations and thermal diffusion further accelerated pinch-off events. Near the bottom wall, at a distance downstream, the breakdown (pinch-off) event accelerated. This was due to the infusion of small eddies into the flamelet regime. These small-scale, semi-detached, burned gases sometimes reach self-ignition temperature and start to burn out or disappear. This is commonly known as a burned-out event Figure 10(ve). The pinch-off and burnout events occurred spontaneously and were controlled by parameters such as pressure, turbulent length scale, and velocity fluctuation. Flame profiles observed in this research work are very thin (thin reaction zone); therefore, all the images presented below are intensified and zoomed in for better interpretation and visualization.

The authors conducted a qualitative analysis of flame behavior in this article. Therefore, the authors combined all the images together and present them here so that an overall understanding of how flame propagates and interacts under different grids and Reynolds numbers could be made. First, thin fluctuating flame profiles are seen in Figures 11–13. Additionally, the flame width increased as it moved downstream. This is referred to as flame broadening. The local flame diffusivity increased due to the enhanced heat and mass transfer imparted to the reaction zone by the turbulent eddies. This caused the flame's core structure (width) to increase. As reported in [37,38], these large-scale eddies and wrinkles should enhance the reactant burning rate, flame burning area, and flame propagation speed. As the velocity increased downstream, high-energy fluid structures formed. This increased the turbulent eddy diffusivity and suppressed the strain rate effect, influencing flame stretching. Broadened flame structures were observed under all grids.

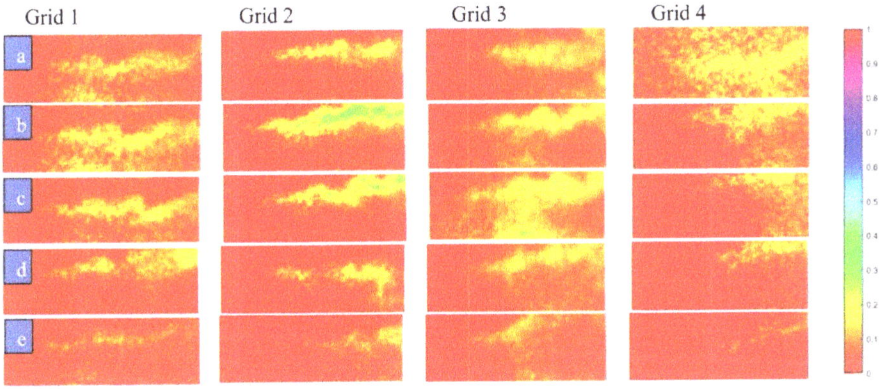

Figure 11. Flame profiles at $\Phi = 0.8$ and $Re = 15{,}000$. Here, (**a**) = initial combustion stage; (**b**) = initiation of wrinkles; (**c**) = formation of more wrinkles; (**d**) = breakdown (pinch-off) of the flamelets; and (**e**) = burnout of the flamelets. The HSV colormap shows flame intensity between 0 and 1. [Flame profiles are zoomed in for better visualization].

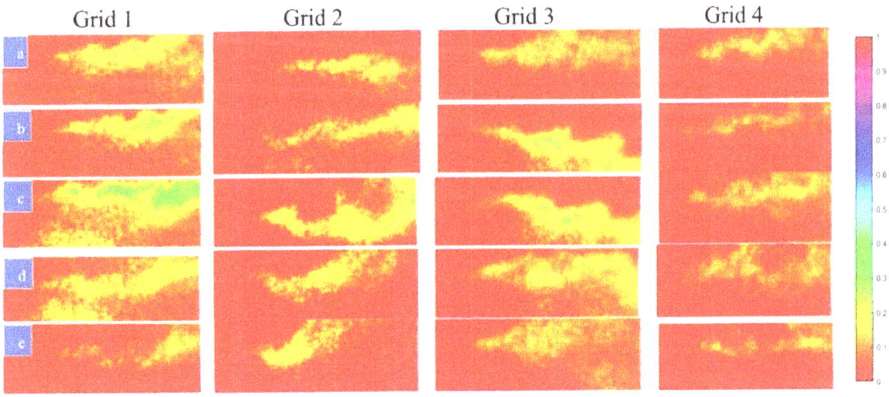

Figure 12. Flame profiles at $\Phi = 1.0$ and $Re = 15{,}000$. Here, (**a**) = initial combustion stage; (**b**) = initiation of wrinkles; (**c**) = formation of more wrinkles; (**d**) = breakdown (pinch-off) of the flamelets; and (**e**) = burnout of the flamelets. [Flame profiles are zoomed in for better visualization].

The effect of grids on flame characteristics has been studied (see Figures 11–13). The flame profiles under Grids 1 and 3 had more wrinkle growths. This was due to the higher change in thermal diffusion and viscosity of the flame under those grids, as also reported in [5]. However, there is less presence of large-scale wrinkles in the flames under those grids. Thus, flame stretching and flame width are less under Grids 1 and 3. The opposite is true for the flame under Grids 2 and 4 (see Figures 11–13).

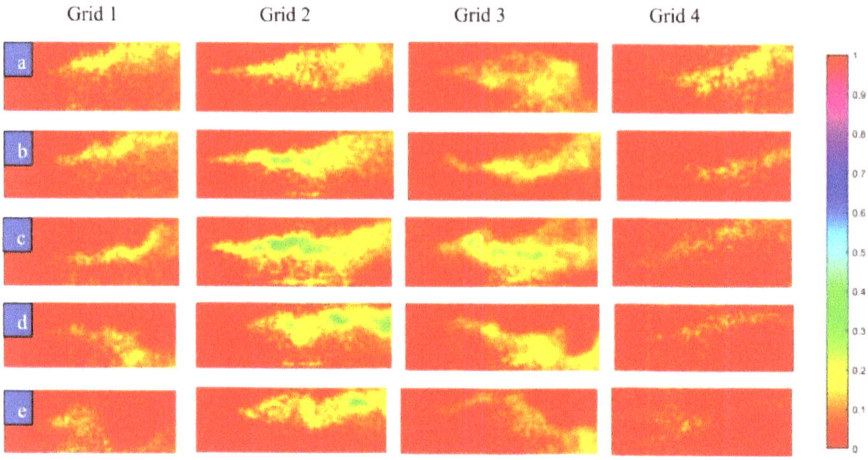

Figure 13. Flame profiles at Φ = 1.2 and *Re* = 15,000. Here, (**a**) = initial combustion stage; (**b**) = initiation of wrinkles; (**c**) = formation of more wrinkles; (**d**) = breakdown (pinch-off) of the flamelets; and (**e**) = burnout of the flamelets. [Flame profiles are zoomed in for better visualization].

The effect of two different Reynold's numbers on flame evolution has been studied. The flame expansion was faster at *Re* = 30,000 (20 m/s) (Figures 14–16) compared to *Re* = 15,000 (10 m/s) (Figures 11–13). The local flame strain rate increased with the increase in flow velocity. This strain tries to suppress flame growth. However, higher-order turbulence at higher *Re* aided in overcoming this challenge. Similar research on flame expansion rate is reported in [39]. Additionally, at *Re* = 30,000, the flow mixture velocity (20 m/s) dominated over turbulent eddy viscosity ($u'l_o$ or $u'\eta$), see Table 6. Therefore, the flame front broadened in both axial and radial directions and allowed the small eddies to infuse into the flame core (see Figures 14–16).

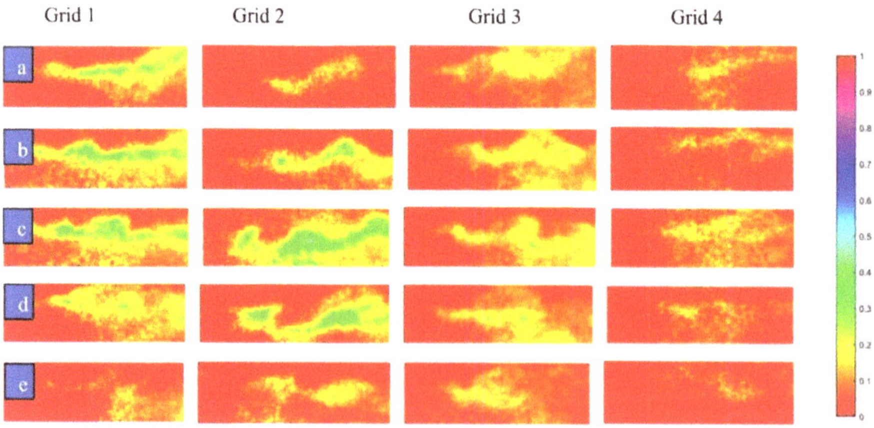

Figure 14. Flame profiles at Φ = 0.8 and *Re* = 30,000. Here, (**a**) = initial combustion stage; (**b**) = initiation of wrinkles; (**c**) = formation of more wrinkles; (**d**) = breakdown (pinch-off) of the flamelets; and (**e**) = burnout of the flamelets. [Flame profiles are zoomed in for better visualization].

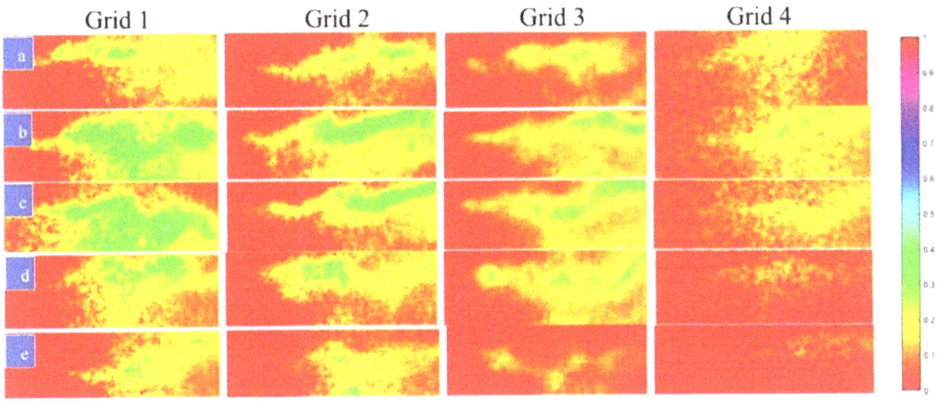

Figure 15. Flame profiles at Φ = 1.0 and *Re* = 30,000. Here, (**a**) = initial combustion stage; (**b**) = initiation of wrinkles; (**c**) = formation of more wrinkles; (**d**) = breakdown (pinch-off) of the flamelets; and (**e**) = burnout of the flamelets. [Flame profiles are zoomed in for better visualization].

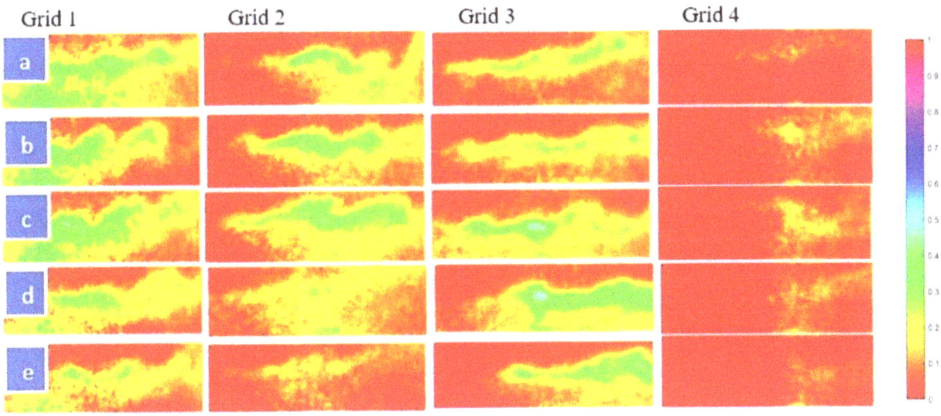

Figure 16. Flame profiles at Φ = 1.2 and *Re* = 30,000. Here, (**a**) = initial combustion stage; (**b**) = initiation of wrinkles; (**c**) = formation of more wrinkles; (**d**) = breakdown (pinch-off) of the flamelets; and (**e**) = burnout of the flamelets. [Flame profiles are zoomed in for better visualization].

The flame wrinkling (irregularly shaped edges) was higher at *Re* = 30,000 compared to *Re* = 15,000. This was because a large number of small-scale eddies and higher flame instabilities (mostly the flame fluctuation) were present at *Re* = 30,000 (see Figures 14–16).

The effect of the equivalence ratio (Φ) on flame propagation was also investigated. The reactant-product pocket formation rate increased with fuel loading as Φ increased. Therefore, at *Re* = 30,000 and Φ > 1, the flame wrinkling rate is maximum. Additionally, the pinch-off and burnout rates were higher at *Re* = 30,000 (Figures 14, 15 and 16d,e). The same finding is also reported in [40]. In addition, the flame stretching rate was found to be higher at *Re* = 30,000. This was due to the large-scale wrinkles that interacted with each other and formed new wrinkles along the flame edges. As a result, the flame stretched across those wrinkles. The opposite is true for *Re* = 15,000.

The flame profiles alone do not offer a comprehensive understanding of flame behavior. To gain more detailed insights into crucial flame characteristics such as flame boundaries and wrinkles, the authors utilized MATLAB image processing tools with in-house coding (see Figure 4) and generated corresponding flame edges (See Figure 17).

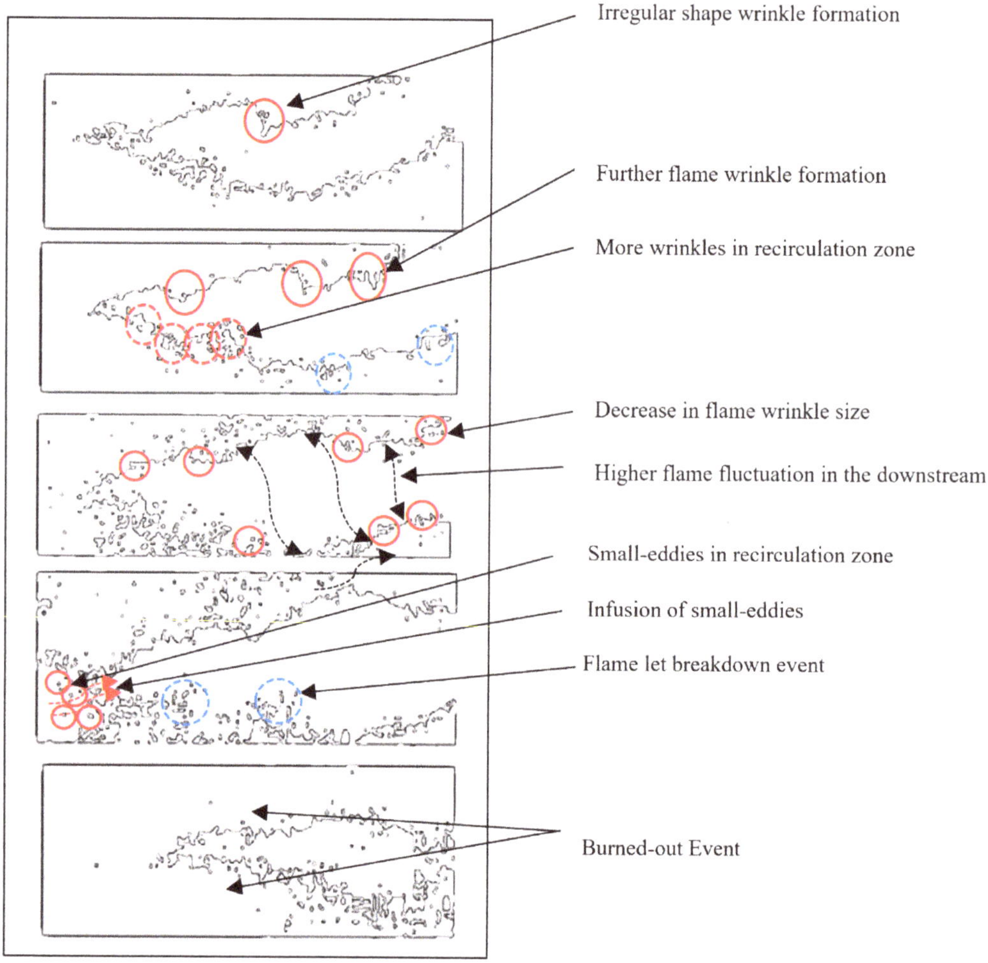

Figure 17. Typical flame edge profiles.

Figure 17 shows how the wrinkle initiated and grew with time. It also shows how the wrinkles changed their shape and size from ignition through the blowout. The wrinkles have an irregular shape with handgrip-like structures. Wrinkle shape does not change, but wrinkle size changes throughout the flame evolution (ignition to burnout). The flame edges show the presence of many small eddies in the recirculation zone. The same behavior is noticed in the PIV analysis. This validates the role of the backward-facing step in flame anchoring.

The flame had more fluctuation downstream (see flame edges under Figures 18–23). Flame growth was mostly controlled by flame wrinkling and turbulence. The flame was continuous and had no flame extinction for the *Re* conditions considered in this paper. Wrinkle formation (Figures 18, 19, 20, 21, 22 and 23b) was adequate to maintain a continuous flame and not too extreme. The flame edges also showed how the small eddies infused into the flame core and the flamelet (small-scale flame eddies) detached and reattached to the flame core. These events were more common at *Re* = 30,000 (see Figures 21, 22 and 23b–d) due to the increase in flame displacement rate. In other words, vorticity-strain rate effects were higher at *Re* = 30,000.

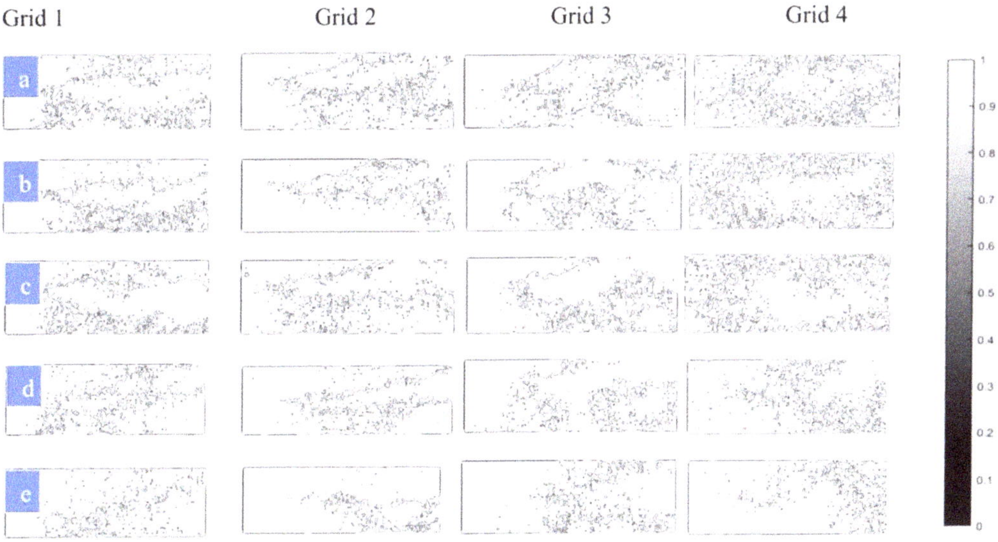

Figure 18. Flame edges at Φ = 0.8 and *Re* = 15,000. Here, (**a**) = initial combustion stage; (**b**) = initiation of wrinkles; (**c**) = formation of more wrinkles; (**d**) = breakdown (pinch-off) of the flamelets; and (**e**) = burnout of the flamelets. [Flame edges are zoomed in for better visualization].

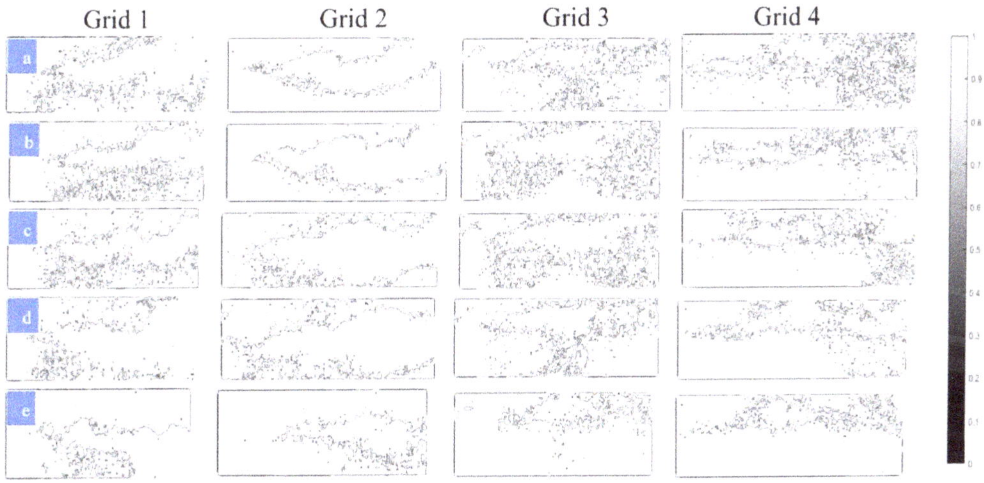

Figure 19. Flame edges at Φ = 1.0 and *Re* = 15,000. Here, (**a**) = initial combustion stage; (**b**) = initiation of wrinkles; (**c**) = formation of more wrinkles; (**d**) = breakdown (pinch-off) of the flamelets; and (**e**) = burnout of the flamelets. [Flame edges are zoomed in for better visualization].

The flame was thinner at *Re* = 30,000 and wider at *Re* = 15,000 (see Figures 18–23). This could be justified by the PIV-derived data presented before. The velocity contours (previously presented in Figure 5) show that the local flow velocity was high and residence time was low across the shear layer at *Re* = 30,000. Consequently, the flame did not have enough time to expand laterally and became thinner. The opposite is true at *Re* = 15,000. The flame edges also show that the flame wrinkling rate is higher under Grids 1 and 3, whereas pinch-off and burnout rates are higher under Grids 2 and 4. The same behavior is noticed in the flame profiles reported earlier in Figures 11–16. A similar observation is reported in [41,42].

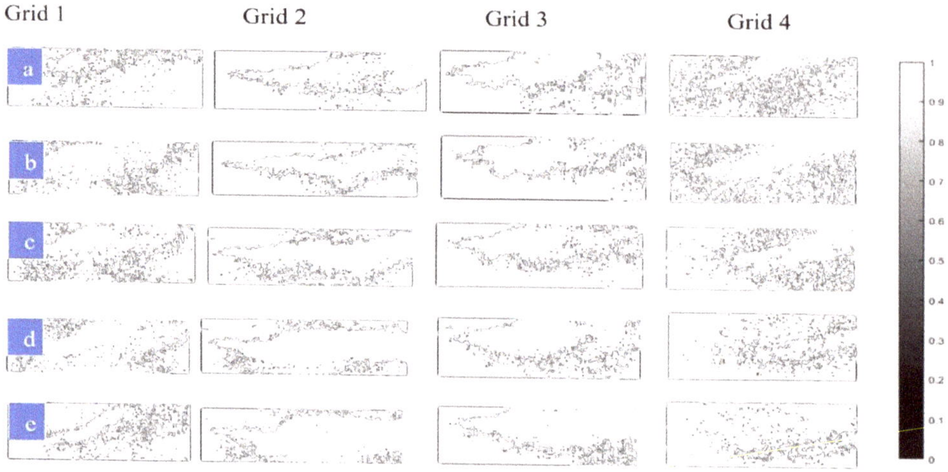

Figure 20. Flame edges at Φ = 1.2 and *Re* = 15,000. Here, (**a**) = initial combustion stage; (**b**) = initiation of wrinkles; (**c**) = formation of more wrinkles; (**d**) = breakdown (pinch-off) of the flamelets; and (**e**) = burnout of the flamelets. [Flame edges are zoomed in for better visualization].

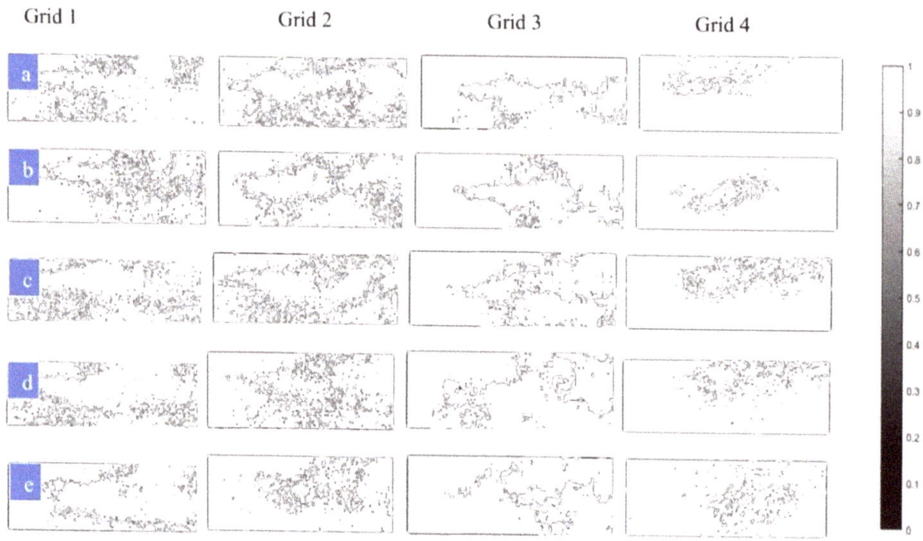

Figure 21. Flame edges at Φ = 0.8 and *Re* = 30,000. Here, (**a**) = initial combustion stage; (**b**) = initiation of wrinkles; (**c**) = formation of more wrinkles; (**d**) = breakdown (pinch-off) of the flamelets; and (**e**) = burnout of the flamelets. [Flame edges are zoomed in for better visualization].

The flame profiles (Figures 11–16) and flame edges (Figures 18–23) presented above show that the CH layer (reaction zone layer) is thinner at both *Re* conditions. However, the CH layer thickness increased slightly downstream. This happens because the turbulence dominated over the strain rate downstream. The CH layer resembled both flamelet-like and diffuse-like structures. The authors believe that the flame signal detected by the C-X CH PLIF was not entirely the CH signal: there might be a presence of some CH * chemiluminescence. This is why in some cases the flame profiles looked diffuse. The authors completed some preliminary iterations on methane–air premixed reacting flames using the CH PLIF. The authors believe that there are still many questions and issues that need to be addressed in future research iterations. For example, the CH filter and the operating parameters of the camera intensifier units such as gate pulse width and gate exposure/opening time need to be re-tuned according to the higher test operating conditions. Additionally, the wavelength tuning for the C-X band of CH PLIF should be further examined. The flame edges showed that the CH signal decreased with the increase in *Re*. The authors will also consider this in the next research iteration. The authors are recommending preheating the reactants at elevated temperatures for higher *Re* combustion tests. The incorporation of higher-order alkaline into the reactants might resolve this issue. However, a series of experimental tests and more theoretical research must be conducted before making any conclusion on this issue.

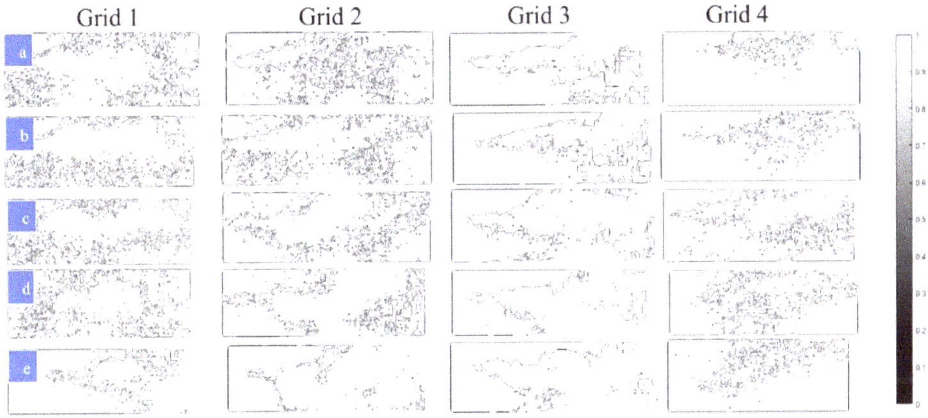

Figure 22. Flame edges at Φ = 1.0 and *Re* = 30,000. Here, (**a**) = initial combustion stage; (**b**) = initiation of wrinkles; (**c**) = formation of more wrinkles; (**d**) = breakdown (pinch-off) of the flamelets; and (**e**) = burnout of the flamelets. [Flame edges are zoomed in for better visualization].

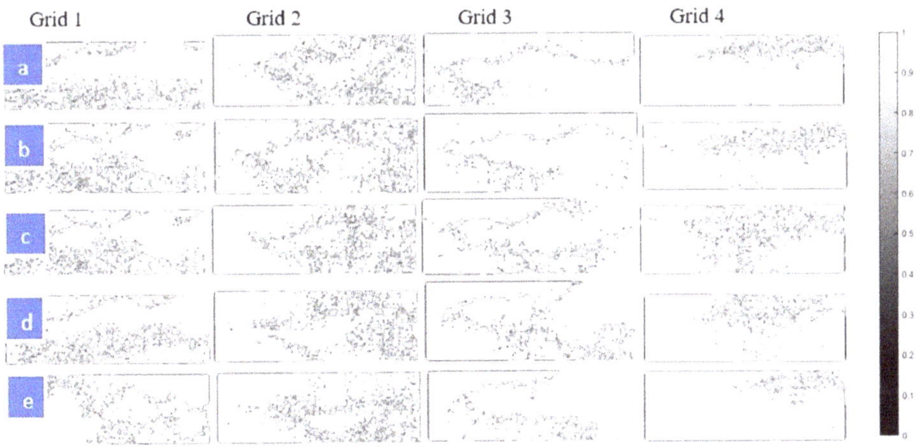

Figure 23. Flame edges at Φ = 1.2 and *Re* = 30,000. Here, (**a**) = initial combustion stage; (**b**) = initiation of wrinkles; (**c**) = formation of more wrinkles; (**d**) = breakdown (pinch-off) of the flamelets; and (**e**) = burnout of the flamelets. [Flame edges are zoomed in for better visualization].

The effect of equivalence ratios (Φ) on flame evolution was investigated using the flame profiles and edges reported above. Data from the temperature sensors (k-type) indicate that the chamber heat release rate increased with the increase in fuel loading, Φ. This is because the combustion radical fraction and their reaction rates increase with Φ. The same behavior is reported in [1,5,43]. The authors used a series of temperature sensors (non-intrusive) on the combustor walls. The temperature sensors sensed the wall face temperature during the combustion. After that, the LabView tool generated the associated heat release rate from the combustor walls, exhaust chamber, and cooling chamber. This information has already been published by two of the co-authors of this paper, [28,32].

The flame reaches the maximum static temperature (~2200 K) at Φ =1.2. Therefore, according to the literature [6], the flame surface consumption rate, flame surface area, and flame length should increase at Φ > 1. Additionally, the flame surface area directly relates to the flame burning velocity and flame front curvature. Thus, at higher Φ, the flame burning area and flame front curvature should increase [3,31,43]. The flame fluctuation (less radical kinetics) is less for the lean mixture (Φ = 0.8). A moderate level of flame fluctuation is seen at Φ = 1.2 (see Figures 18–23). However, this small fluctuation does not affect the flame stability studied in this paper, as also reported in [4]. The authors believe that more tests at variable Φ need to be conducted to conclude these predictions.

3.3. Flame Regime Plot

The laminar flame characteristics such as laminar flame speed and laminar flame thickness used in this paper are extracted from the papers published by two of the co-authors of this article, [28,32]. Information regarding the laminar flame characteristics is also well-reported in the literature [44,45]. The flow characteristics were extracted using the PIV results reported in this paper. All this information was used to locate the flame on the Borghi–Peters diagram (see Figure 24). The PIV results and flame characteristics used to locate the flame regime boundaries are summarized in Tables 6 and 7.

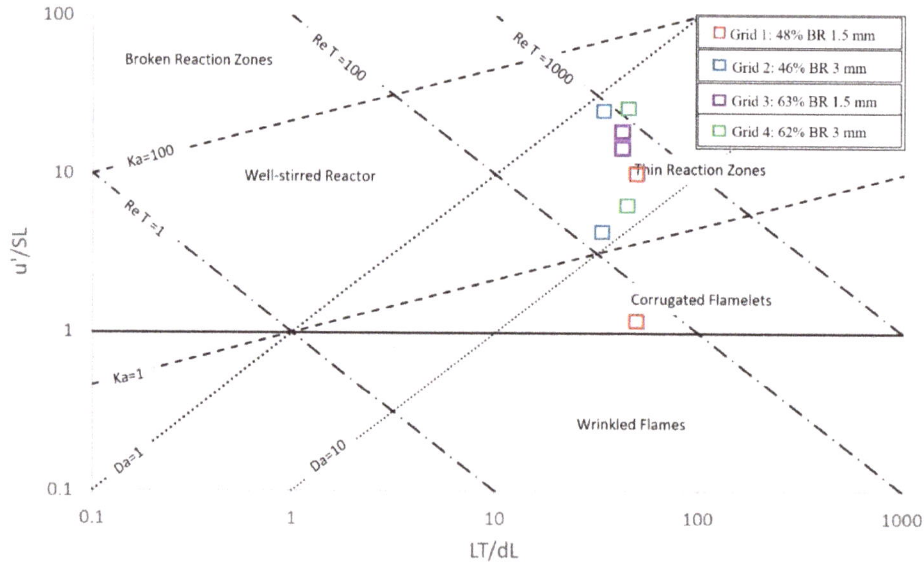

Figure 24. The flame regime location at *Re* = 15,000 and *Re* = 30,000 for all grids.

The thickened flame (thin reaction zone) was observed at all flow conditions except the flame under Grid 1 at *Re* = 15,000 (Figure 24). At this specific condition, the flame falls in the corrugated flamelet regime. This happened because of the velocity fluctuation relative to laminar flame speed ($\frac{u'}{S_L}$) was low under this specific condition. The thickness of small-scale eddies (Kolmogorov scales, η) was less than the flame thickness (δ). The small-scale eddies were infused into the flame and made the reaction zone thinner. Therefore, the reaction zone was much thinner at Re = 30,000 than Re = 15,000. The flame did not fall in the broken reaction zone so there was no flame break up. It is hypothesized that the flame wrinkling was not severe and flame stabilized at both flow conditions.

Table 7. Borghi–Peters diagram parameters.

Flow Rate (CFM)	**Grid 1**		**Grid 2**		**Grid 3**		**Grid 4**	
	L_T/δ_L	u'/S_L	L_T/δ_L	u'/S_L	L_T/δ_L	u'/S_L	L_T/δ_L	u'/S_L
12 (10 m/s, Re = 15,000)	49.371	2.987	33.171	4.405	41.77	14.471	45.57	6.594
25 (20 m/s, Re = 30,000)	49.371	9.285	33.171	25.665	41.77	19.727	45.57	28.875

The flame location on the Borghi–Peter diagram gives some insight into the flame behavior at different turbulence levels. The flame lies on the top of the thin reaction zone under Grids 2 and 4 (see Figure 24). This means that the Kolmogorov time scale is less than the chemical time scale (higher Ka value) and more flamelet breakdown should be expected under Grids 2 and 4. On the other hand, the flame fell at the core of the thin reaction zone under Grids 1 and 3. Therefore, the reaction rate dominated over the diffusion rate (lower Da value) and more flame wrinkling should be seen under Grids 1 and 3. This is exactly what the authors noticed from the flame profile and flame edges reported previously in this paper.

4. Conclusions

The current study reports the flow and flame characteristics in extreme turbulence environments using 10 kHz PIV and 10 kHz PLIF. The premixed methane–air flow inside a small-scale windowed combustor was considered for this study. Different flow conditions (Re = 15,000 (10 m/s), Re = 30,000 (20 m/s)), fuel concentrations (Φ = 0.8, 1.0, and 1.2) and grid turbulators (grids) (BR = 46%, 48%, 62% and 63%) were used for flow and flame study. PIV results show that the flow turbulence increased downstream due to the drastic change in the flow passage after the backward-facing step. The turbulence was higher in the shear layer. Perforated stainless steel (SS) plates, namely grid turbulators (grids), created isotropic homogeneous turbulence inside the combustor. The backward-facing step helped in small eddy formation, flow acceleration, and flame stabilization. The small HD grids (Grids 1 and 3) produced continuous fluid structures and comparatively less turbulence in the flow. On the other hand, large HD grids (Grids 2 and 4) created large-scale fluid structures and higher velocity fluctuation (u') in the flow. The flame front study was carried out using C-X CH PLIF at a wavelength of 314.415 nm. The higher laser absorption and emission coefficients gave outstanding flame profiles at this wavelength. The reaction rate was faster than the diffusion rate under Grids 1 and 3 (lower Da value). This was due to the presence of high-pressure fluctuations and thermal instabilities in the flame. Thus, the flame wrinkling was higher under Grids 1 and 3. The turbulence intensities (I) and flame thermal diffusivity are higher under Grids 2 and 4 (higher Ka value). Accordingly, the pinch-off and burnout rates are higher under these grids. The PIV and PLIF study revealed that the Re, turbulent length scales (L_T), and wrinkling mostly controlled flame evolution. The flame growth rate was faster at Re = 30,000 compared to Re = 15,000 due to the higher velocity fluctuations and thermal instabilities at Re = 30,000. The flame edges showed that the flame had both flamelet and diffuse-like structures. Further, the flame had irregularly shaped wrinkles and the size of those wrinkles decreased with time as the flame progressed downstream. Flame edges justified the infusion of small eddies into the flame core; this eventually helped in flame anchoring. Flame expansion and flame stretching are higher at Re = 30,000 compared to Re = 15,000. The flame mainly fell in the thin reaction zone (thickened flame regime) except for one that fell in the corrugated flame regime. However, the flame was close to the broken reaction zone. For this, the authors are recommending taking extra precautions in future high-speed combustion tests. Additionally, the authors are aware that the flame signal detected by the C-X CH PLIF was not entirely the CH radical signal. Some CH * chemiluminescence might be present in the flame. This is a preliminary study where the authors implemented the CH PLIF in the high-pressure turbulent reacting premixed combustion environment. The authors want to conduct more studies to see how the C-X CH filter and camera intensifier properties such as gate pulse width and gate exposure/opening time play a role in flame image detection.

Author Contributions: Conceptualization, A.C.; methodology, M.A.H., M.N.A.I., M.D.L.T. and A.A.Z.; software, M.A.H. and M.N.A.I.; validation, M.A.H.; formal analysis, M.A.H.; investigation, M.A.H.; resources, A.C.; data curation, M.A.H., M.N.A.I., M.D.L.T. and A.A.Z.; writing—original draft preparation, M.A.H.; writing—review and editing, M.A.H.; visualization, M.A.H., M.N.A.I., M.D.L.T. and A.A.Z.; supervision, A.C.; project administration, A.C.; funding acquisition, A.C. All authors have read and agreed to the published version of the manuscript.

Funding: This research work was supported by the National Aeronautics and Space Administration (NASA) (award No(s) NNX09AV09A) and the Army Research Office (ARO) (award No(s) W911NF-13-1-0156).

Data Availability Statement: Source data and other raw data are available from the corresponding author upon request.

Acknowledgments: The authors would like to thank former graduate research assistants at the Aerospace Center at UTEP: Martin Alejandro De La Torre (currently Engineer, Blue Origin, FL, USA) and Arturo Acosta-Zamora (currently Engineer, Intel, OR, USA) for their contribution to preliminary research work. The authors also would like to thank the faculties and staff of the Aerospace Center at UTEP for their supervision and support.

Conflicts of Interest: The authors declare no conflict of interest.

Nomenclature

PLIF	lanar laser-induced fluorescence
PIV	Particle image velocimetry
HD	Hole diameter
BR	Blockage ratio
CCD	Charged couple device
Φ	Equivalence ratio
x	Mole fraction
$\overline{\omega}$	Vorticity, curl of mean velocity
ω_z	Vorticity in the z-direction, computed from the velocity gradients
$\frac{\partial u}{\partial y}$	X-velocity gradient in the Y direction
u'	Velocity fluctuation
I	Turbulent intensity
Da	Damkohler number
Ka	Karlovitz number
S_L	Laminar flame speed
δ_L	Laminar flame thickness
Re	Reynolds number
Re_T	Turbulent Reynolds number
L_T	Turbulent length scale
u_{RMS}	RMS of u velocity, computed from the velocity variance
v_{RMS}	RMS of v velocity, computed from the velocity variance
FOS	Factor of safety

References

1. Bobba, M.K. *Flame Stabilization and Mixing Characteristics in a Stagnation Point Reverse Flow Combustor*; Georgia Institute of Technology: Atlanta, GA, USA, 2007.
2. Hossain, M.A. *Design of a High Intensity Turbulent Combustion System*; The University of Texas at El Paso: El Paso, TX, USA, 2015.

3. Sadiq, A.M.; Sleiti, A.K.; Ahmed, S.F. Turbulent Flames in Enclosed Combustion Chambers: Characteristics and Visualization—A Review. *J. Energy Resour. Technol.* **2020**, *142*, 080801. [CrossRef]
4. Periagaram, K. *Determination of Flame Characteristics in a Low Swirl Burner at Gas Turbine Conditions through Reaction Zone Imaging*; Georgia Institute Technology: Atlanta, GA, USA, 2012.
5. Wang, Z.; Bai, Z.; Yu, G.; Yelishala, S.; Metghalchi, H. The Critical Pressure at the Onset of Flame Instability of Syngas/Air/Diluent Outwardly Expanding Flame at Different Initial Temperatures and Pressures. *J. Energy Resour. Technol.* **2019**, *141*, 082207. [CrossRef]
6. Allison, P.M. *Experimental Characterization of Combustion Instabilities and Flow-Flame Dynamics in a Partially-Premixed Gas Turbine Model Combustor*; The University of Michigan: Ann Arbor, MI, USA, 2013.
7. Troiani, G.; Lapenna, P.E.; Lamioni, R.; Creta, F. Self-wrinkling induced by Darrieus-Landau instability in turbulent premixed Bunsen flames from low to moderately high Reynolds numbers. *Phys. Rev. Fluids* **2022**, *7*, 053202. [CrossRef]
8. Lapenna, P.E.; Troiani, G.; Lamioni, R.; Creta, F. Mitigation of Darrieus–Landau instability effects on turbulent premixed flames. *Proc. Combust. Inst.* **2021**, *38*, 2885–2892. [CrossRef]
9. Chaudhuri, S.; Saha, A.; Law, C.K. On flame–turbulence interaction in constant-pressure expanding flames. *Proc. Combust. Inst.* **2015**, *35*, 1331–1339. [CrossRef]
10. Lipatnikov, A.N.; Chomiak, J.; Sabelnikov, V.A.; Nishiki, S.; Hasegawa, T. Unburned mixture fingers in premixed turbulent flames. *Proc. Combust. Inst.* **2015**, *35*, 1401–1408. [CrossRef]
11. Chaudhuri, S. Life of flame particles embedded in premixed flames interacting with near isotropic turbulence. *Proc. Combust. Inst.* **2015**, *35*, 1305–1312. [CrossRef]
12. Zhou, R.; Balusamy, S.; Sweeney, M.S.; Barlow, R.S.; Hochgreb, S. Flow field measurements of a series of turbulent premixed and stratified methane/air flames. *Combust. Flame* **2013**, *160*, 2017–2028. [CrossRef]
13. Fan, L.; Chong, C.T.; Tanno, K.; McGrath, D.; Zheng, Y.; Hochgreb, S. Measurement of the effect of water droplets on strained laminar flames using two-phase PIV. *Proc. Combust. Inst.* **2021**, *38*, 3183–3192. [CrossRef]
14. Ruan, C.; Chen, F.; Cai, W.; Qian, Y.; Yu, L.; Lu, X. Principles of non-intrusive diagnostic techniques and their applications for fundamental studies of combustion instabilities in gas turbine combustors: A brief review. *Aerosp. Sci. Technol.* **2019**, *84*, 585–603. [CrossRef]
15. Hedman, T.D.; Cho, K.Y.; Pfeil, M.A.; Satija, A.; Mongia, H.C.; Groven, L.J.; Lucht, R.P.; Son, S.F. High speed OH PLIF applied to multiphase combustion (Review). *Combust. Explos. Shock. Waves* **2016**, *52*, 1–13. [CrossRef]
16. Driscoll, J.F.; Chen, J.H.; Skiba, A.W.; Carter, C.D.; Hawkes, E.R.; Wang, H. Premixed flames subjected to extreme turbulence: Some questions and recent answers. *Prog. Energy Combust. Sci.* **2020**, *76*, 100802. [CrossRef]
17. Farooq, A.; Alquaity, A.B.; Raza, M.; Nasir, E.F.; Yao, S.; Ren, W. Laser sensors for energy systems and process industries: Perspectives and directions. *Prog. Energy Combust. Sci.* **2022**, *91*, 100997. [CrossRef]
18. Skiba, A.W.; Carter, C.D.; Hammack, S.D.; Lee, T. A simplified approach to simultaneous multi-scalar imaging in turbulent flames. *Combust. Flame* **2018**, *189*, 207–211. [CrossRef]
19. Fluorescence, P.L. *Role of Planar Laser-Induced Fluorescence in Combustion Research*; AerospaceLab: Mont-Saint-Guibert, Belgium, 2009; pp. 1–14.
20. Zhou, B.; Kiefer, J.; Zetterberg, J.; Li, Z.; Aldén, M. Strategy for PLIF single-shot HCO imaging in turbulent methane/air flames. *Combust. Flame* **2014**, *161*, 1566–1574. [CrossRef]
21. Baudoin, E.; Bai, X.S.; Yan, B.; Liu, C.; Yu, R.; Lantz, A.; Hosseini, S.M.; Li, B.; Elbaz, A.; Sami, M.; et al. Effect of Partial Premixing on Stabilization and Local Extinction of Turbulent Methane/Air Flames. *Flow Turbul. Combust.* **2013**, *90*, 269–284. [CrossRef]
22. Sjöholm, J.; Rosell, J.; Li, B.; Richter, M.; Li, Z.; Bai, X.-S.; Aldén, M. Simultaneous visualization of OH, CH, CH_2O and toluene PLIF in a methane jet flame with varying degrees of turbulence. *Proc. Combust. Inst.* **2013**, *34*, 1475–1482. [CrossRef]
23. Escofet-Martin, D.; Chien, Y.-C.; Dunn-Rankin, D. PLIF and chemiluminescence in a small laminar coflow methane-air diffusion flame at elevated pressures. *Combust. Flame* **2022**, *243*, 112067. [CrossRef]
24. Driscoll, J.F.; The University of Michigan. Turbulent CombustionExperimental and Fundamental Models. 2016. Available online: https://cefrc.princeton.edu/sites/g/files/toruqf1071/files/combustion-summer-school/lecture-notes/Driscoll_Friday_4M.pdf (accessed on 26 May 2023).
25. Carter, C.D.; Donbar, J.M.; Driscoll, J.F. Simultaneous CH planar laser-induced fluorescence and particle imaging velocimetry in turbulent non-premixed flames. In Proceedings of the 36th AIAA Aerospace Sciences Meeting and Exhibit, Reno, NV, USA, 6–9 January 1997.
26. Carter, C.D.; Hammack, S.; Lee, T. High-speed flamefront imaging in premixed turbulent flames using planar laser-induced fluorescence of the CH C−X band. *Combust. Flame* **2016**, *168*, 66–74. [CrossRef]
27. Tanahashi, M.; Murakami, S.; Choi, G.-M. Simultaneous CH–OH PLIF and stereoscopic PIV measurements of turbulent premixed flames. *Proc. Combust. Inst.* **2005**, *30*, 1665–1672. [CrossRef]
28. Acosta-zamora, A. *Flame front Structure Studies of Highly Turbulent Reacting Flow over a Backward Facing Step Using kHz OH-CH Planar Laser Induced Fluorescence*; The University of Texas at El Paso: El Paso, TX, USA, 2016.
29. Vagelopoulos, C.M.; Frank, J.H. An experimental and numerical study on the adequacy of CH as a flame marker in premixed methane flames. *Proc. Combust. Inst.* **2005**, *30*, 241–249. [CrossRef]

30. Hossain, M.A. Laser Diagnostics of Compressible and High-Intensity Premixed Methane-Air Combustion Inside a Backward Facing Step Combustor Using High-Repetition Rate CH-PLIF and PIV. Ph.D. Thesis, The University of Texas at El Paso, El Paso, TX, USA, 2020.
31. Filatyev, S.A.; Driscoll, J.F.; Carter, C.D.; Donbar, J.M. Measured properties of turbulent premixed flames for model assessment, including burning velocities, stretch rates, and surface densities. *Combust. Flame* **2005**, *141*, 1–21. [CrossRef]
32. De La Torre, M.A. *Characterization of High Intensity Turbulent Flows through Time Resolved Particle Image Velocimetry in a Backward Facing Step Combustor*; University of Texas at El Paso: El Paso, TX, USA, 2016.
33. Hossain, A.; Islam, N.A.; Hossain, W.; Choudhuri, A.R. CH-PLIF Diagnostics of High-Intensity Turbulent Premixed Methane-Air Combustion. In Proceedings of the AIAA Scitech 2020 Forum, Orlando, FL, USA, 6–10 January 2020; pp. 1–14. [CrossRef]
34. Islam, N.A.; Hossain, A.; De La Torre, M.; Acosta-Zamora, A.; Choudhuri, A.R. Laser Diagnostics of Highly Turbulent Premixed Methane-Air Flow over a Backward Facing Step Using PIV and OH-PLIF. In Proceedings of the AIAA Propulsion and Energy Forum, Indianapolis, IN, USA, 19–22 August 2019; pp. 1–12. [CrossRef]
35. Hossain, M.A.; Islam, M.N.A.; Choudhuri, A. Flame Imaging of Highly Turbulent Premixed Methane-Air Combustion Using Planar Laser Induced Fluorescence (PLIF) of CH (C-X). In Proceedings of the ASME 2019 Power Conference, Salt Lake City, UT, USA, 15–18 July 2019; pp. 1–7.
36. Tyagi, A.; Boxx, I.; Peluso, S.; O'Connor, J. Statistics and topology of local flame–flame interactions in turbulent flames. *Combust. Flame* **2019**, *203*, 92–104. [CrossRef]
37. Skiba, A.W.; Carter, C.D.; Hammack, S.D.; Miller, J.D.; Gord, J.R.; Driscoll, J.F. The influence of large eddies on the structure of turbulent premixed flames characterized with stereo-PIV and multi-species PLIF at 20 kHz. *Proc. Combust. Inst.* **2019**, *37*, 2477–2484. [CrossRef]
38. Wang, Z.; Zhou, B.; Yu, S.; Brackmann, C.; Li, Z.; Richter, M.; Aldén, M.; Bai, X.-S. Structure and burning velocity of turbulent premixed methane/air jet flames in thin-reaction zone and distributed reaction zone regimes. *Proc. Combust. Inst.* **2019**, *37*, 2537–2544. [CrossRef]
39. HDave, H.L.; Mohan, A.; Chaudhuri, S. Genesis and evolution of premixed flames in turbulence. *Combust. Flame* **2018**, *196*, 386–399. [CrossRef]
40. Beeri, Z.; Blunsdon, C.A.; Malalasekera, W.M.G.; Dent, J.C. Comprehensive Modeling of Turbulent Flames with the Coherent Flame-Sheet Model—Part II: High-Momentum Reactive Jets. *J. Energy Resour. Technol.* **1996**, *118*, 72–76. [CrossRef]
41. Yoshida, S.; Naka, Y.; Minamoto, Y.; Shimura, M.; Tanahashi, M. Propagation Characteristics of Turbulent Methane-Air Premixed Flames at Elevated Pressure in a Constant Volume Vessel. In Proceedings of the 18th International Symposium on the Application of Laser and Imaging Techniques to Fluid Mechanics, Lisbon, Portugal, 4–7 July 2016; pp. 1–12.
42. OAskari, O.; Metghalchi, H.; Moghaddas, A.; Hannani, S.K.; Ebrahimi, R.; Hemmati, H. Fundamental study of spray and partially premixed combustion of methane/air mixture. *J. Energy Resour. Technol.* **2013**, *135*, 021001. [CrossRef]
43. Mulla, I.A.; Dowlut, A.; Hussain, T.; Nikolaou, Z.M.; Chakravarthy, S.R.; Swaminathan, N.; Balachandran, R. Heat release rate estimation in laminar premixed flames using laser-induced fluorescence of CH_2O and H-atom. *Combust. Flame* **2016**, *165*, 373–383. [CrossRef]
44. Dyakov, I.V.; De Ruyck, J.; Brussel, V.U.; Bosschaart, K.J.; De Goey, P. Measurement of adiabatic burning velocity in methane-oxygen-nitrogen mixtures. *Combust. Sci. Technol.* **2001**, *172*, 81–96. [CrossRef]
45. Gens, J.G.; Mauss, F.; Peters, N. Analytic approximations of burning velocities and flame thicknesses of lean hydrogen, methane, ethylene, ethane, acetylene, and propane flames. In Proceedings of the Twenty-Fourth Symposium (International) on Combustion, Sydney, Australia, 5–10 July 1992; The Combustion Institute: Pittsburgh, PA, USA, 1992; pp. 129–135.

aerospace

MDPI

Article

The Linear Stability of Liquid Film with Oscillatory Gas Velocity

Xiangdong Deng [1], Baolu Shi [1,2,*], Yong Tang [1,2] and Ningfei Wang [1]

1 School of Aerospace Engineering, Beijing Institute of Technology, Beijing 100081, China; 3120205056@bit.edu.cn (X.D.); tangyong@bit.edu.cn (Y.T.)
2 Chongqing Innovation Center, Beijing Institute of Technology, Chongqing 404100, China
* Correspondence: shibaolu@bit.edu.cn

Abstract: The present study theoretically investigated the linear instability of a liquid film sheared by gas flow under acoustic oscillations. In this work, the velocity oscillations of the gas are used to approximately characterize the acoustic oscillations, and the ratio of the conduction heat flux to the evaporation heat flux is used to characterize the heat and mass transfer. Considering the much stronger impact of the heat convection than the heat conduction in practical cases, a correction factor is introduced to satisfy the heat flux ratio within a reasonable range. Because of the oscillatory velocity of gas, several unstable regions, involving the KHI region and the parametric instability (PI) region, appear. The impact of the velocity oscillations on the KHI is related to the forcing frequency. Increasing the oscillatory velocity amplitude promotes the KHI when the forcing frequency is large, while the KHI is restrained with the increase in the oscillatory velocity amplitude when the forcing frequency is small. Since the viscous dissipation is enhanced when the forcing oscillations frequency increases, the PI is suppressed. In addition, when the surface tension decreases, the interfacial instability is also promoted. Increasing the gas–liquid density ratio can destabilize the surface. However, the impact of the heat and mass transfer on the interfacial instability is neglectable as the gas–liquid density ratio is large. Furthermore, the heat and mass transfer have a promoting impact on the PI and KHI, while their destabilizing effect on the indentation between unstable regions is greater. It is significant to note that the location of the maximum growth rate would be in the most unstable region.

Keywords: liquid film; acoustic oscillations; heat and mass transfer; Kelvin–Helmholtz instability

Citation: Deng, X.; Shi, B.; Tang, Y.; Wang, N. The Linear Stability of Liquid Film with Oscillatory Gas Velocity. *Aerospace* **2023**, *10*, 691. https://doi.org/10.3390/aerospace10080691

Academic Editor: Jian Liu

Received: 14 June 2023
Revised: 29 July 2023
Accepted: 31 July 2023
Published: 3 August 2023

1. Introduction

Liquid film cooling is an effective cooling method, which can prevent the wall of the combustor from overheating, oxidation, and destruction. Experimental results have predicted that the detachment of droplets from the liquid layer of the liquid film cooling cannot be neglected [1]. The entrainment is a significant reason for the mass loss of the liquid film, which can result in a decrease in the liquid film cooling efficiency. This entrainment phenomenon is related to the Kelvin–Helmholtz instability (KHI) [2,3], and it can be regarded as the process of droplets being peeled from liquid film, similar to the atomization process, as the liquid film is sheared by the gas stream. As the liquid is surrounded by the gas stream, the velocity of which is higher than that of the liquid, the momentum transfer from the high-speed gas to the liquid dominates the breakup process. Various theoretical and experimental studies [4–9] of the atomization process have been carried out. The typical examples of Lasheras and Hopfinger [10] showed that the interfacial instability was the main factor of the atomization process, and the instability was suppressed by the surface tension. The KHI plays a significant role in the atomization mechanism [11–17]. Rayana et al. [18] extended the theoretical method of Varga et al. [11] to study the process of the droplet peeling from liquid film. It should be noted that there was only gas on the upper side of the liquid film, the lower side of the liquid film was limited by the wall rather than the free liquid film on both sides, and the results obtained by the linear stability theory agreed well with those of the experiments.

In summary, the generation of droplets can still be attributed to the general growth of the perturbation wave, and the KHI has an important impact on the process of air-assisted or air-blast atomization. The process of droplet production under the acoustic oscillations can result in combustion instability. For example, in an F-1 liquid rocket engine, the pulse combustion of liquid fuel peeled from the liquid layer of film cooling under acoustic oscillation was considered to be a main factor of the combustion instability [19]. Baillot et al. [20] considered that the pressure oscillations may reach 10% or even higher, as the combustion instability appears in the combustion chamber, which inevitably affects the atomization process. Even though the fuel injection systems are extremely insensitive to pressure oscillations in the combustion chamber, combustion instability with acoustic oscillations can also occur [21]. The pressure oscillations in the combustion chamber were mainly reflected in oscillations of the gas velocity [22]. Jia et al. [23–26] investigated the instability of coaxial jets and sheets, and the oscillatory gas velocity was used to characterize the acoustic fluctuations. Due to the oscillatory gas velocity, parametric instability (PI) appeared, and the surface waves were also oscillatory [27]. Because the viscous damping effect is enhanced when the frequency of oscillatory gas velocity increases, the PI is restrained. However, the KHI is enhanced when the frequency of oscillatory gas velocity increases. The KHI of the liquid jet is more impressionable to the rotating velocity of gas than the PI [28].

The present study focuses on the K-H instability of viscous liquid film with a co-flowing gas stream under acoustic oscillations. The acoustic oscillations are expressed as the oscillations of gas velocity. In addition, considering that the phase change caused by the temperature difference may affect the growth of the disturbance amplitude [29–31], the heat and mass transfer are also taken into account based on the simplified model of Hsieh [32,33]. The ratio of the conduction heat flux to the evaporation heat flux is used to characterize the heat and mass transfer. Considering that the heat convection is much stronger than the heat conduction in practical cases, a correction factor is introduced to satisfy the heat flux ratio within a reasonable range. The viscous potential flow theory is used to obtain a dispersion relation, which has been used to investigate the KHI in previous studies [34–40]. The dispersion relation between the temporal growth rate and the disturbance wave number is obtained. Furthermore, the effects of the acoustic oscillations, Weber number, heat and mass transfer, as well as other key parameters on the interfacial instability are discussed.

2. Program Formulation

Figure 1 shows that a two-dimensional channel is set horizontally, and the gas is over the liquid. The density, viscosity coefficients, surface tension coefficient, and the thickness of the liquid film are ρ_l, μ_l, σ, and h_l, respectively. The density, viscosity coefficient, and thickness of the gas stream are ρ_g, μ_g, and h_g, respectively. As the gas velocity is higher than that of the liquid, the KHI appears due to the velocity discontinuity. Hence, a coordinate system that moves with the liquid film is considered in this work. The basic velocities of gas and liquid in the x and y directions are $(U_g, 0)$ and $(0, 0)$, respectively. The temperatures at the upper and lower walls are T_1 and T_2, respectively, and T_i represents the temperature at the surface. When the acoustic oscillations are taken into account, the expression of the gas velocity in the x direction is

$$U_g = \left(\overline{U}_g - \overline{U}_l\right) + \Delta\overline{U}_g \cos \omega_s t, \tag{1}$$

where $\overline{U}_g$ is the basic axial velocity, $\Delta\overline{U}_g$ represents the forcing oscillations amplitude, ω_s denotes the forcing oscillations frequency, and t is the time.

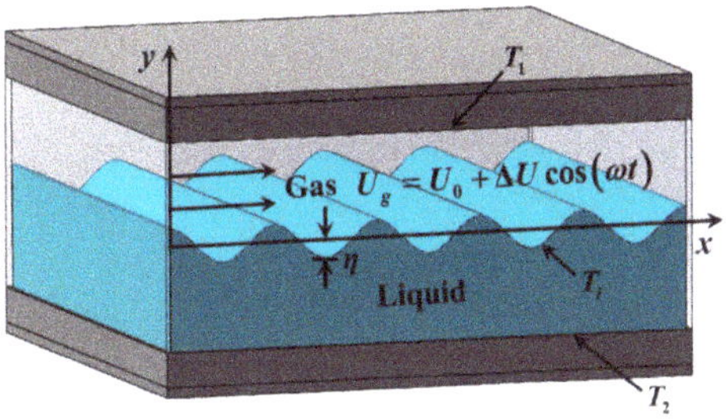

Figure 1. Schematic of the model.

The temperature difference can lead to Marangoni effects [41], and it significantly affects the interfacial instability. Various articles have examined the Marangoni effects on the instability of liquid jets, sheets, and droplets [42–51]. It should be noted that the previous work has focused on the instability behavior dominated by the surface tension, which is named capillary instability or Rayleigh–Plateau instability, and the aerodynamic effect was neglectable. In this case, the Marangoni effects are obvious. However, as the liquid film is sheared by co-flowing gas stream, the instability behavior is driven by the aerodynamic force. In this case, the surface tension restrains the instability. The results of a recent study [52] predicted that the impact of the Marangoni effects would be weak, as the aerodynamic force dominates the instability behavior. In this work, the surface tension coefficient is a constant. Therefore, the Marangoni effects are not taken into account.

The expression of the interfacial displacement as the surface is disturbed in the present study is

$$\eta = D(t)\exp(ikx), \tag{2}$$

where k is the wavenumber.

As $\mathbf{v}_j = \nabla\varphi_j$, and the continuity equation $\mathbf{v}_j = 0$, the perturbation potentials φ_j satisfy the Laplace equation, which is

$$\nabla^2\varphi_j = 0, \tag{3}$$

where the right subscripts $j = g, l$ represent the gas and liquid phases, respectively. The Laplace operator is $\nabla^2 = \partial^2/\partial x^2 + \partial^2/\partial y^2$.

The pressure disturbance can be calculated by the Lagrange integral, which is given by

$$p_l = -\rho_l \frac{\partial\varphi_l}{\partial t}, \tag{4}$$

$$p_g = -\rho_g\left(\frac{\partial\varphi_g}{\partial t} + U_g\frac{\partial\varphi_g}{\partial t}\right). \tag{5}$$

The boundary condition when $y = h_g$ is

$$\frac{\partial\varphi_g}{\partial y} = 0,\ y = h_g. \tag{6}$$

The expressions of the conditions involving heat and mass transfer as $y = 0$, as shown in Appendix A, are

$$\frac{\partial\varphi_l}{\partial y} = \frac{\partial\eta}{\partial t} + \frac{\overline{\Lambda}}{\rho_l}\eta,\ y = 0, \tag{7}$$

$$\frac{\partial\varphi_g}{\partial y} = \frac{\partial\eta}{\partial t} + U_g\frac{\partial\eta}{\partial x} + \frac{\overline{\Lambda}}{\rho_g}\eta,\ y = 0. \tag{8}$$

The boundary condition when $y = -h_l$ is

$$\frac{\partial \varphi_l}{\partial y} = 0,\ y = -h_l. \tag{9}$$

The normal stress must be balanced at $y = 0$, and the boundary condition is given by

$$-p_g + 2\mu_g \frac{\partial^2 \varphi_g}{\partial y^2} - \left(-p_l + 2\mu_l \frac{\partial^2 \varphi_g}{\partial y^2} \right) = -\sigma \frac{\partial^2 \eta}{\partial x^2},\ y = 0. \tag{10}$$

The solutions to Equation (3) are

$$\varphi_l = A_1 \sinh(ky) + A_2 \cosh(ky), \tag{11}$$

$$\varphi_g = B_1 \sinh(ky) + B_2 \cosh(ky), \tag{12}$$

where A_1, A_2, B_1, and B_2 are coefficients, which are determined by the boundary conditions. Substituting Equations (11) and (12) into Equations (6)–(9) yields

$$\varphi_l = \frac{1}{k}\left[\sinh(ky) + \frac{\cosh(kh_l)}{\sinh(kh_l)}\cosh(ky)\right]\left[D'(t) + \frac{\overline{\Lambda}}{\rho_l}D(t)\right]\exp(ikx), \tag{13}$$

$$\varphi_g = \frac{1}{k}\left[\sinh(ky) - \frac{\cosh(kh_g)}{\sinh(kh_g)}\cosh(ky)\right]\left[D'(t) + ikU_g D(t) + \frac{\overline{\Lambda}}{\rho_g}D(t)\right]\exp(ikx). \tag{14}$$

Utilizing Equations (4) and (5), the normal stress balance boundary condition (i.e., Equation (10)) is

$$\rho_g\left(\frac{\partial \varphi_g}{\partial t} + U_g \frac{\partial \varphi_g}{\partial y}\right) + 2\mu_g \frac{\partial^2 \varphi_g}{\partial y^2} - \left(\rho_l \frac{\partial \varphi_l}{\partial t} + 2\mu_l \frac{\partial^2 \varphi_g}{\partial y^2}\right) + \sigma \frac{\partial^2 \eta}{\partial x^2} = 0. \tag{15}$$

Substituting the expressions of the perturbation potentials into the normal stress balance boundary condition yields

$$D''(t) + \frac{m}{f}D'(t) + \frac{n}{f}D(t) = 0, \tag{16}$$

where

$$f = \rho_l \coth(kh_l) + \rho_g \coth(kh_g), \tag{17}$$

$$m = 2\mu_l k^2 \coth(kh_g) + 2ik\rho_g U_g \coth(kh_g) + 2\mu_g k^2 \coth(kh_g) + \overline{\Lambda}\coth(kh_l) + \overline{\Lambda}\coth(kh_g), \tag{18}$$

$$\begin{aligned} n = {} & 2\mu_g ik^3 U_g \coth(kh_g) + 2\mu_g k^2 \tfrac{\overline{\Lambda}}{\rho_g}\coth(kh_g) - \rho_g k^2 U_g^2 \coth(kh_g) \\ & + ikU_g \overline{\Lambda}\coth(kh_g) + 2\mu_l k^2 \tfrac{\overline{\Lambda}}{\rho_l}\coth(kh_l) + ik\rho_g U_g' \coth(kh_g) + \sigma k^3 \end{aligned}. \tag{19}$$

The Hill's equation can be obtained using the following transformation [23,25]:

$$D(t) = F(\tau)\exp\left(-\int \frac{m}{2f}dt\right), \tag{20}$$

$$\tau = 0.5\omega_s t, \tag{21}$$

$$U_g = (\overline{U}_g - \overline{U}_l) + \Delta\overline{U}_g \cos 2\tau. \tag{22}$$

Taking Equations (20)–(22) into Equation (16) (as shown in Appendix B), the result is

$$\frac{d^2F(\tau)}{d\tau^2} + [\theta_0 + 2\theta_2\cos(2\tau) + 2\theta_4\cos(4\tau)]F(\tau) = 0, \tag{23}$$

where

$$\theta_0 = \frac{4}{\omega_s^2}\left\{\begin{array}{l} -\frac{G^2}{4f^2} + \frac{H + 2\mu_g ik^3(\overline{U}_g - \overline{U}_l)\coth(kh_g) + ik(\overline{U}_g - \overline{U}_l)\overline{\Lambda}\coth(kh_g)}{f} \\ -\frac{ik\rho_g G(\overline{U}_g - \overline{U}_l)\coth(kh_g) + k^2\rho_g\rho_l\left[(\overline{U}_g - \overline{U}_l)^2 + \frac{1}{2}\Delta\overline{U}_g^2\right]\coth(kh_g)\coth(kh_l)}{f^2} \end{array}\right\}. \tag{24}$$

$$2\theta_2 = \frac{4}{\omega_s^2}\left[\begin{array}{l} -\frac{ik\rho_g G\Delta\overline{U}_g\coth(kh_g) + 2k^2\rho_g\rho_l(\overline{U}_g - \overline{U}_l)\Delta\overline{U}_g\coth(kh_l)\coth(kh_g)}{f^2} \\ +\frac{ik\Delta\overline{U}_g\overline{\Lambda}\coth(kh_g) + 2\mu_g ik^3\Delta\overline{U}_g\coth(kh_g)}{f} \end{array}\right], \tag{25}$$

$$2\theta_4 = -\frac{2}{\omega_s^2}\frac{k^2\rho_g\rho_l\Delta\overline{U}_g^2\coth(kh_g)\coth(kh_l)}{f^2}, \tag{26}$$

$$G = 2\mu_g k^2\coth(kh_l) + 2\mu_l k^2\coth(kh_g) + [\coth(kh_l) + \coth(kh_g)]\overline{\Lambda}, \tag{27}$$

$$H = 2\mu_g k^2\frac{\overline{\Lambda}}{\rho_g}\coth(kh_g) + 2\mu_l k^2\frac{\overline{\Lambda}}{\rho_l}\coth(kh_l) + \sigma k^3. \tag{28}$$

Considering Equation (20), the expression of the growth rate can be written as

$$\omega_r = 0.5\beta_r\omega_s - \frac{\mu_l k^2\coth(kh_l) + \mu_g k^2\coth(kh_g)}{f} - \frac{\coth(kh_g) + \coth(kh_l)}{2f}\overline{\Lambda}, \tag{29}$$

where β_r is a characteristic exponent, which is introduced to obtain the dispersion relation (i.e., Equation (29)), and the equation to obtain β_r is

$$\cosh(\pi\beta) = 1 - 2\sin^2\left(\frac{\pi\theta_0^{1/2}}{2}\right) - \frac{\pi}{4\theta_0^{1/2}}\left(\frac{\theta_2^2}{1-\theta_0} + \frac{\theta_4^2}{2^2-\theta_0}\right)\sin\left(\pi\theta_0^{1/2}\right). \tag{30}$$

The nondimensionalized dispersion relation is

$$\Omega_r = 0.5\overline{\beta}_r\Omega_s - \frac{K^2\coth(K) + \xi K^2\coth(\alpha K)}{\overline{f}Re} - \frac{\coth(\alpha K) + \coth(K)}{2\overline{f}}\Lambda, \tag{31}$$

$$\cosh(\pi\overline{\beta}) = 1 - 2\sin^2\left(\frac{\pi\overline{\theta}_0^{1/2}}{2}\right) - \frac{\pi}{4\overline{\theta}_0^{1/2}}\left(\frac{\overline{\theta}_2^2}{1-\overline{\theta}_0} + \frac{\overline{\theta}_4^2}{2^2-\overline{\theta}_0}\right)\sin\left(\pi\overline{\theta}_0^{1/2}\right), \tag{32}$$

where

$$\overline{\theta}_0 = \frac{4}{\Omega_s^2}\left[\begin{array}{l} -\frac{\overline{G}^2}{4\overline{f}^2} + \frac{\overline{H} + \frac{2}{Re}\xi iK^3\coth(\alpha K) + iK\Lambda\coth(\alpha K)}{\overline{f}} \\ -\frac{iK\rho\overline{G}\coth(\alpha K) + K^2\rho\left(1 + \frac{1}{2}\Delta\gamma^2\right)\coth(\alpha K)\coth(K)}{\overline{f}^2} \end{array}\right], \tag{33}$$

$$2\overline{\theta}_2 = \frac{4}{\Omega_s^2}\left[\begin{array}{l} -\frac{iK\rho G\Delta\gamma\coth(\alpha K) + 2K^2\rho\Delta\gamma\coth(K)\coth(\alpha K)}{\overline{f}^2} \\ +\frac{\frac{2}{Re}\xi iK^3\Delta\gamma\coth(\alpha K) + iK\Delta\gamma\Lambda\coth(\alpha K)}{\overline{f}} \end{array}\right], \tag{34}$$

$$2\overline{\theta}_4 = -\frac{2}{\Omega_s^2}\frac{K^2\rho\Delta\gamma^2\coth(\alpha K)\coth(K)}{\overline{f}^2}, \quad (35)$$

$$\overline{f} = \coth(K) + \rho\coth(\alpha K), \quad (36)$$

$$\overline{G} = \frac{2}{Re}\xi K^2\coth(\alpha K) + \frac{2}{Re}K^2\coth(K) + [\coth(K) + \coth(\alpha K)]\Lambda, \quad (37)$$

$$\overline{H} = \frac{2}{\rho Re}\xi K^2\Lambda\coth(\alpha K) + \frac{2}{Re}K^2\Lambda\coth(K) + \frac{K^3}{We}, \quad (38)$$

where $K = kh_l$ is a nondimensional wavenumber, $\alpha = h_g/h_l$ is the thickness ratio, $\rho = \rho_g/\rho_l$ denotes the density ratio, $\Omega_r = \omega_r h_l/(\overline{U}_g - \overline{U}_l)$ represents the nondimensional temporal growth rate, $\xi = \mu_g/\mu_l$ is the viscosity ratio between the gas and liquid phases, $\Omega_s = \omega_s h_l/(\overline{U}_g - \overline{U}_l)$ is the nondimensional forcing frequency, $\Delta\gamma = \Delta\overline{U}_g/(\overline{U}_g - \overline{U}_l)$ is the nondimensional oscillations amplitude, $We = \rho_l(\overline{U}_g - \overline{U}_l)^2 h_l/\sigma$ is the Weber number, and $Re = \rho_l(\overline{U}_g - \overline{U}_l)h_l/\mu_l$ denotes the Reynolds number. $\Lambda = C(1+\alpha)/[\rho_l(\overline{U}_g - \overline{U}_l)L]$ is the heat flux ratio between the conduction heat flux from the wall to the interface and the evaporation heat flux at the gas–liquid interface.

3. Results and Discussion

Funada et al. [53] studied the KHI of a liquid film in a channel, and the dispersion relation without gravitational acceleration is

$$\begin{aligned}&\left[\rho_g(\omega + ik\overline{U}_g)^2 + 2\mu_g k^2(\omega + ik\overline{U}_g)\right]\coth(kh_g) + (\rho_l - \rho_g)ak \\ &+\left[\rho_l(\omega + ik\overline{U}_l)^2 + 2\mu_l k^2(\omega + ik\overline{U}_l)\right]\coth(kh_l) + \sigma k^3 = 0\end{aligned}. \quad (39)$$

Equation (39) is obtained in a fixed coordinate system. Under the coordinate system of the present study, it is written as

$$\begin{aligned}&\left\{\rho_g[\omega + ik(\overline{U}_g - \overline{U}_l)]^2 + 2\mu_g k^2[\omega + ik(\overline{U}_g - \overline{U}_l)]\right\}\coth(kh_g) \\ &+(\rho_l\omega^2 + 2\mu_l k^2\omega)\coth(kh_l) + \sigma k^3 = 0\end{aligned}. \quad (40)$$

The nondimensionalized Equation (40) is

$$\left[\rho(\Omega + iK)^2 + \frac{2}{Re}\xi K^2(\Omega + iK)\right]\coth(\alpha K) + \left(\Omega^2 + \frac{2}{Re}K^2\Omega\right)\coth(K) + \frac{K^3}{We} = 0, \quad (41)$$

where $\Omega = \Omega_r + i\Omega_I$ represents the nondimensional complex growth rate, the imaginary part Ω_I predicts the nondimensional disturbance frequency, and the real part Ω_r denotes the nondimensional growth rate. k is the wavenumber, and the wavelength $w = 2\pi/k$. $i = \sqrt{-1}$ is the imaginary unit.

As the phase change and acoustic oscillations are ignored, the liquid layer is sheared by the high-speed gas stream, and this simplified model is similar to that of Funada et al. [53]. Figure 2 shows the results obtained by Equation (41) and the dispersion relation of the present study without the phase change and acoustic oscillations (i.e., $\Delta\gamma = 0$ and $\Lambda = 0$). The curves shown in Figure 2 are normal. Before a maximum value was reached, the curves increased firstly with the wavenumber. After the maximum value, the curves decreased due to the impact of the viscosity of the liquid and the surface tension, and an abscissa-crossing point was reached. The dominant wavenumber is the wavenumber corresponding to the maximum growth rate. In addition, the range of the wavenumber from 0 to the abscissa-crossing point is known the unstable range. It can also be observed that the interfacial instability was promoted as the surface tension decreased. Figure 3 displays the results of this work compared with that of Asthana and Agrawal [34] when the heat and mass transfer were considered. In addition, the results of the simplified model in this work were in good agreement with previous studies, and the correctness of the dispersion relation obtained in this work was verified to a certain degree.

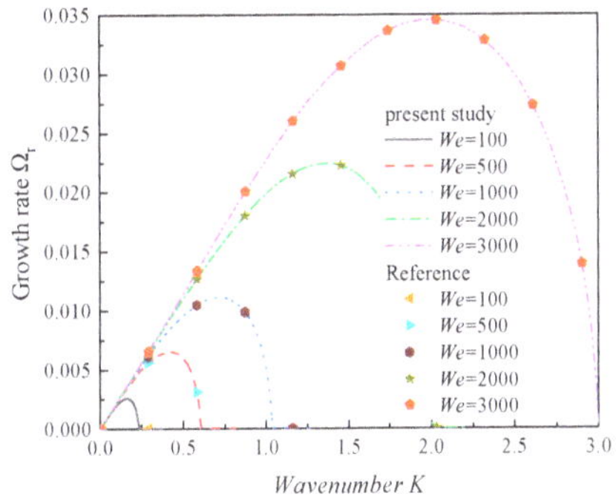

Figure 2. The results compared with Funada et al. [53] when the Weber number varies ($Re = 3000$, $\xi = 0.001$, $\rho = 0.001$, $\alpha = 2$).

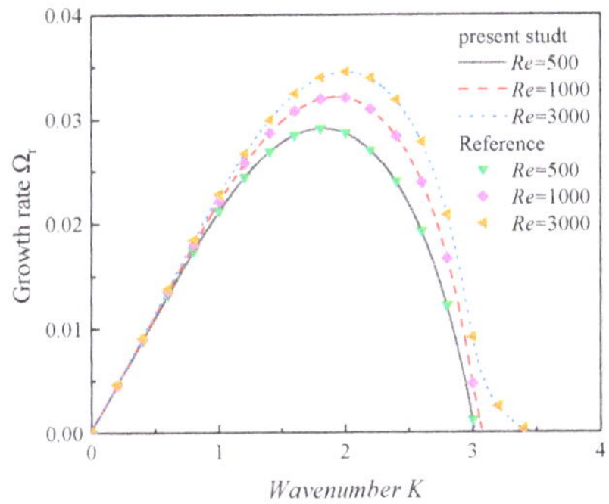

Figure 3. The results compared with Asthana and Agrawal [34] by varying the Reynolds number ($We = 3000$, $\xi = 0.001$, $\Lambda = 0.0001$, $\rho = 0.001$, $\alpha = 2$).

According to previous studies [1,54], the values of the typical physical parameters are shown in the following: the liquid density is $\rho_l = 800\ \mathrm{kg/m^3}$, the gas density is $\rho_g = 1\ \mathrm{kg/m^3}$, $T_1 - T_i = 1000\ \mathrm{K}$, $T_i - T_2 = 200\ \mathrm{K}$, the thermal conductivity coefficient of the liquid is $\kappa_l = 0.08\ \mathrm{W/(m \cdot K)}$, the thermal conductivity coefficient of the gas $\kappa_g = 0.1\ \mathrm{W/(m \cdot K)}$, the latent heat is $L = 250\ \mathrm{kJ/kg}$, the velocity difference between the gas and the liquid is $\overline{U}_g - \overline{U}_l = 5\ \mathrm{m/s}$, the viscosity of the liquid is $\mu_l = 0.008\ \mathrm{Pa \cdot s}$, the surface tension coefficient is $\sigma = 0.02\ \mathrm{N/m}$, the viscosity of the gas is $\mu_g = 0.000008\ \mathrm{Pa \cdot s}$, the thickness of the liquid film is $h_l = 0.001\ \mathrm{m}$, the thickness of the gas stream is $h_g = 0.00625\ \mathrm{m}$, and the frequency of the acoustic oscillations is 1000 Hz. The values of the typical nondimensional parameters are calculated as the reference state in the following: $We = 10,000$, $Re = 5000$, $\rho = 0.00125$, $\alpha = 6.25$, $\Omega_s = 0.2$, $\xi = 0.001$, and the heat flux ratio is $\Lambda = 0.0001856$. In fact, the heat flux on the combustor wall of a liquid rocket can reach $10^6\ \mathrm{W/m^2}$ or higher, and the heat convection is much stronger than heat conduction. In this case, the heat flux ratio $\Lambda \geq 0.001$. Therefore, setting $\Lambda = 0.0001856$ is not reasonable, and a correction factor ψ should be introduced to satisfy Λ within a reasonable range, i.e., $\Lambda = \psi C(1+\alpha)/\left[\rho_l(\overline{U}_g - \overline{U}_l)L\right]$. The heat flux ratio Λ was set to 0.002 in the following.

Figure 4 shows the effect of the nondimensional oscillations amplitude of the gas velocity on the interfacial instability. As shown in Figure 4, only one unstable region occurred as the acoustic oscillations were ignored (i.e., $\Delta\gamma = 0$), without a phase transition. This unstable region is inherent, driven by aerodynamic force, and it is named the KHI region. As the oscillations of the gas velocity were neglected, the expression of the growth rate is $\Omega_r = 0.5\Omega_s\sqrt{\overline{\theta}_0}$. This means that this unstable region is dominated by $\overline{\theta}_0$. However, several unstable regions appeared without a phase transition when the oscillation amplitude of the gas velocity $\Delta\gamma \neq 0$. According to Equations (30) and (31), these unstable regions after the first unstable region, which are named the parametric instability (PI) region, are governed by $\overline{\theta}_2$ and $\overline{\theta}_4$. The full wavelength assumption of Mayer [13] indicted that the unstable disturbed waves in unstable regions can grow generally and break up into droplets. Therefore, the appearance of the PI is beneficial for the formation of smaller droplets. More importantly, Figure 4 shows the most unstable PI region was the second unstable region. Hence, the first and second unstable region only are discussed in the following. Furthermore, the location of the dominant wavenumber would be in the first unstable region as the oscillation amplitude $\Delta\gamma$ is larger than a critical value $\Delta\gamma_{cr}$, and it would be in the second unstable region as $\Delta\gamma > \Delta\gamma_{cr}$. Because the growth rate increases faster in the PI region than in the KHI region, as observed in Figure 4, the PI was more impressionable to the oscillations of the gas velocity than the KHI. Moreover, the results in Figure 4b show that the KHI was enhanced when the oscillatory velocity amplitude increased in the case of $\Omega_s = 0.5$. However, the KHI was restrained with the increase in the oscillatory velocity amplitude when $\Omega_s = 0.2$, as observed in Figure 4a. It can be concluded that the impact of the velocity oscillations on KHI is related to the forcing frequency. The KHI is enhanced with an increasing oscillatory velocity amplitude when the forcing frequency is large, while the KHI is restrained as the oscillatory velocity amplitude increases in the case when the forcing frequency is small.

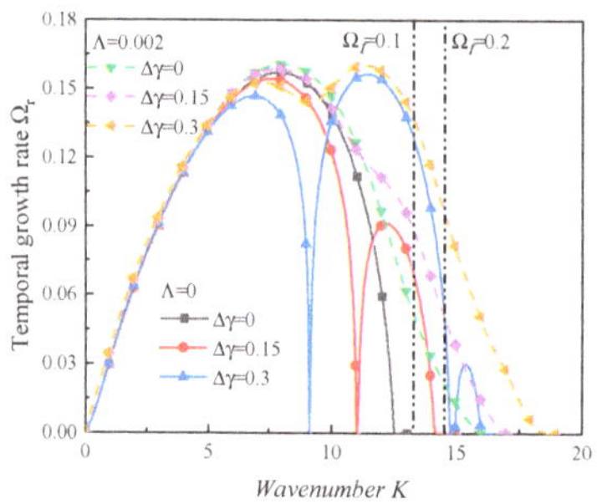

(a)

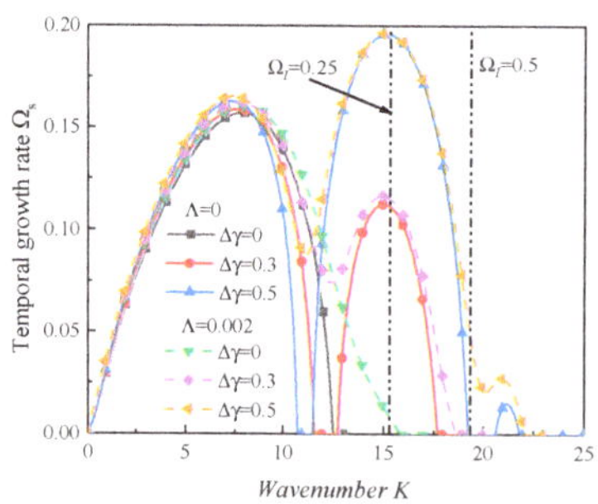

(b)

Figure 4. The dispersion curves with varied oscillations amplitude (We = 10,000, Re = 5000, ρ = 0.00125, α = 6.25, ξ = 0.001, Λ = 0.002). (**a**) Ωs = 0.2. (**b**) Ωs = 0.5.

The inherent frequency can be calculated by Equation (20), the expression of which can be written as

$$\Omega_I = 0.5\Omega_s\sqrt{\bar{\theta}_0}. \tag{42}$$

The results of a previous study [55] predicted that the waves at half or harmonic frequencies were stimulated by the velocity oscillations, and the resonant wavenumber around the parametric instability regions is $\Omega_I = e\Omega_s/2$, where $e = 1, 2 \cdots$. In the present study, the resonant wavenumbers should be 13.26, 14.53, . . . , ·as the forcing frequency is 0.2. When the forcing frequency is 0.5, the resonant wavenumbers should be 15.30, 19.34, . . . , and these wavenumbers have been marked in Figure 4. However, because the variation in the inherent frequency of the system is limited, the parametric instability region would not be located at these resonant wavenumbers [55]. Furthermore, the wavenumbers around which the parametric instability appears are related to the forcing frequency, as can be seen in Figure 5 and Equation (40).

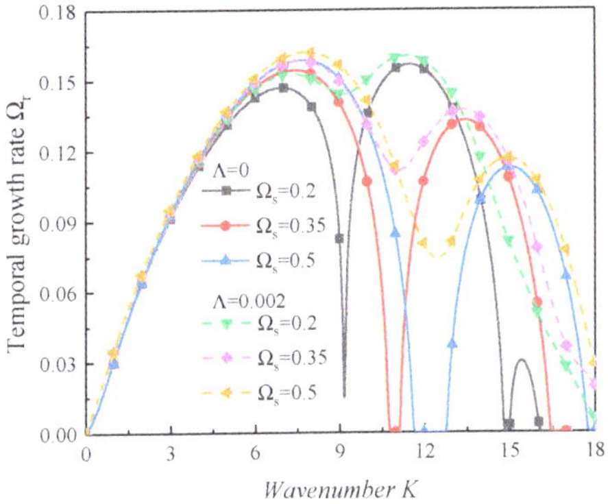

Figure 5. The dispersion curves with varied forcing oscillations frequencies (*We* = 10,000, *Re* = 5000, $\Delta\gamma$ = 0.3, ρ = 0.00125, α = 6.25, ξ =0.001).

Figure 5 exhibits the influence of the forcing oscillations frequency. Obviously, as the forcing oscillations frequency increased, the instability of the surface was enhanced with an increasing forcing oscillations frequency, while the parametric instability was prevented. Meanwhile, the stabilizing effect of the increasing forcing oscillations in the parametric instability region was greater than the stabilizing effect in the KHI region, as observed in Figure 5. According to the dispersion relation equation of the present study (i.e., Equation (30)), the contribution of the viscous dissipation is

$$\Omega_v = -\frac{K^2\coth(K)}{\bar{f}Re} - \frac{\xi K^2\coth(\alpha K)}{\bar{f}Re}. \tag{43}$$

The term $K^2\coth(K)/\bar{f}Re$ in Equation (41) denotes the damping effect of the liquid viscosity on the interfacial instability, and the term $\xi K^2\coth(\alpha K)/\bar{f}Re$ represents the contribution of the gas viscosity. As the wavenumber $K \gg 1$, $\coth(K) \to 1$, the viscous dissipation of the liquid phase can be proportional approximately to K^2/Re, as $\xi \ll 1$. In this case, the viscous dissipation was enhanced with an increasing wavenumber. According to Equation (42), the resonant wavenumber of the PI region increased because of the increase in the forcing oscillations frequency. Therefore, the PI was restrained because of the strongly viscous dissipation, and the regions of the PI moved to the right, as shown in Figure 5. In addition, it can be observed obviously that there was a critical forcing oscillations frequency Ω_{cr}; the location of dominant wavenumber would be in the second unstable region when $\Omega_s < \Omega_{cr}$, and it would be in the first unstable region when $\Omega_s > \Omega_{cr}$.

As can be observed in Figure 6, the KHI as well as the PI were enhanced when the gas–liquid density increased. Meanwhile, the unstable regions expanded when the gas–liquid density ratio increased. Even though the heat and mass transfer were neglected, the gaps between unstable regions disappeared when the gas–liquid density ration was large enough. This suggests that the increases in the gas–liquid density ratio may have accelerated the breakup process of the liquid film. Due to the increase in the gas–liquid density ratio, the aerodynamic force effect was enhanced. Therefore, the interface between the gas and liquid phases tended to be more unstable with an increasing gas–liquid density ratio. In addition, it is worth noting that the growth rate in the PI region increased faster than that in the KHI region when the gas–liquid density ratio rose. It can be concluded that the PI was more susceptible to the gas–liquid density than the KHI. Furthermore, the inherent frequency of the system increased because of the increase in the gas–liquid density ratio, and it led the parametric unstable regions to move to the right, as shown in Figure 6a. Moreover, the dispersion curves with and without the heat and mass transfer were very approximate as the gas–liquid density ratio was large, as observed in Figure 6b. In this case, the effect of the heat and mass transfer could be neglected.

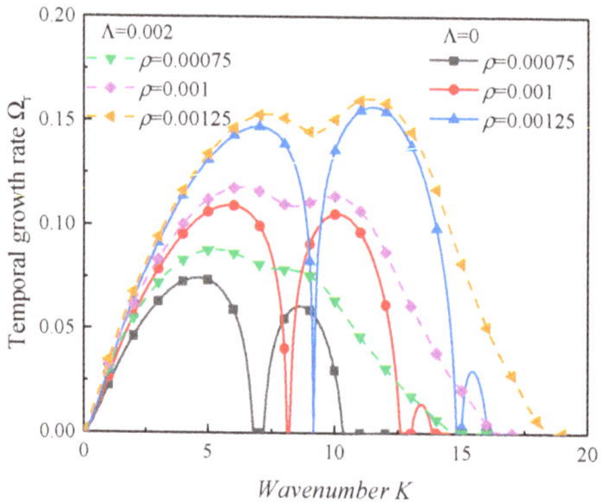

(a)

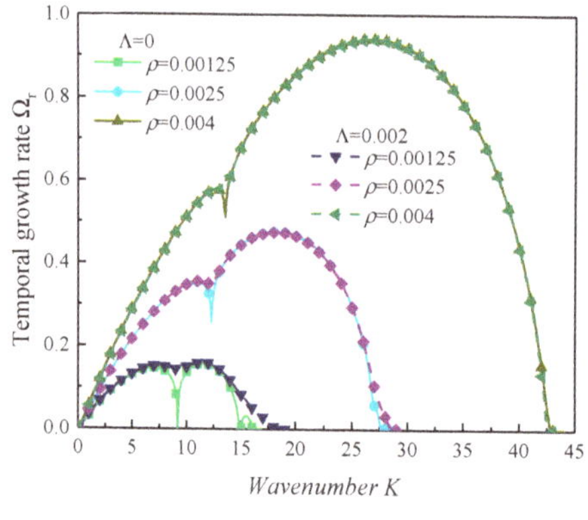

(b)

Figure 6. The dispersion curves with varied gas–liquid density ratios (We = 10,000, Re = 5000, $\Delta\gamma = 0.3$, $\Omega_s = 0.2$, $\alpha = 6.25$, $\xi = 0.001$).

Figure 7 exhibits the influence of the surface tension on the interfacial instability of the viscous liquid film. As observed in Figure 7, with the increase in the Weber number, the maximum growth rate as well as the dominant wavenumber increased. The KHI was dominated by the aerodynamic effect, and it was suppressed by the surface tension. Hence, increasing the Weber number made the gas–liquid interface tend to be more unstable. According to the study of Varga et al. [11,12,14–16], the droplets after the breakup process may become smaller in this condition. In addition, the results in Figure 7 show that the growth rate in the PI region increased faster than that in the KHI region. This predicted that the impact of the oscillatory velocity on the KHI would be weaker than that on the parametric instability. One can also observe, in Equation (32), the term K^3/We, which denotes the effect of surface tension on the interfacial instability. The damping effect of the surface tension would become stronger with an increasing wavenumber K.

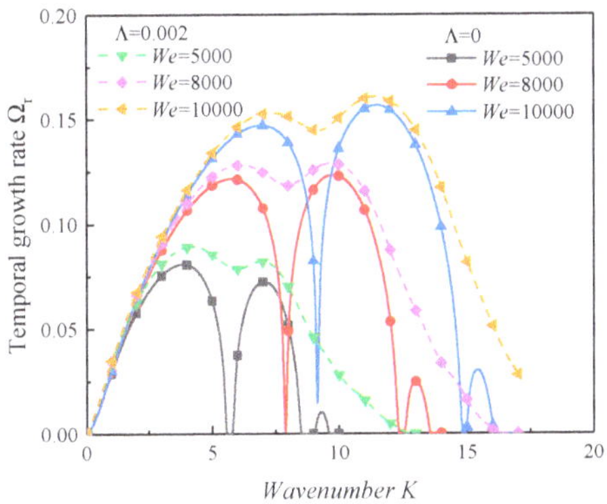

Figure 7. The dispersion curves with varied Weber numbers (Re = 5000, $\Delta\gamma$ = 0.3, Ω_s = 0.2, ρ = 0.00125, α = 6.25, ξ = 0.001).

Then, the effect of the viscosity of the liquid film on the surface instability was examined. It is obvious in Figure 8 that the both the inherent and PI regions increased when the Reynolds number increased. This predicted that the viscosity of the liquid would play a stabilizing role. Jia et al. [23] examined a two-dimensional liquid jet exposed to gas, as the longitudinal acoustic oscillations were considered, and the authors reported that the liquid viscous dissipation was proportional to K^2/Re. According to Equation (40) in this work, the damping effect of the viscosity of the liquid film was proportional approximately to K^2/Re, similar to the study of Kumar [56]. Moreover, it can also be concluded that there is a critical Reynolds number Re_{cr}, the location of the maximum growth rate is in the KHI region as $Re < Re_{cr}$, and it shifts to the second unstable region when $Re > Re_{cr}$, as shown in Figure 8. A larger wavenumber corresponds to a shorter wavelength because of $\lambda = 2\pi/K$. According to recent studies [11,12,14–16], the atomization quality may be improved in the breakup process when the Reynolds number increases.

Figure 9 shows the influence of the viscosity of the gas medium on the interfacial instability of the liquid film. It is clear that the growth rate of the inherent and parametric regions increased as the gas–liquid viscosity increased. The results predicted that increasing the viscosity of the gas would promote the KHI and PI. This conclusion is consistent with that reported by Funada et al. [57], who examined the linear stability of a viscous liquid jet exposed to a viscous gas phase. However, its destabilizing effect is neglectable because the gas–liquid viscosity ratio $\xi \ll 1$.

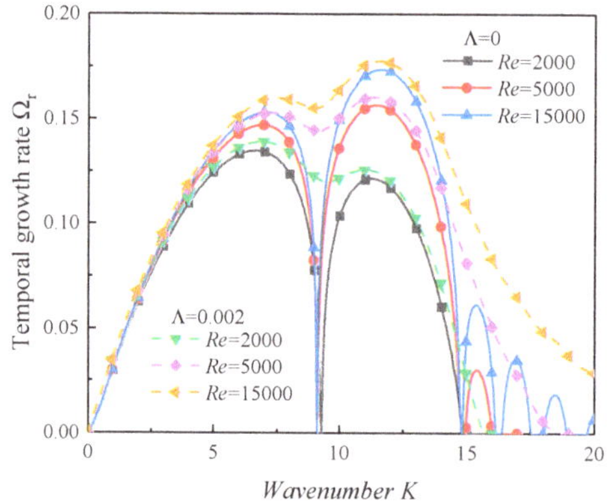

Figure 8. The dispersion curves with varied Reynolds numbers (We = 10,000, $\Delta\gamma = 0.3$, $\Omega_s = 0.2$, $\rho = 0.00125$, $\alpha = 6.25$, $\xi = 0.001$).

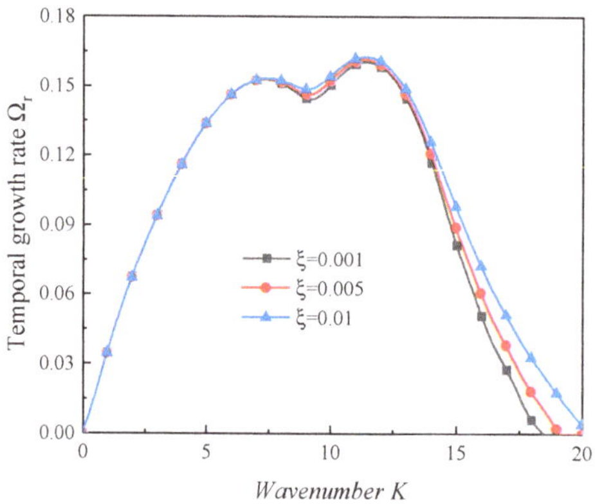

Figure 9. The dispersion curves when the gas–liquid viscosity ratio varies (We = 10,000, Re = 5000, $\Delta\gamma = 0.3$, $\Omega_s = 0.2$, $\rho = 0.00125$, $\alpha = 6.25$, $\Lambda = 0.002$).

Figure 10 shows the variation in the growth rate by varying Λ from 0 to 0.003 as the acoustic oscillations are taken into account. In this work, Λ represents the ratio between the conduction heat flux and the evaporation heat flux, and it is employed to characterize the heat and mass transfer. As displayed in Figure 10, the interfacial instability was promoted as the heat and mass transfer were enhanced. It can be concluded that the heat and mass transfer had a destabilizing effect. Furthermore, the unstable regions expanded obviously as the phase transition appeared. Zeng et al. [58] investigated the primary breakup process numerically with the evaporation effect. The results suggested that the phase change could promote the KHI. Chan et al. [59] examined the mass transfer on the KHI at the gas–liquid interface. The authors concluded that the evaporation enhanced the wave instability. Although the physical model of the present study was somewhat different from those of Zeng et al. [58] and Chan et al. [59], the qualitative tendency was consistent.

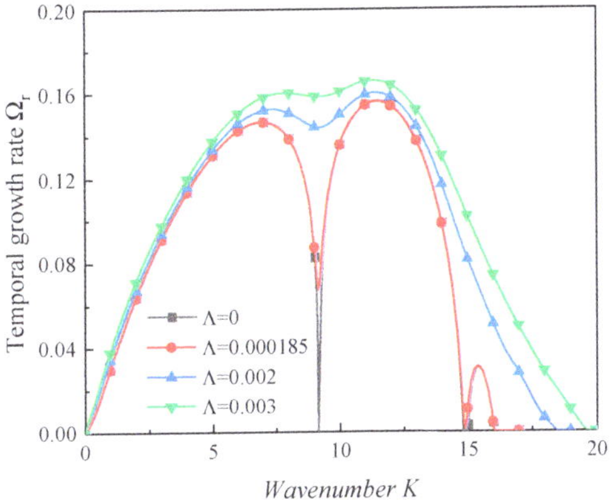

Figure 10. The effect of the heat and mass transfer (We = 10,000, Re = 5000, $\Delta\gamma$ = 0.3, Ω_s = 0.2, ρ = 0.00125, α = 6.25, ξ = 0.001).

4. Conclusions

This study investigated the instability of a planner liquid film theoretically when the phase change and acoustic oscillations were taken into consideration. The gas velocity oscillations were employed to characterize the acoustic oscillations. The results showed that the parametric instability (PI) appeared when the acoustic oscillations were taken into account. The impact of the velocity oscillations on the Kelvin–Helmholtz instability (KHI) was related to the forcing frequency. The results suggest that the KHI was enhanced with an increasing oscillatory velocity amplitude when the forcing frequency was large, while the KHI was restrained as the oscillatory velocity amplitude increased in the case of the small forcing frequency. In addition, the PI was more susceptible to the oscillations of the gas velocity than the KHI. Due to the enhancement of the viscous damping effect, the PI was prevented when the forcing oscillations frequency increased. Furthermore, the resonant wavenumber and frequency varied with the forcing oscillations frequency. Both the KHI and PI were promoted with the enhancement of the heat and mass transfer. More importantly, the inherent frequency of the system increased with an increase in the gas–liquid density ratio, and it led to an increase in the wavenumber of the PI. As the gas–liquid density ratio was large enough, the influence of the heat and mass transfer could be neglected. Moreover, the dominant wavenumber could be transferred from the KHI region to the PI region when the Weber number increased. The liquid viscosity restrained the interfacial instability. Although the gas viscosity destabilized the surface of the liquid film, its destabilizing impact was neglectable because of the low viscosity. Considering Mayer's full wavelength assumption [13], which indicates that the unstable disturbed waves in unstable regions can grow generally and break up into droplets, the acoustic oscillations are beneficial for obtaining smaller droplets. In addition, increasing the Weber number, Reynolds number, and gas–liquid density ratio not only accelerates the breakup process of the liquid film but also improves the atomization quality.

Author Contributions: X.D.: Methodology; Formal Analysis; Writing—Original Draft Preparation. B.S.: Validation; Writing—Review and Editing. Y.T.: Validation; Writing—Review and Editing. N.W.: Validation; Writing—Review and Editing. All authors have read and agreed to the published version of the manuscript.

Funding: This work was funded by the National Natural Science Foundation of China (Grant No. 52106129) and the Science Center for Gas Turbine Project (Grant No. P2022-A-II-006-002).

Data Availability Statement: Not applicable.

Conflicts of Interest: The authors declare no conflict of interest.

Nomenclature

A_1	undetermined coefficients
A_2	undetermined coefficient
B_1	undetermined coefficient
B_2	undetermined coefficient
C	undetermined coefficient
D	a function of time
h_g	thickness of gas
h_l	thickness of liquid
k	wavenumber
K	nondimensional wavenumber
p	pressure disturbance
p_g	disturbed pressure of gas
p_l	disturbed pressure of liquid
Re	Reynolds number
t	time
U_g	axial gas velocity in coordinate system moving with liquid jet
$U_{g,0}$	axial gas velocity in fixed coordinate system
U_l	axial velocity of liquid
U_g	axial velocity of gas
$\overline{U}_l$	axial basic velocity of liquid
$\overline{U}_g$	axial basic velocity of gas
$\Delta\overline{U}_g$	oscillations amplitude of gas velocity
$\mathbf{v}$	velocity vector
w	wavelength
We	Weber number
α	thickness ratio between gas and liquid
φ_g	perturbation potential of gas
φ_l	perturbation potential of liquid
$\Delta\gamma$	nondimensional oscillations amplitude
η	disturbed amplitude
μ_g	gas viscosity
μ_l	liquid viscosity
ρ	gas–liquid density ratio
ρ_g	gas density
ρ_l	liquid density
σ	surface tension coefficient
Λ	heat flux ratio
Ω	nondimensional complex growth rate
Ω_I	nondimensional inherent frequency
Ω_i	nondimensional disturbance frequency
Ω_r	nondimensional temporal growth rate
Ω_s	nondimensional forcing oscillation frequency
Ω_v	contribution of viscosity to the instability
ω	complex growth rate
ω_i	complex disturbance frequency
ω_r	temporal growth rate
ω_s	forcing oscillations frequency
ξ	gas–liquid viscosity ratio

Appendix A

The mass transfer, which occurs with the heat transfer, changes the amplitude of the disturbance at the interface through the phase change between gas and liquid. To obtain the kinematic boundary conditions of the disturbance wave including the effect of the mass transfer, Heish [32,33] built the mass transfer balance and heat transfer balance conditions. The mass transfer is approximated by the ratio of the conduction heat flux and the evaporation heat flux in kinematic boundary conditions. As in the studies of Hsieh [32,33], the mass transfer flux balance is

$$\rho_l\left(\frac{\partial\varphi_l}{\partial r}-\frac{\partial\eta}{\partial t}\right)=\rho_g\left(\frac{\partial\varphi_g}{\partial r}-\frac{\partial\eta}{\partial t}-U_g\frac{\partial\eta}{\partial x}\right). \quad \text{(A1)}$$

When the phase transition appears, the heat absorbed by the phase transition [32,33] is

$$L\rho_l\left(\frac{\partial\varphi_l}{\partial r}-\frac{\partial\eta}{\partial t}\right)=S(\eta), \quad \text{(A2)}$$

where L in Equation (A2) is the latent heat released by evaporation, and $S(\eta)$ represents the heat flux. In the simplified physical model of Hsieh et al. [32,33], T_1 represents the temperature of the upper wall, T_2 denotes the temperature of the lower wall, T_i represents the temperature of the surface, and T_i should be the saturation temperature when the phase transition appears. We assume that the heat transfer from the upper wall to the surface is by heat conduction, and the heat transfer from the surface to the lower wall is by heat conduction. Thus, the heat flux $S(\eta)$ at the disturbed interface is

$$S(\eta)=\frac{\kappa_g(T_i-T_1)}{h_g-\eta}+\frac{\kappa_l(T_i-T_2)}{h_l+\eta}, \quad \text{(A3)}$$

where κ_l and κ_g are the thermal conductivity coefficients of liquid and gas, respectively. Obviously, Equation (A3) is not a linear expression of the disturbed amplitude η. Therefore, Equation (A3) should be linearized by the Taylor series. Expanding the $S(\eta)$ in the Taylor series at $\eta=0$, the following equation can be obtained:

$$S(\eta)=S(0)+S'(0)\eta+\frac{1}{2}S''(0)\eta^2+\cdots. \quad \text{(A4)}$$

To represent $S(\eta)$ as a function of the disturbed amplitude η, the case of $\eta=0$ was assumed as an equilibrium state in previous work [32,33], and the heat flux should be zero.

$$S(0)=\frac{\kappa_g(T_i-T_1)}{h_g}+\frac{\kappa_l(T_i-T_2)}{h_l}=0. \quad \text{(A5)}$$

It can be obtained from Equation (A5) as follows

$$\frac{\kappa_g(T_i-T_1)}{h_g}=-\frac{\kappa_l(T_i-T_2)}{h_l}=C, \quad \text{(A6)}$$

where C is a constant. Taking Equations (A3)–(A5) into Equation (A6), the linearized expression is

$$S(\eta)=\frac{\kappa_g(T_i-T_1)}{h_g^2}-\frac{\kappa_l(T_i-T_2)}{h_l^2}=C\left(\frac{1}{h_g}+\frac{1}{h_l}\right)\eta. \quad \text{(A7)}$$

Utilizing the Equations (A1), (A2), and (A7), the result is

$$\frac{\partial\varphi_l}{\partial y}=\frac{\partial\eta}{\partial t}+\frac{\overline{\Lambda}}{\rho_l}\eta, \quad \text{(A8)}$$

$$\frac{\partial \varphi_g}{\partial y} = \frac{\partial \eta}{\partial t} + U_g \frac{\partial \eta}{\partial x} + \frac{\overline{\Lambda}}{\rho_g}\eta, \tag{A9}$$

where $\overline{\Lambda} = C(1/h_g + 1/h_l)/L$, and the Equations (A8) and (A9) are the boundary conditions at the disturbed surface.

Appendix B

Taking Equations (20) and (21) into Equation (16) yields

$$\frac{\partial^2 F(\tau)}{\partial \tau^2} + \frac{4}{\omega_s^2}\left(-\frac{m'}{2f} - \frac{m^2}{4f^2} + \frac{n}{f}\right)F(\tau) = 0. \tag{A10}$$

Considering that

$$G = 2\mu_g k^2 \coth(kh_l) + 2\mu_l k^2 \coth(kh_g) + [\coth(kh_l) + \coth(kh_g)]\overline{\Lambda} \tag{A11}$$

$$H = 2\mu_g k^2 \frac{\overline{\Lambda}}{\rho_g}\coth(kh_g) + 2\mu_l k^2 \frac{\overline{\Lambda}}{\rho_l}\coth(kh_l) + \sigma k^3, \tag{A12}$$

the expressions of m', m^2, and n are

$$m' = 2ik\rho_g U_g' \coth(kh_g), \tag{A13}$$

$$\begin{aligned} m^2 &= [G + 2ik\rho_g U_g \coth(kh_g)]^2 = G^2 + 4Gik\rho_g U_g \coth(kh_g) - 4k^2\rho_g^2 U_g^2 \coth^2(kh_g) \\ &= G^2 + 4Gik\rho_g \coth(kh_g)[(\overline{U}_g - \overline{U}_l) + \Delta\overline{U}_g \cos 2\tau] \\ &\quad -4k^2\rho_g^2 \coth^2(kh_g)[(\overline{U}_g - \overline{U}_l) + \Delta\overline{U}_g \cos 2\tau]^2 \end{aligned}, \tag{A14}$$

$$\begin{aligned} n &= 2\mu_g ik^3 U_g \coth(kh_g) - \rho_g k^2 U_g^2 \coth(kh_g) + ikU_g\overline{\Lambda}\coth(kh_g) + H \\ &= H + 2\mu_g ik^3[(\overline{U}_g - \overline{U}_l) + \Delta\overline{U}_g \cos 2\tau]\coth(kh_g) + ik\rho_g U_g' \coth(kh_g) \\ &\quad -\rho_g k^2[(\overline{U}_g - \overline{U}_l) + \Delta\overline{U}_g \cos 2\tau]^2 \coth(kh_g) + ik[(\overline{U}_g - \overline{U}_l) + \Delta\overline{U}_g \cos 2\tau]\overline{\Lambda}\coth(kh_g) \end{aligned}. \tag{A15}$$

For the cosine function, there is

$$\cos 4\tau = \cos^2 2\tau - \sin^2 2\tau = 2\cos^2 2\tau - 1, \tag{A16}$$

Taking Equations (A13)–(A16) into Equation (A10), the results are

$$\frac{d^2F(\tau)}{d\tau^2} + [\theta_0 + 2\theta_2 \cos(2\tau) + 2\theta_4 \cos(4\tau)]F(\tau) = 0, \tag{A17}$$

where

$$\theta_0 = \frac{4}{\omega_s^2}\left\{ \begin{array}{l} -\frac{G^2}{4f^2} + \frac{H + 2\mu_g ik^3(\overline{U}_g - \overline{U}_l)\coth(kh_g) + ik(\overline{U}_g - \overline{U}_l)\overline{\Lambda}\coth(kh_g)}{f} \\ -\frac{ik\rho_g G(\overline{U}_g - \overline{U}_l)\coth(kh_g) + k^2\rho_g\rho_l\left[(\overline{U}_g - \overline{U}_l)^2 + \frac{1}{2}\Delta\overline{U}_g^2\right]\coth(kh_g)\coth(kh_l)}{f^2} \end{array} \right\}, \tag{A18}$$

$$2\theta_2 = \frac{4}{\omega_s^2}\left[\begin{array}{l} -\frac{ik\rho_g G\Delta\overline{U}_g \coth(kh_g) + 2k^2\rho_g\rho_l(\overline{U}_g - \overline{U}_l)\Delta\overline{U}_g \coth(kh_l)\coth(kh_g)}{f^2} \\ +\frac{ik\Delta\overline{U}_g\overline{\Lambda}\coth(kh_g) + 2\mu_g ik^3\Delta\overline{U}_g \coth(kh_g)}{f} \end{array} \right], \tag{A19}$$

$$2\theta_4 = -\frac{2}{\omega_s^2}\frac{k^2\rho_g\rho_l\Delta\overline{U}_g^2\coth(kh_g)\coth(kh_l)}{f^2}. \tag{A20}$$

References

1. Gater, R.A.; L'Ecuyer, M.R. A Fundament Investigation of the Phenomena That Characterize Liquid-Film Cooling. *Int. J. Heat Mass Transf.* **1970**, *13*, 1925–1939. [CrossRef]
2. Petrarolo, A.; Kobald, M.; Schlechtriem, S. Understanding Kelvin–Helmholtz instability in paraffin-based hybrid rocket fuels. *Exp. Fluids* **2018**, *59*, 62. [CrossRef]
3. Petrarolo, A.; Kobald, M.; Schlechtriem, S. Optical analysis of the liquid layer combustion of paraffin-based hybrid rocket fuels. *Acta Astronaut.* **2019**, *158*, 313–322. [CrossRef]
4. Yang, L.J.; Fu, Q.F.; Qu, Y.Y.; Gu, B.; Zhang, M.Z. Breakup of a power-law liquid sheet formed by an impinging jet injector. *Int. J. Multiph. Flow* **2012**, *39*, 37–44. [CrossRef]
5. Fu, Q.F.; Yang, L.J.; Qu, Y.Y. Measurement of annular liquid film thickness in an open-end swirl injector. *Aerosp. Sci. Technol.* **2011**, *15*, 117–124. [CrossRef]
6. Fu, Q.F.; Yang, L.J.; Qu, Y.Y.; Gu, B. Linear Stability Analysis of a Conical Liquid Sheet. *J. Propul. Power* **2010**, *26*, 955–968. [CrossRef]
7. Fu, Q.F.; Yang, L.J.; Wang, X.D. Theoretical and Experimental Study of the Dynamics of a Liquid Swirl Injector. *J. Propul. Power* **2010**, *26*, 94–101. [CrossRef]
8. Yang, L.J.; Qu, Y.Y.; Fu, Q.F.; Gu, B. Linear Stability Analysis of a Non-Newtonian Liquid Sheet. *J. Propul. Power* **2010**, *26*, 1212–1224. [CrossRef]
9. Yang, L.J.; Fu, Q.F.; Qu, Y.Y.; Zhang, W.; Du, M.L.; Xu, B.R. Spray characteristics of gelled propellants in swirl injectors. *Fuel* **2012**, *97*, 253–261. [CrossRef]
10. Lasheras, J.C.; Hopfinger, E.J. Liquid Jet Instability and Atomization in a Coaxial Gas Stream. *Annu. Rev. Fluid Mech.* **2000**, *32*, 275–308. [CrossRef]
11. Varga, C.M.; Lasheras, J.C.; Hopfinger, E.J. Initial breakup of a small-diameter liquid jet by a high-speed gas stream. *J. Fluid Mech.* **2003**, *497*, 405–434. [CrossRef]
12. Aliseda, A.; Hopfinger, E.J.; Lasheras, J.C.; Kremer, D.M.; Berchielli, A.; Connolly, E.K. Atomization of viscous and non-Newtonian liquids by a coaxial, highspeed gas jet. Experiments and droplet size modeling. *Int. J. Multiphase Flow.* **2008**, *34*, 161–175. [CrossRef]
13. Mayer, E. Theory of liquid atomization in high velocity gas streams. *ARS J.* **1961**, *31*, 1783–1785.
14. Qin, L.Z.; Yi, R.; Yang, L.J. Theoretical breakup model in the planar liquid sheets exposed to high-speed gas and droplet size prediction. *Int. J. Multiph. Flow* **2018**, *98*, 158–167. [CrossRef]
15. Fu, Q.F.; Yao, M.W.; Yang, L.J.; Xie, L. Atomization Model of Liquid Jets Exposed to Subsonic Crossflows. *AIAA J.* **2020**, *58*, 2347–2351. [CrossRef]
16. Yang, L.J.; Gao, Y.P.; Li, J.X.; Fu, Q.F. Theoretical atomization model of a coaxial gas–liquid jet. *Phys. Fluids* **2020**, *32*, 124108. [CrossRef]
17. Yang, L.J.; Fu, Q.F. Stability of Confined Gas–Liquid Shear Flows in Recessed Shear Coaxial Injectors. *J. Propul. Power* **2012**, *28*, 1413–1424. [CrossRef]
18. Rayana, F.B.; Cartellier, A.; Hopfinger, E.J. Assisted Atomization of a Liquid Layer: Investigation of the Parameters Affecting the Mean Drop Size Prediction. In Proceedings of the International Conference on Liquid Atomization and Spray Systems (ICLASS), Kyoto, Japan, 27 September 2006; pp. 1–8.
19. Anderson, W.E.; Yang, V. Liquid Rocket Engine Combustion Instability. In *Progress in Astronautics and Aeronautics*; AIAA: Washington, DC, USA, 1995; pp. 1–78.
20. Baillot, F.; Blaisot, J.B.; Boisdron, G.; Dumouchel, C. Behaviour of an air-assisted jet submitted to a transverse high-frequency acoustic field. *J. Fluid Mech.* **2003**, *640*, 305–342. [CrossRef]
21. Ćosić, B.; Moeck, J.P.; Paschereit, C.O. Nonlinear Instability Analysis for Partially Premixed Swirl Flames. *Combust. Sci. Technol.* **2014**, *186*, 713–736. [CrossRef]
22. Kheirkhaha, S.; Cirtwill, J.D.M.; Saini, P.; Venkatesanb, K.; Steinberga, A.M. Dynamics and mechanisms of pressure, heat release rate, and fuel spray coupling during intermittent thermoacoustic oscillations in a model aeronautical combustor at elevated pressure. *Combust. Flame* **2017**, *185*, 319–334. [CrossRef]
23. Jia, B.Q.; Xie, L.; Cui, X.; Yang, L.J.; Fu, Q.F. Linear Stability of Confined Coaxial Jets in the Presence of Gas Velocity Oscillations with Heat and Mass Transfer. *Phys. Fluids* **2019**, *31*, 092101. [CrossRef]
24. Jia, B.Q.; Xie, L.; Cui, X.; Yang, L.J.; Fu, Q.F. Linear instability of viscoelastic planar liquid sheets in the presence of gas velocity oscillations. *J. Non Newton. Fluid Mech.* **2019**, *273*, 104169. [CrossRef]
25. Jia, B.Q.; Yang, L.J.; Xie, L.; Fu, Q.F.; Cui, X. Linear stability of confined swirling annular liquid layers in the presence of gas velocity oscillations with heat and mass transfer. *Int. J. Heat Mass Transf.* **2019**, *138*, 117–125. [CrossRef]

26. Guan, X.Y.; Jia, B.Q.; Yang, L.J.; Fu, Q.F. Linear instability of an annular liquid jet with gas velocity oscillations. *Phys. Fluids* **2021**, *33*, 054110. [CrossRef]
27. Deng, X.D.; Jia, B.Q.; Cui, X.; Wang, N.F.; Shi, B.L. Temporal Instability of Liquid Jet in Swirling Gas with Axial Velocity Oscillations. *AIAA J.* **2022**, *60*, 3852–3862. [CrossRef]
28. Deng, X.D.; Wang, H.R.; Cui, X.; Xie, L.; Jia, B.Q. Temporal instability of confined three-dimensional liquid jet with heat and mass transfer under longitudinal acoustic oscillations. *Phys. Fluids* **2022**, *34*, 102107. [CrossRef]
29. Fu, Q.F.; Deng, X.D.; Yang, L.J. Kelvin–Helmholtz Instability of Confined Oldroyd-B Liquid Film with Heat and Mass Transfer. *J. Non-Newton. Fluid Mech.* **2019**, *267*, 28–34. [CrossRef]
30. Fu, Q.F.; Deng, X.D.; Jia, B.Q.; Yang, L.J. Temporal instability of a confined liquid film with heat and mass transfer. *AIAA J.* **2018**, *56*, 2615–2622. [CrossRef]
31. Mohanta, L.; Cheung, F.B.; Bajorek, S.M. Stability of coaxial jets confined in a tube with heat and mass transfer. *Phys. A* **2016**, *443*, 333–346. [CrossRef]
32. Hsieh, D.Y. Effects of heat and mass transfer on Rayleigh-Taylor instability. *J. Fluid Eng. T. ASME* **1972**, *94*, 156–160. [CrossRef]
33. Hsieh, D.Y. Interfacial stability with mass and heat transfer. *Phys. Fluids* **1978**, *21*, 745–748. [CrossRef]
34. Asthana, R.; Agrawal, G.S. Viscous potential flow analysis of Kelvin–Helmholtz instability with mass transfer and vaporization. *Phys. A* **2007**, *382*, 389–404. [CrossRef]
35. Asthana, R.; Agrawal, G.S. Viscous potential flow analysis of electrohydrodynamic Kelvin–Helmholtz instability with heat and mass transfer. *Int. J. Eng. Sci.* **2010**, *48*, 1925–1936. [CrossRef]
36. Awasthi, M.K.; Asthana, R.; Agrawal, G.S. Pressure corrections for the potential flow analysis of Kelvin–Helmholtz instability with heat and mass transfer. *Int. J. Heat Mass Transf.* **2012**, *55*, 2345–2352. [CrossRef]
37. Awasthi, M.K.; Asthana, R.; Agrawal, G.S. Viscous correction for the viscous potential flow analysis of Kelvin–Helmholtz instability of cylindrical flow with heat and mass transfer. *Int. J. Heat Mass Transf.* **2014**, *78*, 251–259. [CrossRef]
38. Awasthi, M.K. Kelvin-Helmholtz instability of viscoelastic liquid-viscous gas interface with heat and mass transfer. *Int. J. Therm. Sci.* **2021**, *161*, 106710. [CrossRef]
39. Moatimid, G.M.; Obied Allah, M.H.; Hassan, M.A. Kelvin-Helmholtz instability for flow in porous media under the influence of oblique magnetic fields: A viscous potential flow analysis. *Phys. Plasmas* **2013**, *20*, 102111. [CrossRef]
40. Awasthi, M.K.; Asthana, R.; Uddin, Z. Nonlinear Study of Kelvin-Helmholtz instability of cylindrical flow with mass and heat transfer. *Int. J. Commun. Heat Mass* **2016**, *71*, 216–224. [CrossRef]
41. Scriven, L.E.; Sternling, V.S. The Marangoni effects. *Nature* **1960**, *187*, 186–788. [CrossRef]
42. Oron, A.; Deissler, R.T.; Duh, J.C. Marangoni instability in a liquid sheet. *Adv. Space Res.* **1995**, *16*, 83–86. [CrossRef]
43. Dávalos-Orozco, L.A. Thermocapillar instability of liquid sheets in motion. *Colloids Surf. A* **1999**, *157*, 223–233. [CrossRef]
44. Funada, T. Marangoni instability of thin liquid sheet. *J. Phys. Soc. Jpn.* **1986**, *55*, 2191–2202. [CrossRef]
45. Tong, M.X.; Yang, L.J.; Fu, Q.F. Thermocapillar instability of a two-dimensional viscoelastic planar liquid sheet in surrounding gas. *Phys. Fluid* **2014**, *26*, 033105. [CrossRef]
46. Zhang, S.; Lan, X.D.; Zhou, M. Thermocapillary instability of a liquid sheet with centrifugal force. *J. Braz. Soc. Mech. Sci.* **2018**, *47*, 40–47. [CrossRef]
47. Hu, K.X.; He, M.; Chen, Q.S. Instability of thermocapillary liquid layers for Oldroyd-B fluid. *Phys. Fluid* **2016**, *28*, 033105. [CrossRef]
48. Hu, K.X.; He, M.; Chen, Q.S.; Liu, R. Linear stability of thermocapillary liquid layers of a shear-thinning fluid. *Phys. Fluid* **2017**, *29*, 073101. [CrossRef]
49. Chandrasekhar, S. *Hydrodynamic and Hydro-Magnetic Stability*; Dover publications, Inc.: New York, NY, USA, 1961.
50. Hasegawa, K.; Manzaki, Y. Marangoni fireworks: Atomization dynamics of binary droplets on an oil pool. *Phys. Fluids* **2021**, *33*, 034124. [CrossRef]
51. Moezzi, M.; Sajjadi, M.; Hossein Hejazi, S. Thermally driven Marangoni effects on the spreading dynamics of droplets. *Int. J. Multiph. Flow* **2023**, *159*, 104335. [CrossRef]
52. Cui, X.; Jia, B.Q. Thermal Effect on the Instability of Annular Liquid Jet. *Aerospace* **2021**, *8*, 382. [CrossRef]
53. Funada, T.; Joseph, D.D. Viscous potential flow analysis of Kelvin-Helmholtz instability in a channel. *J. Fluid Mech.* **2001**, *445*, 263–283. [CrossRef]
54. Yang, L.; Zhang, Q.; Zhang, H.Q.; Wang, B. Numerical investigation on the performance of internal flow and atomization in the recessed gas-centered swirl coaxial injectors. *Aerosp. Sci. Technol.* **2022**, *129*, 107858. [CrossRef]
55. Benjamin, T.B.; Ursell, F. The stability of the plane free surface of a liquid in vertical periodic motion. *Proc. R. Soc. Lond. Ser. A* **1954**, *225*, 505–515. [CrossRef]
56. Kumar, K. Linear Theory of Faraday Instability in Viscous Liquids. *Proc. R. Soc. A* **1996**, *452*, 1113–1126. [CrossRef]
57. Funada, T.; Joseph, D.D.; Yamashita, S. Stability of a Liquid Jet into Incompressible Gases and Liquids. *Int. J. Multiph. Flow* **2004**, *30*, 1279–1310. [CrossRef]

58. Zeng, P.; Sarholz, S.; Iwainsky, C.; Binninger, B.; Peters, N.; Herrmann, M. Simulation of Primary Breakup for Diesel Spray with Phase Transition. In *Recent Advances in Parallel Virtual Machine and Message Passing Interface: 16th European PVM/MPI Users' Group Meeting, Espoo, Finland, 7–10 September 2009*; Springer: Berlin/Heidelberg, Germany, 2009.
59. Chan, S.H.; Wang, Y.S.; Tan, C.C. The effect of mass transfer on Kelvin-Helmholtz instability at the gas-liquid interface of a sonic reacting and non-reacting gas jet submerged in a liquid. *Int. J. Heat Mass Transf.* **1994**, *37*, 1123–1132. [CrossRef]

aerospace

MDPI

Article

Liquid Nitrogen Flow Boiling Critical Heat Flux in Additively Manufactured Cooling Channels

Debra Ortega, Alejandro Amador, Mohiuddin Ahmad, Ahsan Choudhuri and Md Mahamudur Rahman *

Aerospace Center, Department of Aerospace and Mechanical Engineering, The University of Texas at El Paso, El Paso, TX 79968, USA
* Correspondence: mrahman15@utep.edu; Tel.: +1-915-747-8305

Abstract: This paper presents an experimental characterization of liquid nitrogen (LN_2) flow boiling in additively manufactured minichannels. There is a pressing need of concerted efforts from the space exploration and thermal transport communities to design high-performance rocket engine cooling channels. A close observation of the literature gaps warrants a systematic cryogenic flow boiling characterization of asymmetrically heated small (<3 mm) non-circular channels fabricated with advanced manufacturing technologies at mass flux > 3000 kg/m^{2}s and pressure > 1 MPa. As such, this work presents the LN_2 flow boiling results for three asymmetrically heated additively manufactured GR-Cop42 channels of 1.8 mm, 2.3 mm, and 2.5 mm hydraulic diameters. Twenty different tests have been performed at mass flux~3805–14,295 kg/m^{2}s, pressures~1.38 and 1.59 MPa, and subcooling~0 and 5 K. A maximum departure from nucleate boiling (DNB)-type critical heat flux (CHF) of 768 kW/m^2 has been achieved for the 1.8 mm channel. The experimental results show that CHF increases with increasing LN_2 flow rate (337–459 kW/m^2 at 25–57 cm^3/s for 2.3 mm channel) and decreasing channel size (307–768 kW/m^2 for 2.5–1.8 mm channel). Finally, an experimental DNB correlation has been developed with 10.68% mean absolute error.

Keywords: cryogenic flow boiling; critical heat flux; asymmetric heating; non-circular channels; additive manufacturing; GR-Cop42; departure from nucleate boiling correlation

Citation: Ortega, D.; Amador, A.; Ahmad, M.; Choudhuri, A.; Rahman, M.M. Liquid Nitrogen Flow Boiling Critical Heat Flux in Additively Manufactured Cooling Channels. *Aerospace* **2023**, *10*, 499. https://doi.org/10.3390/aerospace10060499

Academic Editor: Yiheng Tong

Received: 27 April 2023
Revised: 19 May 2023
Accepted: 24 May 2023
Published: 25 May 2023

1. Introduction

Cryogenic fluids exhibit some unique characteristics, such as low surface tension, low liquid viscosity, low latent heat of vaporization, small density difference between the liquid and gas phases [1], and large compressibility [2]. These unique thermophysical properties of cryogenic fluids have contributed to its application in the multitude of terrestrial and space systems including cooling of high-temperature superconductors [2,3], cryosurgery [4], metal processing [5], air separation units, chilling down of telescopes and satellites [1], regenerative cooling of liquid rocket engines [6,7] and so on. In regeneratively cooled rocket engines, the propellant or the oxidizer flows through the cooling channels surrounding the combustion chamber and nozzle walls to absorb heat and maintain the structural integrity of the thrust chamber [8,9]. During regenerative cooling, four different heat flux regimes appear to exist to control the wall temperature [10]. At low heat flux values, single-phase forced convection forms a liquid boundary layer to absorb heat from the hot wall. As the imposed heat flux increases, the wall temperature increases significantly to sustain the nucleation of vapor bubbles. The vapor bubbles grow, depart from the wall generating local turbulence, and finally collapse in the bulk cooler liquid. Due to bubble dynamics, a large amount of heat is absorbed via latent heat stored within the coolant in this regime. This regime is known as nucleate boiling and it usually appears in the nozzle throat area. Further increase in heat flux results in the competition between vapor generated at the wall and the replenishing liquid until the hot wall reaches critical heat flux (CHF). Two types of CHF events are encountered for cryogenic cooling, known as departure from nucleate boiling (DNB) type CHF and dry-out type CHF [1]. During DNB, the vapor bubbles from the wall are unable to penetrate the highly sub-cooled liquid core flow at high mass flux resulting in a localized vapor blanket. Conversely, during dry-out type CHF, annular liquid film appears in the cooling channels at low degrees of subcooling and low mass flow rates. However, regardless of the type of CHF event, it identifies the maximum feasible heat transfer rate for the combination of given fluid–surface interface conditions. If the applied heat flux exceeds the CHF limit, an unstable vapor film acting as an insulation layer significantly increases the wall temperature. A further increase in wall temperature is associated with the radiation heat transfer regime due to a drastic increase in vapor film temperature. However, the cooling channel design aims to avoid both the unstable gas film and radiation heat transfer regimes to achieve efficient cooling and structural integrity. Therefore, it is essential to identify the CHF limits in the design of two-phase systems.

A variety of publications have reported flow boiling CHF values for cryogenic liquids including liquid hydrogen (LH_2), helium (LHe), nitrogen (LN_2), and methane (LCH_4). In 2021, Ganesan et al. [1] consolidated 2312 CHF data points from world cryogenic literature dating back to 1959. The database contains single component cryogens flowing vertically upward or downward, and horizontal, and symmetrically heated straight stationary circular channels. Some consolidated results are presented in Figure 1. Figure 1a presents the number of available CHF data points as a function of mass velocity. It can be seen that more data are available up to mass flux value of 3000 kg/m^2 s. However, Cho et al. [8] performed regenerative cooling heat transfer analysis using SUPERTRAPP code for F-1 and RS-27A engines at 30,000 kg/m^2s mass flux. Boysan et al. [11] performed numerical simulations to investigate the effect of cooling channel geometry for liquid propellant rocket engines using ~6000 kg/m^2s liquid hydrogen mass flux. Moreover, Wang et al. [12] developed a multidisciplinary, computational methodology using a combination of CFD/thermal models to predict the hot-gas-side and coolant-side heat transfer for regeneratively cooled liquid rocket engine combustors where ~11,000 kg/m^2 s mass flux was used for liquid hydrogen coolant. Therefore, it is essential to measure the cryogenic flow boiling CHF at large mass flux values.

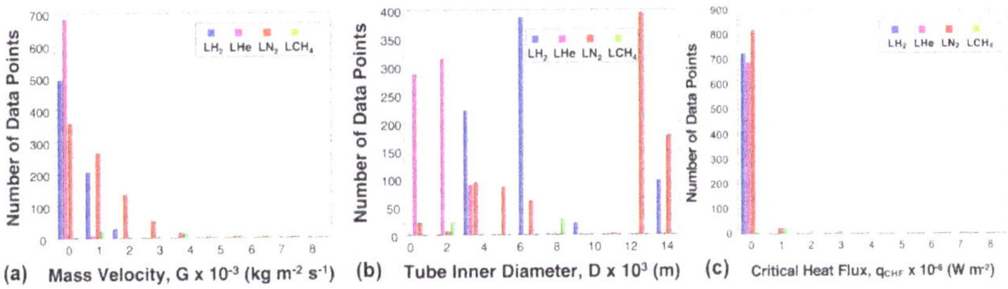

Figure 1. Distribution of data in CHF database relative to (**a**) mass velocity, (**b**) tube diameter, and (**c**) measured CHF [1].

Figure 1b presents the number of available CHF data points as a function of cylindrical tube diameter. Albeit there exists a significant number of CHF data points for small diameter channels (D < 3 mm) using Lhe, the flow boiling results below 3 mm channel diameter are limited for LN_2, LH_2, and LCH_4. However, it is well-established that CHF increases with decreasing channel diameter. The slope of the velocity profile in the two-phase boundary layer increases with decreasing channel diameter. This effect with increased mass flux facilitates the bubble departure from the heated wall and increases condensation in the liquid core [13]. Wadel [14] performed analytical investigations on seven coolant channel designs using a coupling of the Rocket Thermal Evaluation code and the Two-Dimensional Kinetics code. The high aspect ratio cooling channel designs with hydraulic diameters varying from 0.49 mm to 0.94 mm showed hot combustion gas-side wall temperature reduction up to 22% with coolant pressure drop increasing to 7.5%. Van Noord et al. [15] conducted sub-cooled LCH_4 boiling tests with electrically heated Inconel tubes of 0.66 mm,

1.4 mm, and 2.1 mm diameters and achieved ~10 MW/m^2 CHF value for the 1.4 mm diameter channel. Therefore, it is essential to experimentally investigate the cryogenic flow boiling CHF using small channels.

Figure 1c presents the number of available CHF data points as a function of measured CHF values. It is evident that most of the measured CHF values are below 0.1 MW/m^2, more specifically for LH_2, Lhe, LN_2, and fewer reported above for LCH_4. Moreover, rocket engine thrust chamber walls are subjected to high heat transfer rates of 0.8–82 MW/m^2 [16]. As such, cooling channel designs with high heat transfer capabilities appear to be a significant design consideration for rocket engines.

In addition to high coolant mass flux, small channel cross-sectional area, and high heat transfer rates, other factors, such as channel material, channel geometry, fabrication method and imposed heat flux direction, play significant roles in cooling channel design for liquid rocket engines. Significant experimental research has been reported on cryogenic flow boiling in symmetrically heated circular channels [1]. However, rocket engine cooling channels experience asymmetric heat flux from the combustion chamber hot wall [17–20]. Moreover, the effect of high aspect ratio cooling channels on increasing regenerative cooling performance and improving pressure drop is well-documented [18]. Wadel showed hot combustion gas-side wall temperature reduction up to 22% with coolant pressure drop increasing to 7.5% for high aspect ratio cooling channels [14]. Santhosh and Kuzhiveli performed two-dimensional heat transfer analysis for high aspect ratio channels and concluded that maximum cooling and minimum pressure drop can be achieved for an optimum aspect ratio value [21]. Wadel and Meyer showed a 27% reduction in pressure drop with 13% reduction in hot-gas-side wall temperature using high aspect ratio channel at low coolant flow rate as compared to conventional channel at full-scale mass flow test [22]. Carlile and Quentmeyer experimentally demonstrated that the combustion chamber wall temperature decreases by 30% as compared to the baseline low-aspect-ratio channel for the same coolant pressure drop [23]. Yet, a limited amount of research has addressed the experimental cryogenic flow boiling using non-circular geometries, such as square and high aspect ratio rectangular channels, more specifically asymmetrically heated non-circular cooling channels. Additionally, a vast majority of the cryogenic flow boiling tests have been conducted using stainless steel tubes while some researchers utilized nickel alloys and fewer employed copper and silver tubes [1]. However, Glenn Research Copper (GR-Cop) alloys are the preferred combustion chamber materials due to their high thermal conductivity, low creep and low-cycle fatigue, oxidation and blanching resistance, and high material strength at elevated temperatures with sustained operations [24,25]. Moreover, the National Aeronautics and Space Administration (NASA) has observed significant potential with additive manufacturing (AM) that offers reduced lead time and cost of fabrication for complex high-performance liquid rocket engine components. However, there is a lack of literature on the characterization of fuel film cooling in additively manufactured GR-Cop cooling channels [25]. Therefore, it is essential to investigate the cryogenic flow boiling heat transfer performance for the optimization of high-performance engine systems.

Motivated by the above-mentioned research gaps of cryogenic flow boiling tests in small asymmetrically heated non-circular channels of high heat transfer capability fabricated using advanced manufacturing technologies, this article aims to present an experimental characterization of cryogenic flow boiling heat transfer in additively manufactured channels. Here, GR-Cop square, and high aspect ratio rectangular channels have been used during sub-cooled and saturated LN_2 flow boiling. The tests have been performed at a wide variety of mass flux values and different inlet pressure conditions up to CHF. The experimental CHF results are compared with experimental CHF correlations from the literature and, a modified CHF correlation has been developed for LN_2 DNB-type CHF in asymmetrically heated non-circular cooling channels.

2. Materials and Methods

2.1. High Heat Flux Test Facility (HHFTF)

Cryogenic flow boiling experiments are performed using a high heat flux test facility (HHFTF), shown in Figure 2, capable of operating up to 4 MPa pressure and 755 K temperature using liquid nitrogen (LN_2) and liquid methane (LCH_4) as the working fluids [26–28]. The HHFTF is a thermal concentrator system based on conduction that asymmetrically transfers heat to a horizontal cooling channel resembling regenerative cooling during combustion. Tests are performed inside a vacuum chamber (Figure 2a) to reduce convective heat leaks during the tests and potential oxidation of components at high temperatures. A scroll vacuum pump is used to achieve a pressure of 6 Pa inside the vacuum chamber and remove fluid residuals and non-condensable gases from the flow lines. Vacuum pressure inside the chamber is measured using a convection-enhanced Pirani sensor with a Pirani gauge controller. A 316L stainless steel test stand is used to structurally support the test assembly and thermal concentrator inside the vacuum chamber. The custom-built test rig, heat flux sensor and thermal concentrator are compressed using a 6061-aluminum cradle to provide good thermal contacts. Figure 2b shows the schematic of test assembly with aluminum cradle inside the test stand. Nitrogen gas (GN_2) is used to pressurize LN_2 inside the high-pressure liquid cryogen dewar to pump LN_2 to the test section. A pressure regulator is used to control the GN_2 flow to the LN_2 dewar. LN_2 flow rate is adjusted using a precision needle valve. A turbine flowmeter is used to measure the flow rate. Static pressure transducers with thermal standoffs are used to measure pressure along the LN_2 flow line at the inlet of the vacuum chamber, and at the inlet of the cooling channel. Two ungrounded sheathed type-E thermocouples are utilized to measure LN_2 temperatures at the inlet and outlet of the cooling channel. A National Instrument Data Acquisition (NI-DAQ) system is used to monitor and record the measurements from all the sensors. Figure 2c shows the picture of the high heat flux test facility.

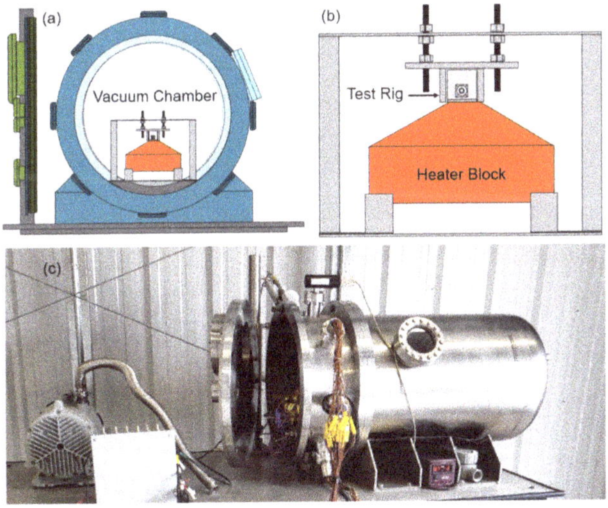

Figure 2. Cryogenic flow boiling test facility: (**a**) schematic of test assembly inside the vacuum chamber, (**b**) schematic of the test stand, test rig and heater block assembly, and (**c**) picture of the high heat flux test facility.

2.2. Thermal Concentrator

To provide asymmetric heating to the cooling channel, the HHFTF consists of a thermal concentrator, as shown in Figure 3. The thermal concentrator is a trapezoidal C12200 copper block that contains 17 cartridge heaters capable of generating ~7.2 kW power.

The cartridge heaters are controlled using a variable transformer (VARIAC) of adjustable voltage ranging from 0 to 130 V. The safe operation of the cartridge heaters is monitored using 17 type-K thermocouples (Figure 3, $T_{16} - T_{32}$) inserted into the concentrator block near the base of the cartridge heaters. The thermal concentrator top has a surface area of 2.5 cm width × 5 cm length, resulting in ~5.7 MW/m^2 heat flux capability of the HHFTF. The thermal concentrator assembly also contains a custom-built heat flux sensor to measure the given heat flux to the cooling channel during flow boiling tests. The heat flux sensor consists of a copper wafer of 2.5 cm width, 5 cm length and 6.35 mm thickness that contains six slots for thermocouples. The distance between the top three thermocouples (Figure 3, $T_{10} - T_{12}$) and bottom three thermocouples (Figure 3, $T_{13} - T_{15}$) is ~3 mm. Type-T thermocouples are inserted into the copper wafer to measure heat flux using the Fourier conduction law. Thermally conductive ceramic paste is used between the copper wafer and thermal concentrator and compressed with aluminum cradle, as shown in Figure 2b, to provide good thermal contact. Here, $T_{L,in}$ and $T_{L,out}$ are the liquid nitrogen inlet and exit temperatures, respectively, and P_{in} and P_{out} are the inlet and exit pressures, respectively.

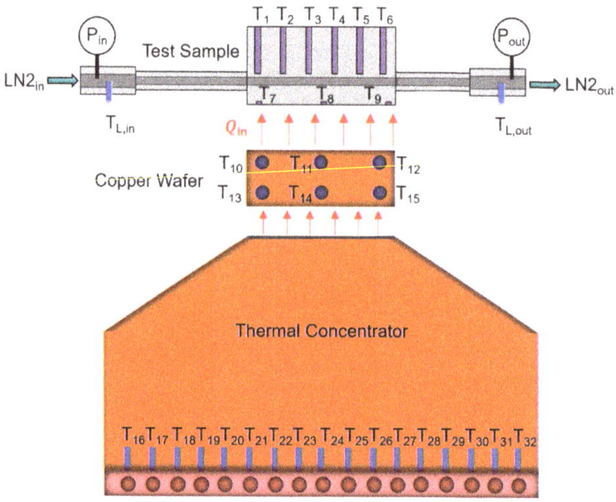

Figure 3. Schematic of heater block, wafer, and test section assembly.

2.3. Test Channels

Three different Glenn Research Copper (GR-Cop42), a copper-chrome-niobium alloy, and cooling channel designs have been fabricated and tested during the sub-cooled and saturated liquid nitrogen flow boiling tests in this paper. The test sections have been designed by the authors and fabricated by the National Aeronautics and Space Administration (NASA) Marshall Space Flight Center (MSFC). Powder Bed Fusion (PBF) advanced additive manufacturing has been used to fabricate the samples. To accommodate the feasibility of various cooling channel tests, the test rigs have been designed such that different cooling channels can be simply substituted for testing without affecting the wafer and thermal concentrator. The test section has a surface area of 2.5 cm width × 5 cm length to match the top surface area of copper wafer and thermal concentrator. Figure 4 shows the design of the test section. Figure 4a shows the cross-sectional view of the test section where w is the width, and h is the height of the cooling channel. Figure 4b presents the side view of the test section showing the cooling channel height, length, and thermocouple locations.

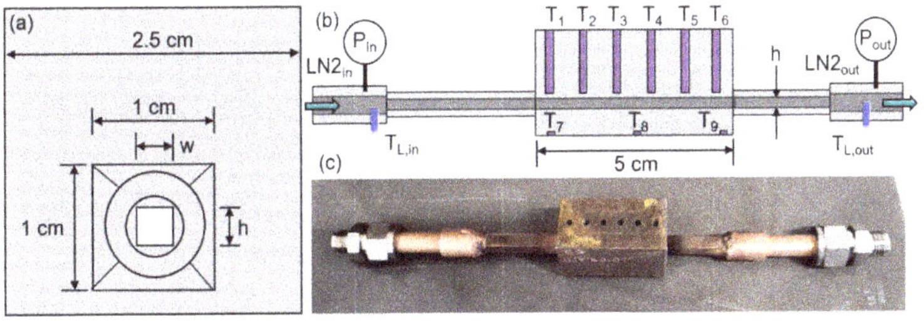

Figure 4. Cryogenic flow boiling test section: (**a**) schematic of the cross-sectional view of the test channel, (**b**) schematic of the side view of the test channel, and (**c**) picture of the GR-Cop42 test section with 2.5 mm hydraulic diameter.

Table 1 presents the dimensions of the various cooling channels tested in this work. The variations in cooling channel designs include two square channels with hydraulic diameters of 1.8 mm (1.8 mm × 1.8 mm), and 2.3 mm (2.3 mm × 2.3 mm), and one high aspect ratio rectangular channel with hydraulic diameter of 2.5 mm (1.8 mm × 4.1 mm). All the channels have a heated length of 5 cm. Figure 4c shows the picture of the test section with the high aspect ratio 2.5 mm hydraulic diameter channel. The test section consists of an entry-length section ($\sim 10D_h$, where D_h is the hydraulic diameter of the colling channel) to allow fully developed flow and exit-length section. The test section also includes six thermocouples located at the top of the cooling channel (Figure 4b, $T_1 - T_6$), three thermocouples underneath the cooling channel (Figure 4b, $T_7 - T_9$), one thermocouple at the LN_2 inlet (Figure 4b, $T_{L,in}$), and one thermocouple at the LN_2 outlet (Figure 4b, $T_{L,out}$) along the flow direction. The distance between the tip of the thermocouple and channel wall is ~1.27 mm.

Table 1. List of different cooling channels tested during LN_2 flow boiling.

Channel	Width, w	Height, h	Length	Hydraulic Diameter, D_h
GR-Cop42	1.8 mm	1.8 mm	5 cm	1.8 mm
GR-Cop42	2.3 mm	2.3 mm	5 cm	2.3 mm
GR-Cop42	1.8 mm	4.1 mm	5 cm	2.5 mm

2.4. Test Procedure

LN_2 flow boiling tests are performed in horizontal orientation starting with achieving ~25 Pa pressure inside the vacuum chamber. Initially, the tube lines and the test channel are cooled using a continuous flow of LN_2 until the system reaches the desired temperature. This is an open flow system where the exiting nitrogen is released to the vent. Three different operating conditions are used in this paper: (i) 1.38 Mpa pressure and corresponding saturation temperature of 108.97 K, (ii) 1.59 Mpa pressure and corresponding saturation temperature of 111.42 K, and (iii) 1.59 Mpa pressure and sub-cooled temperature of 106.42 K (5 K subcooling). Once the desired temperature is reached, the LN2 flow rate is set to the desired value using the precision needle valve. Five various volumetric flow rate values are used in this paper: 25 cm^3/s, 31 cm^3/s, 38 cm^3/s, 47 cm^3/s, and 57 cm^3/s. Then, the test section is heated by powering the cartridge heaters in the thermal concentrator at small increments. The steady-state data points are recorded at each increment of heater power when the temperatures of wafer thermocouples ($T_{10} - T_{15}$) do not change more than 0.5 K for 1 min. A slow and gradual heat input is given to the test channels until a drastic increase in wafer temperatures is observed ensuring that the channel has reached a film boiling regime. This significant change in the nucleate boiling slope is identified as the critical heat flux (CHF) point for cryogenic boiling heat transfer [15]. Since the tests start at low temperatures (109 K or 113 K), system failure is not a concern during these experiments.

2.5. Data Analysis

Total heat applied to the copper wafer has been calculated using the Fourier heat conduction law, shown in Equation (1). Here, Q is the total heat transfer rate given to the copper wafer, k_{wafer} is the thermal conductivity, A_{wafer} is the cross-sectional area that is 12.5 cm^2, ΔT_{wafer} is the temperature difference among the top and bottom thermocouples in the copper wafer, and L_{wafer} is the distance among the top and bottom thermocouples that is approximately 3 mm center-to-center distance.

$$Q = k_{wafer} A_{wafer} \frac{\Delta T_{wafer}}{L_{wafer}} \tag{1}$$

The average temperature difference between the top and bottom thermocouples has been calculated using Equation (2). Here, T_{10}, T_{11}, and T_{12} are the top three thermocouples and T_{13}, T_{14}, and T_{15} are the bottom three thermocouples, as shown in Figure 3.

$$\Delta T_{wafer} = \frac{1}{3}[(T_{13} + T_{14} + T_{15}) - (T_{10} + T_{11} + T_{12})] \tag{2}$$

Heat flux (q'') to the cooling channel has been calculated using the total heat transfer rate across the copper wafer calculated using Equation (1) divided by the wetted area of the cooling channel, as given in Equation (3). Here, A_w is the wetted area of the channel, D_h is the hydraulic diameter and L is the length of the channel.

$$q'' = \frac{Q}{A_w} = \frac{Q}{\pi D_h L} \tag{3}$$

The average wall temperature of the channel at a given heat flux has been calculated from the average temperature of the top six thermocouples ($T_1 - T_6$, Figure 4b) and average of the bottom three thermocouples ($T_7 - T_9$, Figure 4b) of the test section, as shown in Equation (4).

$$T_{wall} = \frac{1}{2}\left[\left(\frac{T_1 + T_2 + T_3 + T_4 + T_5 + T_6}{6}\right) + \left(\frac{T_7 + T_8 + T_9}{3}\right)\right] \tag{4}$$

Finally, the heat transfer coefficient (HTC) of each recorded data point has been calculated using Equation (5), where ΔT_{wall} is the wall superheat that is the temperature difference between the average wall temperature (T_{wall}) calculated from Equation (3) and fluid saturation temperature (T_{sat}) at the corresponding pressure.

$$HTC = \frac{q''}{\Delta T_{wall}} = \frac{q''}{T_{wall} - T_{sat}} \tag{5}$$

The mass flux (G) is calculated from the volumetric flow rate ($\dot{\forall}$) measured using the turbine flow meter as given by Equation (6). Here, ρ_l is the density of liquid nitrogen at saturation temperature, and A_c is the channel cross-sectional area defined by $\pi D_h^2/4$.

$$G = \frac{\rho_l \dot{\forall}}{A_c} = \frac{4\rho_l \dot{\forall}}{\pi D_h^2} \tag{6}$$

2.6. Experimental Uncertainty

The experimental uncertainty associated with the heat flux, wall superheat and heat transfer coefficient are calculated based on the propagation of error analysis that include thermocouple accuracy and uncertainty in various physical dimensions. The propagations of these uncertainties (U) are given in Equations (7)–(9) [29]. Here, $U_{q''}$, $U_{\Delta T_{wall}}$, and U_{HTC} are the uncertainties associated with heat flux, wall superheat and heat transfer coefficient, respectively. $U_{T_{avg,T10-T12}}$ is the uncertainty associated with the average temperature of the wafer top thermocouples calculated from the T_{10} to T_{12} thermocouples ($T_{avg,T10-T12}$), as shown in Figure 3. $U_{T_{avg,T13-T15}}$ is the uncertainty associated with the average temperature of the wafer bottom thermocouples calculated from the T_{13} to T_{15} thermocouples ($T_{avg,T13-T15}$). $U_{T_{avg,T1-T6}}$ is the uncertainty associated with the average temperature of the channel top surface calculated from the T_1 to T_6 ($T_{avg,T1-T6}$) thermocouples, as shown in Figure 4. $U_{avg,T7-T9}$ is the uncertainty associated with the average temperature of the channel bottom surface calculated from the T_7 to T_9 thermocouples ($T_{avg,T7-T9}$). $U_{L_{wafer}}$ is the uncertainty associated with the thermocouple locations in the copper wafer. T_{sat} is the saturation temperature of liquid nitrogen at inlet pressure, and $U_{T_{sat}}$ is the uncertainty associated with the saturation temperature determined from the REFPROP.

$$\frac{U_{q''}}{q''} = \left[\left(\frac{U_{T_{avg,T10-T12}}}{T_{avg,T10-T2}}\right)^2 + \left(\frac{U_{T_{avg,T13-T15}}}{T_{avg,T13-T15}}\right)^2 + \left(\frac{U_{L_{wafer}}}{L_{wafer}}\right)^2\right]^{1/2} \quad (7)$$

$$\frac{U_{\Delta T_{wall}}}{\Delta T_{wall}} = \left[\left(\frac{U_{T_{avg,T1-T6}}}{T_{avg,T1-T6}}\right)^2 + \left(\frac{U_{T_{avg,T7-T9}}}{T_{avg,T7-T9}}\right)^2 + \left(\frac{U_{T_{sat}}}{T_{sat}}\right)^2\right]^{1/2} \quad (8)$$

$$\frac{U_{HTC}}{HTC} = \left[\left(\frac{U_{q''}}{q''}\right)^2 + \left(\frac{U_{\Delta T_{wall}}}{\Delta T_{wall}}\right)^2\right]^{\frac{1}{2}} \quad (9)$$

The uncertainty of the omega type-T thermocouple measurement is ±1%. The uncertainty in the thermocouple locations in the copper wafer is approximated to be a standard machining error of 0.125 mm. Therefore, for the cryogenic flow boiling setup utilized in this study, the maximum uncertainty in the heat flux is calculated to be approximately ±2.52%. The uncertainty of the saturation temperature determined from the REFPROP is associated with the inlet pressure measurement. The error of the static pressure transducer used in this work at the LN_2 inlet is ±0.25%. Therefore, the maximum uncertainty in the wall superheat and heat transfer coefficient are calculated to be approximately ±1.44% and ±2.9%, respectively. Additionally, the uncertainty associated with the volumetric flow rate measurements using turbine flow meter is approximately ±1.23%. Table 2 presents the summary of the experimental uncertainty associated with the heat flux, wall superheat and heat transfer coefficient.

Table 2. List of experimental uncertainty associated with the LN_2 flow boiling.

Parameter	Relative Uncertainty
Thermocouple, T (°C)	±1%
Distance between thermocouples, L_{wafer} (mm)	±2.08%
Pressure transducer, P_{in} (Mpa)	±0.25%
Turbine flow meter, $\forall$ (cm^3/s)	±1.23%
Wall superheat, ΔT_{wall} (K)	±1.44%
Heat flux, q'' (kW/m^2)	±2.52%
Heat transfer coefficient, HTC (kW/m^{2}K)	±2.9%

3. Results and Discussions

To investigate the cryogenic flow boiling in additively manufactured GR-Cop42 cooling channels, 20 tests have been performed at two different inlet pressures, two different degrees of subcooling and five different LN_2 flowrates. The following sections present the boiling results showing the boiling repeatability, effect of mass flux, effect of inlet pressure and degree of subcooling, effect of channel size, and finally an experimental CHF correlation for asymmetrically heated additively manufactured LN_2 cooling channels.

3.1. Repeatability

To characterize the repeatability of the flow boiling performance, the 1.8 mm hydraulic diameter channel (1.8 mm × 1.8 mm square) design has been tested several times. Figure 5 shows the repeated boiling results for two similar channels at two different LN_2 flow rates of 47 cm^3/s, and 57 cm^3/s. X-axis represents the wall superheat, i.e., the difference between the channel average surface temperature and liquid nitrogen saturation temperature (~108.97 K) at 1.38 MPa inlet pressure. Y-axis represents the heat transfer rate per unit wetted area of the cooling channel. The arrows represent CHF point.

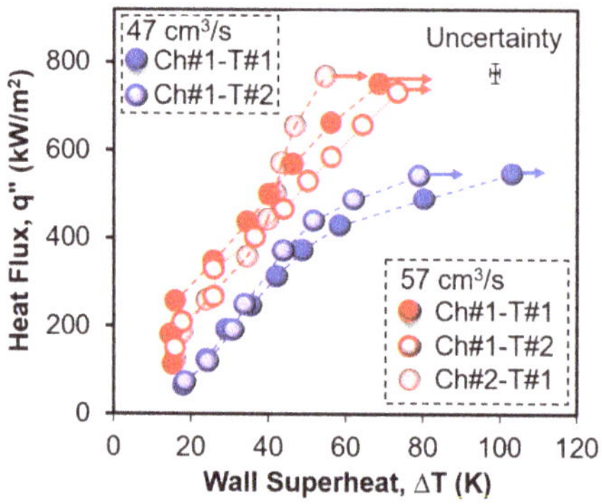

Figure 5. Repeatability tests for two 1.8 mm hydraulic diameter channels at 1.38 MPa inlet pressure, saturation temperature and 57 cm^3/s LN_2 flowrate.

The red circles represent the 1.8 mm channel with 57 cm^3/s LN_2 flow rate, and blue circles represent the 1.8 mm channel with 47 cm^3/s LN_2 flow rate. The dark red and blue filled circles represent the first-time test results for the first sample (Ch#1), the shaded red and blue circles represent the repeated boiling results for the first sample, and the light red circles represent the first-time test result for the similar 1.8 mm channel (Ch#2). A maximum CHF value of 752 kW/m^2 has been achieved for the channel#1 test#1 at 57 cm^3/s flow rate. The second test of the same channel shows a CHF value of 732 kW/m^2. However, the first test of the second channel shows a higher CHF of ~767 kW/m^2. For the 47 cm^3/s flow rate, the first test shows a CHF of 548 kW/m^2 and the second test shows a CHF of 543 kW/m^2. A shift in the boiling curves is present with a maximum variation in CHF of approximately ±1.1% for the 57 cm^3/s LN_2 flow tests, and approximately ±0.5% for the 47 cm^3/s flow tests. This is negligible compared to the experimental uncertainty of ±2.52% for the heat flux results. However, the variation in wall superheat is significant that can be attributed to the fouling, handling of the test samples, test sample manufacturing and boiling itself [30].

3.2. Effect of Mass Flux

During flow boiling, single-phase forced convection and bubble dynamics through nucleation, growth and departure-enabled efficient heat removal from the hot surfaces. Mass flux is one of the strongest parameters that affects the flow boiling heat transfer performance through increased single-phase forced convection. At low mass flux conditions, gravity effect becomes important for horizontal channels [31]. In this work, the flow boiling tests have been performed in horizontal flow orientation. However, the calculated modified Froude number $\left(Fr^* = G\cos\theta/\rho_l\sqrt{gD_h(\rho_l-\rho_v)/\rho_v}\right)$ value varies from 12.95 to 53.77 for the flow boiling tests conducted in this work. Here, G is the LN_2 mass flux (kg/m^2s), ρ_l is the density of liquid nitrogen and ρ_v is the density of nitrogen gas at the saturation temperature of corresponding pressure. θ is the flow orientation (here, $\theta = 0$), and D_h is the hydraulic diameter of the cooling channels. Since the modified Froude number value is significantly greater than 6, the cryogenic liquid flow and heat transfer performance of the cooling channels tested in this work are independent of flow orientation relative to Earth's gravity [1].

To characterize the effect of mass flow rate on LN_2 flow boiling, a series of flow boiling tests have been performed at various LN_2 flow rates using a 2.3 mm hydraulic diameter channel (2.3 mm × 2.3 mm) at 1.38 Mpa pressure and saturation temperature. The results are presented in Figure 6. Figure 6a shows the boiling curves at various volumetric flow rates. The orange, green, blue, and red squares represent 25 cm^3/s, 38 cm^3/s, 47 cm^3/s, and 57 cm^3/s volumetric flow rates, respectively. Solid squares represent test#1 whereas light squares represent test#2. It is evident from Figure 6a that with increasing LN_2 flow rates, the boiling curves are shifted to the left in general, i.e., increasing HTC. A similar trend has also been observed by Ribatski et al. [32], Agostini et al. [33], and Wang and Sefiance [34] and it has been attributed to the transition from partial to fully developed nucleate boiling. During flow boiling, heat transfer is governed by the combination of nucleate boiling and forced convective evaporation. This can be represented by $HTC = S(HTC)_{nb} + F(HTC)_{cov}$ [35]. Here, $(HTC)_{nb}$ is the heat transfer coefficient for nucleate boiling, S is the suppression factor for nucleate boiling, $(HTC)_{cov}$ is the heat transfer coefficient for forced convective evaporation, and F is the forced convective evaporation enhancement factor. As the liquid flow rate increases, the contribution of nucleate boiling (S) reduces due to the suppression of nucleation sites, and the contribution of forced convective evaporation (F) increases for flow boiling. Therefore, for the transition from 25 cm^3/s flow rate (orange squares) to 38 cm^3/s flow rate (dark green squares) in Figure 6a, HTC increases due to the enhanced contribution from forced convective evaporation (F) in addition to the contribution from nucleate boiling (S). However, with a further increase in LN_2 flow rate at 47 cm^3/s (dark blue squares), the boiling curve can be seen shifting to the right as compared to the 38 cm^3/s flow rate (dark green squares) curve. Zhang et al. [36] have observed the collapse of boiling curves at various mass flux values in a vertical mini channel for LN_2 due to the significance of nucleate boiling through the consumption of latent heat. This has also been observed in mini channels by other researchers, such as Balasubramanian et al. [37] and Bertsch et al. [38]. Furthermore, Zhang et al. [36] have shown that the HTC decreases with increasing mass flux due to the suppression of nucleate boiling in the intermediate mass flux regimes. Therefore, the HTC decreasing trend of boiling curve for the dark blue squares as compared to the dark green squares can be attributed to the suppression of augmented nucleate boiling, however, there exist the contribution from enhanced forced convective evaporation. Although the LN_2 flow rate increases from 47 cm^3/s to 57 cm^3/s and the contribution from nucleate boiling decreases, the significant increase in the contribution from forced convective evaporation eventually increases the HTC that can be seen in the red squares in Figure 6a. It can also be seen that the second tests for 38 cm^3/s flow rate (light green squares) and 47 cm^3/s flow rate (light blue squares) are flipped, i.e., the second test boiling curve shifts to the right as compared to the first test for 38 cm^3/s. Since heat transfer at 38 cm^3/s flow rate has a significant contribution from nucleate boiling, the decreasing HTC in the second test can be attributed to the reduction in entrapped non-condensable gases that facilitates the nucleation on boiling surface [39] and aging phenomena [40].

The effect of mass flux on CHF (q''_{CHF}) is presented in Figure 6b where mass flux is calculated from measured volumetric flow rates using Equation (6). A maximum CHF of 459 kW/m^2 has been achieved for the 57 cm^3/s flow rate. The CHF for 47 cm^3/s flow rate is 426 kW/m^2 and 38 cm^3/s flow rate tests show a variation in CHF of 387 kW/m^2 and 401 kW/m^2 between repeated tests. CHF value of 337 kW/m^2 has been achieved for the 25 cm^3/s flow rate which is the minimum among all four flow rates. The results show that CHF increases in a linear trend as the mass flux increases, which has also been reported by Agostini et al. [41], Roday et al. [42], and Ong and Thome [43]. However, the opposite trend has also been observed by other researchers. Qi et al. [44] and Callizo [45] have shown that CHF decreases with increasing mass flux due to the shear-enhanced droplet entrainment effect in vertical channels. However, for horizontal channels, the droplet entrainment effect reduces the flow instability with increasing mass flux [31]. Therefore, the CHF enhancement can be attributed to the reducing influence of flow instability and gravity effects with increasing LN_2 mass flux resulting in an increased forced convective evaporation in this work.

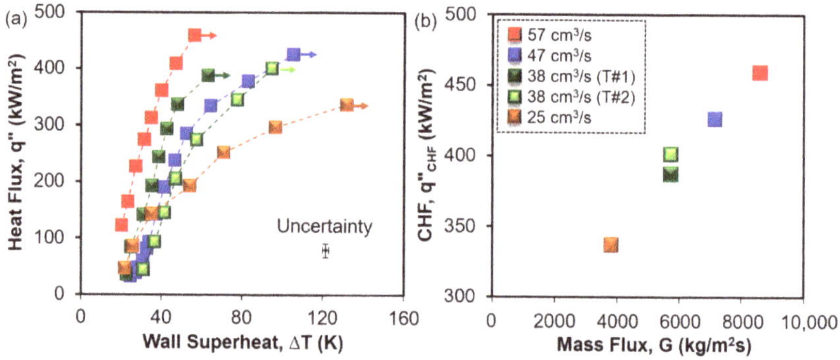

Figure 6. Effect of mass flux on LN_2 flow boiling CHF for 2.3 mm hydraulic diameter GR-Cop42 channel at 1.38 MPa inlet pressure and saturation temperature: (**a**) heat flux as a function of wall-superheat, and (**b**) CHF as a function of mass flux.

3.3. *Effect of Inlet Pressure and Subcooling*

Figure 7 presents the effect of inlet pressure and degree of subcooling on LN_2 flow boiling tests in cooling channels. The purple color represents the mass flow rate of 31 cm^3/s. Circles represent the 1.8 mm hydraulic diameter channel and triangular dots represent the 2.5 mm hydraulic diameter channel. Dark solid purple markers represent boiling results for 1.38 MPa inlet pressure and saturation temperature (test#1), light purple markers represent results for 1.38 MPa inlet pressure and saturation temperature (test#2), open purple represent results from 1.59 MPa inlet pressure and saturation temperature, and cyan filled purple represent boiling results for 1.59 MPa inlet pressure and 5 degrees of subcooling. Figure 7a presents the boiling heat flux up to CHF in Y-axis as a function of channel average wall superheat in X-axis. Figure 7b presents the CHF (Y-axis) as a function of inlet pressure (X-axis) at various degrees of subcooling. It is evident from Figure 7 that for both channels CHF decreases with increasing pressure. The 1.8 mm channel shows a CHF value of ~468 kW/m^2 at 1.38 MPa pressure and zero-degree subcooling. The CHF value decreases to 407 kW/m^2 at 1.59 MPa pressure and corresponding saturation temperature. This represents an approximately 13% reduction in CHF with the increase in pressure. Similarly, the 2.5 mm channel shows the CHF values of 221 kW/m^2 and 218 kW/m^2 between the two tests at 1.38 MPa pressure and saturation temperature. However, the CHF value decreases to 188 kW/m^2 at 1.59 MPa pressure, representing approximately 14% decrease in CHF. Therefore, it is evident that both channels show a similar reduction in CHF with a similar increase in operating pressure. Finally, it can be seen from Figure 7b that by increasing the pressure from 1.38 MPa to 1.59 MPa and decreasing the inlet fluid temperature (saturation to 5 degree of subcooling), the 1.8 mm channel achieves a CHF of 474 kW/m^2, highlighted by the arrow. This is comparable to the boiling CHF of 468 kW/m^2 at 1.38 MPa pressure and saturation temperature. However, Figure 7a shows a significant shift of the boiling curve to the left, representing HTC increase for the sub-cooled boiling of 1.8 mm channel as compared to the saturation boiling.

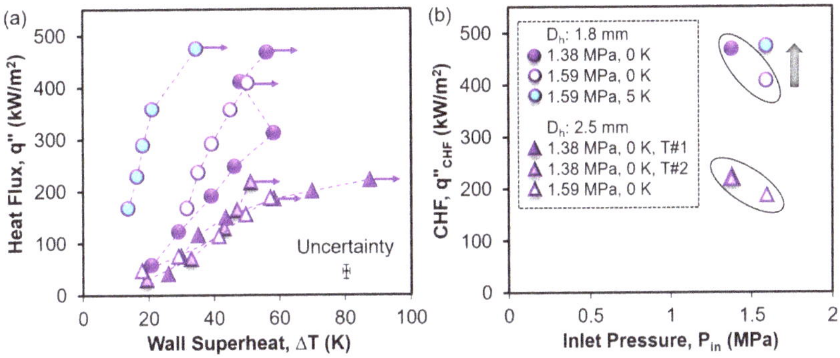

Figure 7. Effect of inlet pressure and degree of subcooling on LN_2 flow boiling CHF at 31 cm^3/s flow rate: (**a**) heat flux as a function of wall superheat, and (**b**) CHF as a function of inlet pressure at 0 K and 5 K degree of subcooling.

The effect of pressure on internal forced convective boiling is complex due to the integrated influence of liquid and vapor density of the fluid, surface tension, flow orientation and gravity. Wu et al. [46] have shown that CHF increases with increasing pressure for water flow boiling in vertical annular mini channels. They have attributed this enhancement to the reduction in the ratio of liquid and vapor density, corresponding velocity ratio, and droplet entrainment effect. Donniacuo et al. [47] have reached the same conclusion for 3 mm diameter horizontal channels and attributed the enhancement to the increase in liquid film thickness and influence of gravity. On the contrary, Charnay et al. [48,49] have presented that high pressure leads to increase in liquid film thickness and decrease in vapor quality resulting in a decrease in CHF. Furthermore, Zhang et al. [36] have shown that heat transfer deteriorates with increasing inlet pressure due to the reduction in surface tension resulting in the earlier appearance of an unsteady annular flow regime. Additionally, the standard thermophysical properties of LN_2 shows that the latent heat of vaporization also decreases with increasing operating pressure, and phase change heat transfer is a strong function of the latent heat of vaporization. Therefore, the decrease in CHF with increasing pressure in this work can be attributed to the reduction in surface tension and latent heat of vaporization of LN_2. However, the increase in CHF with increasing degree of subcooling, as shown in Figure 7b, is well-established in the literature. Celata et al. [50], Vandervort et al. [51], and Inasaka and Nariai [52] have observed an increase in CHF with increasing the degree of subcooling that is consistent with the result presented in this work. Moreover, Bao et al. [53] and Qi et al. [44] have shown that with the increase in pressure, bubble size becomes smaller and bubble departure frequency increases leading to an increase in heat transfer coefficient. Therefore, the shift in boiling curve for the 1.8 mm channel at 1.38 MPa pressure and 5-degree subcooling can be attributed to the bubble size reduction and bubble departure frequency enhancement.

3.4. *Effect of Channel Size*

The effect of channel size on LN2 flow boiling heat transfer is presented in Figure 8. Figure 8a presents the boiling curves of CHF for all three channels (1.8 mm, 2.3 mm, and 2.5 mm hydraulic diameter channels) at 57 cm^3/s and 47 cm^3/s flow rates, 1.38 MPa pressure and corresponding saturation temperature. Figure 8b presents the heat transfer coefficient (HTC) as a function of heat flux for the results from Figure 8a. The circles represent the 1.8 mm channel, squares represent the 2.3 mm channel and triangles represent the 2.5 mm channel. Red color represents the boiling results for 57 cm^3/s flow rate, and blue color represents the boiling results for 47 cm^3/s flow rate. Dark fill color represents the first test and light fill color represents the second test for the same channel. It is evident from the results that the boiling curves shift to the left with decreasing channel size. Figure 8b shows the HTC enhancement with decreasing channel size (triangle to square to circle) as well as increasing mass flow rate (blue to red markers).

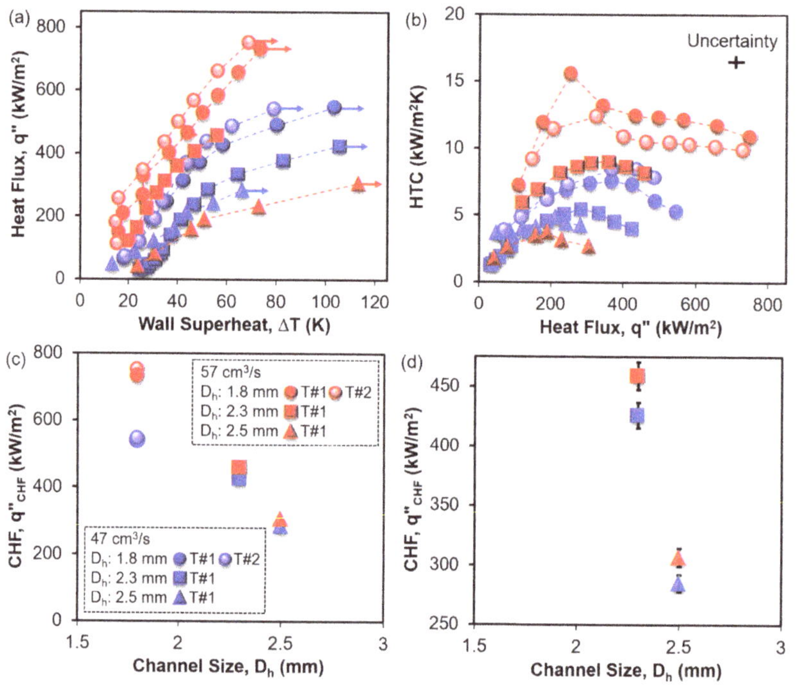

Figure 8. Effect of channel size on LN_2 flow boiling CHF at 1.38 MPa pressure and saturation temperature: (**a**) heat flux as a function of wall superheat, (**b**) HTC as a function of heat flux, (**c**) CHF as a function of channel size for all three channels, and (**d**) CHF as a function of channel size for 2.3 mm and 2.5 mm channels including error bars.

Since the flow boiling heat transfer varies significantly from small channels to microchannels due to the influence of surface tension, time-dependent flow patterns, and bubble dynamics [31,54], it is therefore essential to first identify the classification of cooling channels utilized in this work. According to Kandlikar and Grande [54], channels are classified into seven different types based on the hydraulic diameter: conventional channels (D_h > 3 mm), minichannels (200 µm < D_h ≤ 3 mm), microchannels (10 µm < D_h ≤ 200 µm), transitional microchannels (1 µm < D_h ≤ 10 µm), transitional nanochannels (0.1 µm < D_h ≤ 1 µm), and molecular nanochannels (D_h ≤ 0.1 µm). In this work, the smallest channel hydraulic diameter is 1.8 mm, and the largest channel hydraulic diameter is 2.5 mm. Therefore, the channels used in this work are considered minichannels.

For minichannels, two mechanisms dominate the flow boiling heat transfer: nucleate boiling and forced convective evaporation [55]. The heat transfer coefficient is affected by the heat flux where nucleate boiling is dominant. Tran et al. [56] have shown that HTC increases with increasing heat flux but is independent of mass flux and vapor quality for 2.4 mm square and 2.46 mm circular channels using R12 as the working fluid. This is typical where nucleate boiling is dominant, and HTC is affected by the heat flux [57,58]. However, HTC is affected by the mass flux and vapor quality where the forced convective evaporation is dominant [55]. The thermodynamic exit vapor quality at CHF, calculated from $Q/(\rho_l \dot{\forall} h_{lv})$, where h_{lv} is the latent heat of vaporization, varies approximately from 0.024 to 0.069. Therefore, the influence of vapor exit quality on HTC can be considered insignificant for this work. Indeed, it can be seen from Figure 8a,b that the high aspect ratio

2.5 mm channel does not show a major difference in HTC with increasing mass flux. This is inverse from the observation reported by Wang et al. [34] that can be attributed to the low aspect ratio (AR~2.27) channel utilized in this work as compared to their high aspect ratio channels (AR~10 and 20). However, it is evident from Figure 8b that increasing mass flux (blue to red) significantly increases the HTC for the 1.8 mm channel (circle) and 2.3 mm channel (square). Therefore, it can be implied that the high aspect ratio 2.5 mm channel (1.8 mm × 4.1 mm) exhibits nucleate boiling dominating HTC trend, and that both the 1.8 mm and 2.3 mm channels exhibit the forced convective evaporation dominating HTC enhancement. Furthermore, the boiling results in Figure 8b show that average heat transfer coefficient initially increases and then decreases with increasing heat flux. Wang et al. [34] have attributed this heat transfer deterioration phenomenon to partial or intermittent dry-out. Additionally, for a given mass flux, HTC has been observed to increase with decreasing channel diameter. This increase can be attributed to the decrease in liquid film thickness on the channel walls that increases the contribution of liquid film evaporation for flow boiling heat transfer.

The effect of channel size on CHF is presented in Figure 8c. Due to the high range of CHF values, specifically for the 1.8 mm channel, CHF as a function of channel size for 2.3 mm and 2.5 mm channels is separately presented in Figure 8d with error bars showing the variation in CHF with mass flux. It is clear from Figure 8c,d that CHF increases (307 kW/m^2 to 752 kW/m^2) with decreasing channel size (2.5 mm to 1.8 mm, respectively) and increasing mass flux (548 kw/m^2 at 47 cm^3/s to 752 kW/m^2 at 57 cm^3/s for 1.8 mm channel) in a non-linear fashion. The increasing trend of CHF with decreasing channel diameter is consistent with other researchers, such as Ong and Thome [43], Vandervort et al. [51], and Celata et al. [59]. However, a variation in observations also exists in the literature. Qi et al. [44] have performed LN_2 flow boiling in 0.531 mm, 0.834 mm, 1.042 mm, and 1.931 mm channels and shown that CHF increases with increasing tube diameter. They have attributed this CHF enhancement to the appearance of nucleate boiling. Vandervort et al. [51] have performed sub-cooled flow boiling of water in 0.3 to 2.7 mm tubes up to 40,000 kg/m^{2}s mass flux and observed that CHF increases up to the length-to-diameter ratio of 10. Moreover, Celata et al. [59] have identified a threshold ratio up to 30. The length-to-diameter ratios are 20, 22, and 27 in this work, showing a consistent CHF enhancement above the length-to-diameter ratio of 10 and below 30. Ong and Thome [43] have also shown that CHF values for R134a reach peak value up to a threshold diameter lying between the Confinement number (Co) of 0.78–0.99 for 0.51 mm to 3.04 mm channels. The confinement number (Co) varies from 0.25 to 0.37 in this work, calculated by $\sqrt{\sigma/g(\rho_l-\rho_v)D_h^2}$, where σ is the surface tension, and g is the gravitational acceleration. Since Co is inversely proportional to channel size, a decrease in channel size below 1.8 mm appears to increase CHF up to the threshold diameter (up to Co~0.99) of approximately 0.73 mm at 1.38 MPa pressure and saturation temperature. This increase in CHF with decreasing channel size can be attributed to the increase in the contribution of forced convective evaporation associated with the high mass flow rates and explained through hydrodynamic boundary conditions [44].

3.5. CHF Correlation

Two types of CHF are encountered in cooling channel flow boiling susceptible to the degree of subcooling, mass flux and wall heat flux. Dry-out type CHF is associated with the complete evaporation of the liquid film from the heated wall at low mass flux and low degree of subcooling, and high critical length-to-diameter ratios. On the contrary, departure from nucleate boiling (DNB) type CHF is encountered when a vapor blanket is formed on the heated wall despite the availability of abundant liquid core at high mass flow rates, high degrees of subcooling, and high heat flux values. A significant number of correlations have been proposed by researchers, such as Katto and Kurata, 1980 [60], Katto and Ohno, 1984 [61], Katto and Yokoya, 1987 [62], Shah, 1987 [63], Mudawar and Maddox, 1990 [64], Hall and Mudwar, 2000 [65], Darr, 2019 [66], and Ganesan, 2021 [1], aiming to enable accurate prediction of flow boiling CHF. However, the correlations are strongly dependent on the CHF type as the two-phase flow physics and CHF fundamental mechanisms are significantly different. Therefore, it is imperative to segregate the CHF type for utilizing the appropriate correlation and comparing the test results from this work.

According to Ganesan et al. [1], there exist threshold values of critical void fraction ($\alpha_{CHF} < 0.6$ for LN_2) and modified Boiling number ($Bo^* < 0.33$ for LN_2) that demarcate DNB-type CHF from dry-out type CHF. The critical void fraction has been calculated using Zivi's correlation [67] given by Equation (10).

$$\alpha_{CHF} = \left[1 + \left(\frac{1 - x_{e,CHF}}{x_{e,CHF}}\right)\left(\frac{\rho_l}{\rho_v}\right)^{2/3}\right]^{-1} \tag{10}$$

Here, $x_{e,CHF}$ is the thermodynamic equilibrium quality at the CHF location [1] that has been calculated using Equation (11). Boiling number is defined as $Bo = q''_{CHF}/(Gh_{lv})$, where L_{CHF} is the axial location where CHF occurs.

$$x_{e,CHF} = x_{e,in} + 4Bo\frac{L_{CHF}}{D_h} \tag{11}$$

In this work, L_{CHF} is considered as half of the heated length, i.e., 2.5 cm. The occurrence of CHF near the middle of the asymmetrically heated cooling channel can be observed from the temperature distribution underneath the channel. Figure 9 shows an example of the temperature profile in the bottom three thermocouples (T_7–T_9) for the 2.3 mm channel at 38 cm^3/s flow rate, 1.38 MPa inlet pressure and saturation temperature. The Y-axis represents the average heat flux and X-axis represents the surface temperature. It is evident from Figure 9 that at CHF~387 kW/m^2 the middle thermocouple (T_8) underneath the channel wall shows a significant increase in temperature as compared to the thermocouples near the inlet and exit of the cooling channel. Therefore, half of the heated length has been considered as the CHF length from the channel inlet.

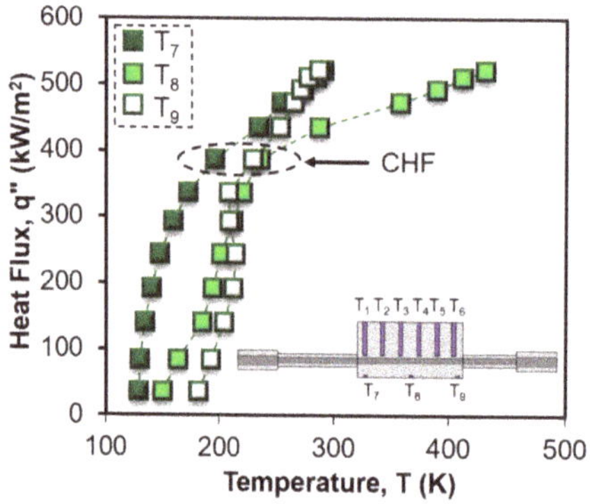

Figure 9. Heat flux as a function of channel base temperature for 2.3 mm channel at 38 cm^3/s flow rate, 1.38 MPa inlet pressure and saturation temperature.

The modified Boiling number (Bo^*) has been calculated from Equation (12). Here, $x_{e,in}$ is the thermodynamic equilibrium quality at the channel inlet defined by $x_{e,in} = (h_{in} - h_{l,in})/h_{lv,in}$, where h_{in} is the enthalpy of fluid at inlet temperature, $h_{l,in}$ is the enthalpy of saturated liquid at inlet pressure, and $h_{lv,in}$ is the latent heat of vaporization at the inlet pressure.

$$Bo^* = \frac{4Bo\frac{L_{CHF}}{D_h}}{1 - x_{e,in}} \tag{12}$$

The modified Boiling number as a function of critical void fraction for all three channels, five different flow rates, two different pressures, and two degrees of subcooling is presented in Figure 10. As can be seen, the maximum modified Boiling number encountered in this work is ~0.028 which is much smaller than the threshold modified Boiling number ($Bo^* \ll 0.33$) for LN_2 DNB-type CHF. Additionally, the maximum critical void fraction encountered in this work is ~0.12, which is also much smaller than the threshold critical void fraction ($\alpha_{CHF} \ll 0.6$) for LN_2 DNB-type CHF. Therefore, it can be postulated that the critical heat flux values associated with this work are DNB-type CHF which is consistent with the literature at high mass flow rates and high heat flux conditions.

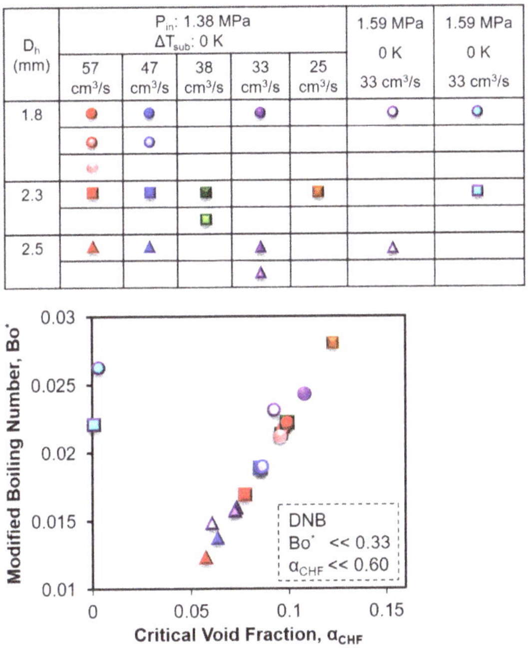

Figure 10. Modified boiling number as a function of critical void fraction.

Although a significant number of CHF correlations are available for flow boiling in cooling channels, a close inspection will reveal that there exists a lack of errorfree, reliable, and simple universal CHF correlation for cryogenic fluids. Recently, Ganesan et al. [1] have proposed a universal CHF correlation for the flow boiling of cryogenic liquids, more specifically for liquid hydrogen, liquid helium, liquid nitrogen, and liquid methane by amassing 2312 CHF data points. The proposed experimental correlations have been shown to work for both horizontal and vertical orientation of flow, and for both DNB and dry-out type CHF. Therefore, this work compares the experimental CHF results with the correlation presented in Ganesan et al. [1]. Equation (13) has been used to predict the CHF values ($q''_{CHF,P}$) for all three cooling channels, five different LN_2 flow rates, two different inlet pressures and two different degrees of subcooling. Here, Weber number ($We_{l,D}$) is based on channel hydraulic diameter and has been calculated from $We_{l,D} = G^2 D_h/(\rho_l \sigma)$, where σ is the liquid–vapor surface tension at the saturation condition. Since this work presents the CHF results for horizontal flow and DNB-type CHF, therefore, the constant values are considered as $c_1 = 0.32$, $c_2 = -0.24$, $c_3 = -0.60$, $c_4 = 0.48$, and $c_5 = 0.69$.

$$q''_{CHF,P} = 0.25\, c_1 We_{l,D}^{c_2} \left(\frac{\rho_v}{\rho_l}\right)^{c_3} (1 - x_{e,in})^{c_4+1} \left(\frac{L_{CHF}}{D_h}\right)^{c_5-1} G\, h_{lv} \quad (13)$$

The LN_2 CHF results from this work are compared with the literature correlation in Figure 11 where all the legends are held the same as in Figure 10. Y-axis represents the predicted CHF and X-axis represents the measured CHF results. Here, the MAE and RMS are calculated using Equations (14) and (15), respectively. $q''_{CHF,P-G}$ is the predicted CHF using Ganesan et al.'s [1] universal correlation, $q''_{CHF,P}$ is the predicted CHF in this work and $q''_{CHF,m}$ is the experimentally measured CHF in this work.

$$MAE = \frac{1}{N}\sum \frac{|q''_{CHF,P} - q''_{CHF,m}|}{q''_{CHF,m}} \times 100\% \quad (14)$$

$$RMS = \left[\frac{1}{N}\sum\left(\frac{q''_{CHF,P} - q''_{CHF,m}}{q''_{CHF,m}}\right)^2\right]^{1/2} \times 100\% \quad (15)$$

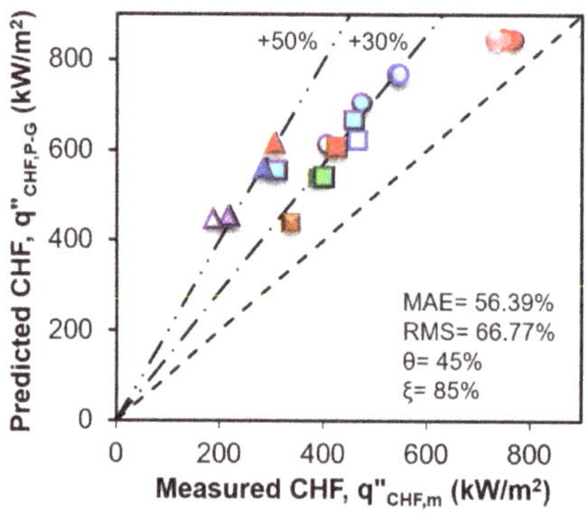

Figure 11. Comparison of all 20 DNB data for LN_2 in additively manufactured channels against Ganesan et al.'s [1] universal correlation for uniformly heated cooling channels.

It is evident from the comparison in Figure 11 that the model overpredicts the DNB results with 56.39% mean absolute error (MAE) and 66.77% root mean squared (RMS) error. Additionally, the percentage of CHF result prediction within ±30% (θ) is 45% and within ±50% (ξ) is 85%. This overprediction can be attributed to the asymmetrically heated non-circular cooling channels tested in this work where Ganesan et al.'s [1] universal correlation has been developed for the uniformly heated tubes. Moreover, the development of analytical models and computational fluid dynamics (CFD) models require intense parametric tests using flow measurement and visualization for validation which is limited for cryogenic experiments. As such, there exists a pressing need to develop an experimental correlation for asymmetrically heated non-circular cooling channels. To develop the correlation for LN_2 flow boiling in asymmetrically heated non-circular channels in this work, the generalized reduced gradient (GRG) solver in Microsoft Excel has been used to identify the constant factors for Equation (13). The calculated values are as follows: $c_1 = 0.0015$, $c_2 = -0.17$, $c_3 = -0.38$, $c_4 = 1.09$, and $c_5 = 1.43$. Utilizing these curve fitting parameters in Equation (13), the CHF values have been calculated for all the 20 test conditions and are compared with the experimental CHF results in Figure 12. As seen from the comparison, the new correlation predicts the LN_2 DNB results with 10.68% mean absolute error (MAE) and 12.67% root mean squared (RMS) error. Additionally, the percentage of CHF result prediction within ±30% (θ) is 100%.

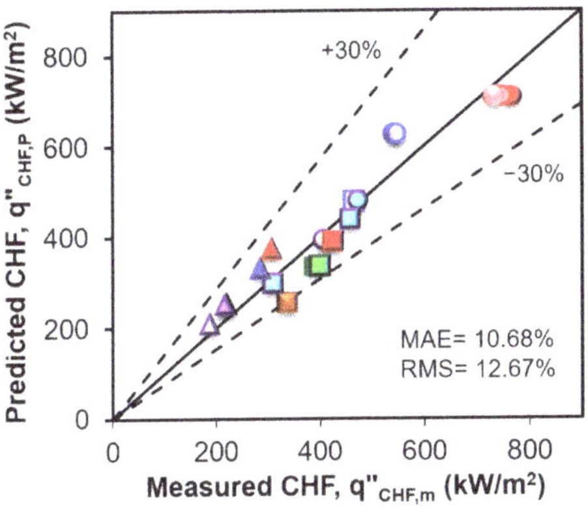

Figure 12. Comparison of experimental DNB results for LN_2 against new correlation for asymmetrically heated non-circular additively manufactured horizontal cooling channels.

However, the new correlation is developed based on the experimental database presented in this work due to the lack of literature on cryogenic flow boiling in asymmetrically heated non-circular cooling channels at high mass flux and pressure values. Moreover, the present work has focused only on additively manufactured channels and LN_2 as the working fluid. Therefore, it is essential to perform a parametric study on the effect of influential parameters including a wide range of hydraulic diameters (micro- to macro-channels), mass flow rates, pressures and corresponding saturation temperatures, degrees of subcooling, vapor quality, channel materials [68], micro/nano-scale structures [69], conventional channels, and various cryogenic fluids, such as LCH_4, LH_2 and LHe, to develop a universal correlation for regenerative cooling channels.

4. Conclusions

Additively manufactured asymmetrically heated non-circular channels have been shown to suggest a new DNB correlation for LN_2 flow boiling. Three different GR-Cop42 channels have been tested at five different flow rates, two inlet pressures and two degrees of subcooling. The repeatability of flow boiling performance has been conducted using the 1.8 mm × 1.8 mm cooling channel showing a maximum CHF variation of ±1.1% at 57 cm^3/s and ±0.5% at 47 cm^3/s LN_2 flow tests which are significantly lower than the measurement uncertainty of ±2.52%. The effect of LN_2 flow rate on flow boiling HTC and CHF has been characterized using a 2.3 mm × 2.3 mm channel (2.3 mm hydraulic diameter). HTC enhancement has been observed with increasing mass flux in general for the tested range of 3824 kg/m^2s to 8605 kg/m^2s. An increase in HTC has been observed with increasing the mass flow rate from 25 cm^3/s to 38 cm^3/s due to the contribution of enhanced nucleate boiling. However, with the increase in LN_2 flow rate to 47 cm^3/s, HTC has been observed to decrease which is attributed to the suppression of nucleate boiling. A further increase in flow rate to 57 cm^3/s, and the significant contribution from forced convective evaporation achieved a significant increase in HTC. However, CHF has been observed to increase linearly as the mass flux increases due to the reducing influence of flow instability and gravitation effect and increasing forced convective evaporation. CHF values of 337 kW/m^2 and 459 kW/m^2 have been achieved at 25 cm^3/s and 57 cm^3/s LN_2 flow rates for the 2.3 mm channel, respectively.

The effect of system pressure on CHF and HTC has been evaluated for the 1.8 mm × 1.8 mm (1.8 mm hydraulic diameter) and 1.8 mm × 4.1 mm (2.5 mm hydraulic diameter) channels at 1.38 MPa and 1.59 MPa pressures, respectively. Increasing system pressure has shown to negatively affect CHF due to the decrease in surface tension and latent heat of vaporization of LN_2. CHF values have been observed to decrease by 13% for the 1.8 mm channel and 14% for the 2.5 mm channel with the increase in system pressure. However,

HTC has been seen to increase with increasing pressure, which is attributed to the reducing bubble size and increasing bubble departure frequency. Moreover, among the three different channels, enhancements in both HTC and CHF have been achieved with decreasing channel size. The experimental results show that CHF increases from 307 kW/m^2 to 752 kW/m^2 with decreasing channel size from 2.5 mm to 1.8 mm, respectively, in a non-linear fashion due to the significant enhancement in the contribution of forced convective evaporation. CHF enhancement from 548 kw/m^2 to 752 kW/m^2 has also been observed for the 1.8 mm channel with the increasing LN_2 flow rate from 47 cm^3/s to 57 cm^3/s.

Through the comparison of modified boiling number and critical void fraction, this work shows the DNB-type CHF for minichannels at high mass flux values and high inlet pressures during LN_2 flow boiling. Finally, it has been found that the literature universal DNB model for uniformly heated tubes overpredicts the DNB heat flux for asymmetrically heated non-circular cooling channels with 56.39% mean absolute error (MAE) and 66.77% root mean squared (RMS) error. Therefore, a new modified DNB correlation for LN_2 has been developed in this work with 10.68% MAE, 12.67% RMS error, and 100% prediction capability within $\pm$30% uncertainty. This work aims to provide a fundamental cryogenic flow boiling heat transfer performance of additively manufactured asymmetrically heated non-circular minichannels at high mass flux values and pressures and bridge the gaps in the literature for high-performance regeneratively cooled rocket engines.

Author Contributions: Conceptualization, M.M.R. and A.C.; methodology, M.M.R. and D.O.; software, D.O. and A.A.; validation, D.O. and A.A.; formal analysis, D.O., M.M.R. and M.A.; investigation, D.O. and A.A.; resources, A.C. and M.M.R.; data curation, D.O., A.A. and M.M.R.; writing—original draft preparation, D.O. and M.M.R.; writing—review and editing, M.A., A.A. and A.C.; visualization, D.O.; supervision, M.M.R.; project administration, M.M.R. and A.C.; funding acquisition, M.M.R. and A.C. All authors have read and agreed to the published version of the manuscript.

Funding: This research was funded by the National Aeronautics and Space Administration (NASA) under Cooperative Agreement number 80NSSC19P1100 Mod. #P00005 and under NASA MIRO Cooperative Agreement number 80NSSC20M0237.

Data Availability Statement: Source data and other raw data are available from the corresponding author upon request.

Conflicts of Interest: The authors declare no conflict of interest.

References

1. Ganesan, V.; Patel, R.; Hartwig, J.; Mudawar, I. Universal Critical Heat Flux (CHF) Correlations for Cryogenic Flow Boiling in Uniformly Heated Tubes. *Int. J. Heat Mass Transf.* **2021**, *166*, 120678. [CrossRef]
2. Gubaidullin, D.A.; Snigerev, B.A. Numerical Simulation of Heat Transfer During Boiling Flow of Cryogenic Fluid in Vertical Tube. *Lobachevskii J. Math.* **2020**, *41*, 1210–1215. [CrossRef]
3. Kwon, D.W.; Sedwick, R.J. Cryogenic heat pipe for cooling high temperature superconductors. *Cryogenics* **2009**, *49*, 514–523. [CrossRef]
4. Marcushamer, M.; King, D.L.; Ruano, N.S. Cryosurgery in the management of mucoceles in children. *Pediatr. Dent.* **1997**, *19*, 292–293.
5. Yuan, K. Cryogenic Boiling and Two-Phase Chilldown Process Under Terrestrial and Microgravity Conditions. Ph.D. Thesis, University of Florida, Gainesville, FL, USA, 2006.
6. Zhou, C.; Yu, N.; Wang, S.; Han, S.; Gong, H.; Cai, G.; Wang, J. The Influence of Thrust Chamber Structure Parameters on Regenerative Cooling Effect with Hydrogen Peroxide as Coolant in Liquid Rocket Engines. *Aerospace* **2023**, *10*, 65. [CrossRef]
7. Ferraiuolo, M.; Leo, M.; Citarella, R. On the Adoption of Global/Local Approaches for the Thermomechanical Analysis and Design of Liquid Rocket Engines. *Appl. Sci.* **2020**, *10*, 7664. [CrossRef]
8. Cho, W.K.; Seol, W.S.; Son, M.; Seo, M.K.; Koo, J. Development of preliminary design program for combustor of regenerative cooled liquid rocket engine. *J. Therm. Sci.* **2011**, *20*, 467–473. [CrossRef]
9. Van Huff, N.E.; Fairchild, D.A. *Liquid Rocket Engine Fluid-Cooled Combustion Chambers*; NASA Lewis Research Center: Cleveland, OH, USA, 1972.
10. Sutton, G.P.; Biblarz, O. *Rocket Propulsion Elements*, 9th ed.; John Wiley & Sons Inc.: Hoboken, NJ, USA, 2017.
11. Boysan, M.E.; Ulas, A.; Toker, K.A.; Seckin, B. Comparison of Different Aspect Ratio Cooling Channel Designs for a Liquid Propellant Rocket Engine. In Proceedings of the 2007 3rd International Conference on Recent Advances in Space Technologies, Istanbul, Turkey, 14–16 June 2007.

12. Wang, T.-S.; Luong, V. Hot-gas-side and coolant-side heat transfer in liquid rocket engine combustors. *J. Thermophys. Heat Transf.* **1994**, *8*, 524–530. [CrossRef]
13. Celata, G.P.; Mariani, A. *Critical Heat Flux, Post Dry-Out and Their Augmentation*; ENEA-RT-ERG-98-10; ENEA: Rome, Italy, 1999.
14. Wadel, M. Comparison of high aspect ratio cooling channel designs for a rocket combustion chamber. In Proceedings of the 33rd Joint Propulsion Conference and Exhibit, Joint Propulsion Conferences, Seattle, WA, USA, 6–9 July 1997; American Institute of Aeronautics and Astronautics: Reston, VA, USA, 1998.
15. Van Noord, J.L.; Center, N.G.R.; Meeting, J.P. *A Heat Transfer Investigation of Liquid and Two-Phase Methane*; National Aeronautics and Space Administration, Glenn Research Center: Cleveland, OH, USA, 2010.
16. Huzel, D.K.; Huang, D.H. *Design of Liquid Propellant Rocket Engines*, 2nd ed.; Scientific and Technical Information Office, National Aeronautics and Space Administration: Washington, DC, USA, 1967.
17. Pizzarelli, M. Modeling of Cooling Channel Flow in Liquid-Propellant Rocket Engines. Ph.D. Thesis, University of Rome, Rome, Italy, 2007.
18. Denies, L. Regenerative Cooling Analysis of Oxygen/Methane Rocket Engines. Master's Thesis, TU Delft, Mekelweg, The Netherlands, 2015.
19. Waxenegger-Wilfing, G.; Dresia, K.; Deeken, J.C.; Oschwald, M. Heat Transfer Prediction for Methane in Regenerative Cooling Channels with Neural Networks. *J. Thermophys. Heat Transf.* **2020**, *34*, 347–357. [CrossRef]
20. Song, J.; Liang, T.; Li, Q.; Cheng, P.; Zhang, D.; Cui, P.; Sun, J. Study on the heat transfer characteristics of regenerative cooling for LOX/LCH4 variable thrust rocket engine. *Case Stud. Therm. Eng.* **2021**, *28*, 101664. [CrossRef]
21. Santhosh, T.; Kuzhiveli, B.T. Heat transfer aspects and analysis of regenerative cooling in semi-cryogenic thrust chamber with fixed and variable aspect ratio coolant channels. In Proceedings of the IOP Conference Series: Materials Science and Engineering, Kazimierz Dolny, Poland, 21–23 November 2019.
22. Wadel, M.; Meyer, M. Validation of high aspect ratio cooling in a 89 kN (20,000 lbf) thrust combustion chamber. In Proceedings of the 32nd Joint Propulsion Conference and Exhibit, Lake Buena Vista, FL, USA, 1–3 July 1996; American Institute of Aeronautics and Astronautics: Reston, VA, USA, 1996.
23. Carlile, J.; Quentmeyer, R. An experimental investigation of high-aspect-ratio cooling passages. In Proceedings of the 28th Joint Propulsion Conference and Exhibit, Nashville, TN, USA, 6–8 July 1992; American Institute of Aeronautics and Astronautics: Reston, VA, USA, 2012.
24. Ellis, D. *GRCop-84: A High-Temperature Copper Alloy for High-Heat-Flux Applications*; NASA Glenn Research Center: Cleveland, OH, USA, 2005.
25. Gradl, P.R.; Protz, C.S.; Cooper, K.; Ellis, D.; Evans, L.J.; Garcia, C. GRCop-42 Development and Hot-fire Testing Using Additive Manufacturing Powder Bed Fusion for Channel-cooled Combustion Chambers. In Proceedings of the AIAA Propulsion and Energy 2019 Forum, Indianapolis, IN, USA, 19–22 August 2019; American Institute of Aeronautics and Astronautics: Reston, VA, USA, 2019.
26. Ortega, D.; Amador, A.; Silva, A.; Choudhuri, A.R.; Rahman, M.M. Effect of Pressure on Liquid Nitrogen Flow Boiling in Additively Manufactured Rocket Engine Cooling Channels. In Proceedings of the AIAA SciTech Forum, National Harbor, MD, USA, 23–27 January 2023; American Institute of Aeronautics and Astronautics: Reston, VA, USA, 2023.
27. Ortega, D.J.; Amador, A.; Choudhuri, A.R.; Rahman, M.M. Experimental Characterization of Critical Heat Flux and Minimum Film Boiling Heat Flux for Additively Manufactured Cooling Channels for Liquid Nitrogen Saturated Flow Boiling. In Proceedings of the ASME International Mechanical Engineering Congress, and Exposition, Columbus, OH, USA, 30 October–3 November 2022.
28. Hernandez, L.; Palacios, R.; Ortega, D.; Adams, J.; Bugarin, L.I.; Rahman, M.D.M.; Choudhuri, A.R. The Effect of Surface Roughness on LCH4 Boiling Heat Transfer Performance of Conventionally and Additively Manufactured Rocket Engine Regenerative Cooling Channels. In Proceedings of the AIAA Propulsion and Energy 2019 Forum, Indianapolis, IN, USA, 19–22 August 2019; American Institute of Aeronautics and Astronautics: Reston, VA, USA, 2019.
29. Moffat, R.J. Describing the uncertainties in experimental results. *Exp. Therm. Fluid Sci.* **1988**, *1*, 3–17. [CrossRef]
30. Rahman, M.M.; Ölçeroğlu, E.; McCarthy, M. Scalable Nanomanufacturing of Virus-templated Coatings for Enhanced Boiling. *Adv. Mater. Interfaces* **2014**, *1*, 1300107. [CrossRef]
31. Liu, X.; Chen, X.; Zhang, Q.; Wang, S.; Hou, Y.; Chen, L. Investigation on CHF of saturated liquid nitrogen flow boiling in a horizontal small channel. *Appl. Therm. Eng.* **2017**, *125*, 1025–1036. [CrossRef]
32. Ribatski, G.; Wojtan, L.; Thome, J.R. An analysis of experimental data and prediction methods for two-phase frictional pressure drop and flow boiling heat transfer in micro-scale channels. *Exp. Therm. Fluid Sci.* **2006**, *31*, 1–19. [CrossRef]
33. Agostini, B.; Thome, J.R.; Fabbri, M.; Michel, B.; Calmi, D.; Kloter, U. High heat flux flow boiling in silicon multi-microchannels—Part I: Heat transfer characteristics of refrigerant R236fa. *Int. J. Heat Mass Transf.* **2008**, *51*, 5400–5414. [CrossRef]
34. Wang, Y.; Sefiane, K. Effects of heat flux, vapour quality, channel hydraulic diameter on flow boiling heat transfer in variable aspect ratio micro-channels using transparent heating. *Int. J. Heat Mass Transf.* **2012**, *55*, 2235–2243. [CrossRef]
35. Lim, T.W.; Kim, J.H. An experimental investigation of heat transfer in forced convective boiling of R134a, R123 and R134a/R123 in a horizontal tube. *KSME Int. J.* **2004**, *18*, 513–525. [CrossRef]
36. Zhang, B.C.; Li, Q.; Wang, Y.; Zhang, J.Q.; Song, J.; Zhuang, F.C. Experimental investigation of nitrogen flow boiling heat transfer in a single mini channel. *J. Zhejiang Univ. Sci. A* **2020**, *21*, 147–166. [CrossRef]

aerospace

MDPI

Article

Performance Analysis and Design of Direct Ammonia Fuel Tubular Solid Oxide Fuel Cell for Shipborne Unmanned Aerial Vehicles

Zhe Wang [1,2], Fan Zhao [1], Yue Ma [1], Rui Xia [1] and Fenghui Han [1,2,*]

[1] Marine Engineering College, Dalian Maritime University, Dalian 116026, China; wang.zhe@dlmu.edu.cn (Z.W.)

[2] National Center for International Research of Subsea Engineering Technology and Equipment, Dalian Maritime University, Dalian 116026, China

* Correspondence: fh.han@dlmu.edu.cn

Abstract: Ammonia is being considered as a promising alternative to hydrogen fuel in solid oxide fuel cells (SOFCs) due to its stability and ease of storage and transportation. This study investigates the feasibility of using ammonia fuel in a tubular SOFC for shipborne unmanned aerial vehicles (UAVs). The paper develops a 3D model of a tubular-anode-supported SOFC single cell and conducts numerical simulations to analyze the impact of different operating conditions on SOFC performance. The study optimizes the SOFC's performance by adjusting its working parameters and overall structure, revealing that increasing temperature and porosity enhance performance, but excessively high values can cause deterioration and instability in the cell. The study also finds that the cathode-supported (CS)-SOFC outperforms the anode-supported (AS)-SOFC, mainly due to its thicker cathode layer, providing better sealing and oxygen supply, resulting in a more uniform current density distribution. The paper provides valuable insights into the potential use of ammonia fuel for shipborne UAVs and offers a foundation for future research and development in the field of SOFCs. The results indicate that increasing the temperature and porosity of the SOFC can enhance battery performance, but excessive values can cause deterioration and instability in the cell. The study also highlights the impact of different operating conditions on SOFC performance, with a significant performance improvement observed in the range of 0.6–0.8 V. Additionally, the CS-SOFC outperforms the AS-SOFC due to its thicker cathode layer, but both have significant potential for development.

Keywords: solid oxide fuel cell; tubular structure; ammonia fuel; multiphysics field simulation; shipborne UAV applications

check for updates

Citation: Wang, Z.; Zhao, F.; Ma, Y.; Xia, R.; Han, F. Performance Analysis and Design of Direct Ammonia Fuel Tubular Solid Oxide Fuel Cell for Shipborne Unmanned Aerial Vehicles. *Aerospace* **2023**, *10*, 397. https://doi.org/10.3390/aerospace10050397

Academic Editors: Jian Liu, Qingfei Hu, Yiheng Tong and Chaoyang Liu

Received: 2 March 2023
Revised: 17 April 2023
Accepted: 18 April 2023
Published: 25 April 2023

1. Introduction

In recent years, the global focus on sustainable development has increased, resulting in a growing demand for carbon-neutral technologies [1]. The development of efficient and clean power systems is not only crucial for shipborne unmanned aerial vehicles (UAVs), but also a key component of transitioning towards a low-carbon economy [2]. Moreover, the availability and sustainability of fuel sources are critical factors in determining the feasibility and long-term viability of a power system for shipborne UAVs [3]. As fossil fuels are gradually depleting and concerns over their environmental impact are increasing, exploring and utilizing alternative fuels has become an urgent and necessary task. Additionally, shipborne UAVs have a broad range of applications, such as maritime surveillance, search and rescue, and ocean monitoring, where long endurance, high reliability, and flexibility are essential requirements [4]. However, traditional power supply systems based on internal combustion engines (ICEs) or batteries cannot meet these requirements. ICEs are noisy, emit pollutants, and require frequent maintenance, while batteries have limited endurance and require frequent recharging [5]. Therefore, the development of a power system that can provide high efficiency, long endurance, and environmental friendliness is critical for the successful deployment and operation of shipborne UAVs in these applications. In this regard, solid oxide fuel cells (SOFCs) have emerged as a promising technology for powering shipborne UAVs [6]. SOFCs can operate on a variety of fuels, including ammonia, which is a clean and renewable fuel source. Ammonia is produced from hydrogen and nitrogen, both of which are readily available and can be produced sustainably through renewable energy sources [7]. Furthermore, the use of ammonia as a fuel for SOFCs does not emit greenhouse gases or other pollutants, making it a highly attractive alternative to traditional fuels [8–10]. In summary, the development of an efficient and clean power system for shipborne UAVs is crucial not only for meeting the demands of various applications but also for achieving sustainability and environmental protection goals. The use of SOFCs fueled by ammonia is a promising solution that can address these challenges and offer long-term benefits.

The demand for shipborne UAVs with long endurance, high reliability, and environmental protection has increased, leading to growing interest in developing efficient and clean power systems that contribute to the goal of carbon neutrality [11,12]. SOFCs are a promising technology for powering shipborne UAVs due to their fuel flexibility, high energy efficiency, and low emissions compared to traditional ICEs or battery power supply systems [13,14]. Ammonia is a promising alternative fuel for SOFCs because of its high energy density, low toxicity, and low cost, and it can be produced sustainably from renewable sources [15,16]. However, further research and development is required to optimize the performance design of ammonia-fueled SOFCs for shipborne UAVs, particularly in terms of operating conditions, durability, stability, and compatibility. Such investigations can contribute to the development of efficient and clean power systems for shipborne UAVs and pave the way for the wider application of ammonia fuel cells in transportation sectors. In the era of high energy consumption, SOFCs are being developed as a new and efficient power generation device to meet high standards for energy conservation and environmental protection. Ammonia, as a stable hydrogen carrier, has the potential to become a highly efficient zero-carbon fuel for SOFCs.

Hydrogen fuel cells are a promising technology for high-efficiency and low-emission electricity generation [17]. However, current regulations and standards for hydrogen production, storage, and transport need improvement, posing a significant challenge for implementing a safe and efficient hydrogen infrastructure [18]. To overcome this, on-site hydrogen generation can be used as an alternative solution to provide fuel for ship-based fuel cells, which could greatly accelerate the commercialization of ship-based fuel cell technology [19,20]. Compared to proton exchange membrane fuel cells, SOFCs operate at much higher temperatures, enabling them to produce high-quality waste heat and achieve higher fuel utilization efficiency [21]. Furthermore, all-solid-state SOFCs offer greater flexibility in installation, facilitating the modular design of ship-based power and propulsion systems and making maintenance and upkeep easier [22,23]. Additionally, tubular SOFCs possess excellent sealing properties that can protect the fuel and prevent leakage, while also meeting the requirements for moisture, dust, and salt spray resistance necessary for marine power machinery. Both atmospheric and pressurized SOFC power systems are well suited for modern large ships and their auxiliary power systems, underscoring their potential for widespread adoption in the shipping industry. The adoption of SOFCs in ship-based power generation and propulsion systems could lead to significant reductions in greenhouse gas emissions and improve overall energy efficiency, thereby contributing to a more sustainable future for the marine transportation industry.

Fuel cells that use on-site hydrogen production are the focus of development for fuel cells in marine UAVs [24,25]. SOFCs that generate hydrogen through direct ammonia cracking can achieve a reasonable combination of fuel supply systems and fuel cells with a simple structure. SOFCs are also suitable for use in ship power plants, making direct-ammonia-cracking SOFCs a potential development direction for marine fuel cells [26,27]. In SOFCs fueled by pure hydrogen, hydrogen and oxygen are directly converted into water, and heat is released. However, for a single SOFC with direct ammonia cracking, the reaction is more complicated. First, the ammonia gas enters the high-temperature fuel channel for a cracking reaction to generate nitrogen and hydrogen. Then, the hydrogen produced electrochemically reacts with the oxygen transported by the air channel to realize energy conversion [28,29]. The heat required for ammonia pyrolysis, as an endothermic reaction, can be provided by ambient temperature. Due to the complexity of the reaction and many undetermined details, most studies in the current literature focus on modeling SOFCs fueled by hydrogen [30]. The reaction in SOFCs is strongly temperature dependent, like most chemical reactions. The distribution of the temperature gradient in the fuel cell directly affects the reaction rate and, therefore, the performance of the fuel cell. Conversely, chemical reactions accompanied by exothermic and endothermic reactions also change the distribution of temperature gradients.

Barelli et al. [31] conducted an experimental test and designed a system for operating an SOFC-based power system using ammonia as a fuel. The study demonstrates the feasibility and potential of using ammonia as a fuel for SOFCs, which could have implications for the development of sustainable energy systems. Khodabandeh et al. [32] investigated the impact of using oil/multiwalled carbon nanotube nanofluids and different cooling channel geometries on the cooling performance of SOFCs. The study shows that using nanofluids can enhance the heat transfer performance of SOFC cooling systems, while the geometrical design of the cooling channels can also significantly affect the system's thermal behavior. Rathore et al. [33] provide a comprehensive review of the progress and prospects of direct ammonia solid oxide fuel cells, highlighting recent advancements in catalyst development, cell design, and system integration, and discussing the challenges and opportunities associated with the technology. Farhad et al. [34] conducted experiments on anode-supported tubular SOFCs and found that the ammonia cracking reaction can suppress temperature increases. They reported that 0.8 L of ammonia could generate 100 W of electricity for 10 h at 0.73 V and 800 °C, indicating that the chemical reactions occurring within the fuel cell have a significant impact on its thermal load. Tariq Nawas et al. [35] established a thermal shock model of a single direct methane steam reforming SOFC and concluded that the methane steam reforming reaction plays a dominant role in generating the thermal load of the cell, highlighting the non-negligible effect of chemical reactions on its thermal load. Buttler et al. [36] observed that every electrochemical reaction has a corresponding thermal neutral voltage. They found that when a fuel cell operates at the thermal neutral voltage, it does not generate or consume more heat. However, operating above or below the thermal neutral voltage consumes more electricity or cools the cell, resulting in temperature gradients inside the cell. Kishimoto et al. [37] developed a 2D mathematical model of ammonia-fueled SOFCs, coupling temperature, material distribution, material flow, and electrochemical properties. They concluded that direct ammonia reforming fuel cells have similar performance to ammonia pre-reforming fuel cells. Kee et al. [38] reported that fuel cell energy density largely depends on the gas composition, particularly the hydrogen content in the mixture. Tan et al. [39] established a three-dimensional model of plate SOFCs with internal ammonia cracking and found that the uneven distribution of current density was mainly caused by the uneven distribution of gas on the air electrode. Hajimolana et al. [40] developed a kinetic model of ammonia tubular SOFCs and found that the internal flow channel size of the fuel cell had the greatest impact on fuel cell efficiency, while the porosity of the cathode had a greater impact on fuel cell performance than the anode. Overall, previous studies have shown that various factors, including chemical reactions, fuel composition, electrode materials, geometric shape, and microstructure parameters, can affect the performance and thermal load of SOFCs.

This paper focuses on the use of ammonia as fuel in tubular SOFCs for shipborne UAVs. A three-dimensional model of a tubular anode-supported SOFC single cell was established in COMSOL Multiphysics [41] using steady-state analysis and multiphysics field coupling. Based on the theoretical principles of tubular SOFCs, numerical simulations and analyses of the working conditions of ammonia-fueled SOFCs were conducted. By adjusting the working parameters or overall structure using a controlled variable method, this paper analyzes and optimizes some factors that affect the performance of the cell. Specifically, it describes the process of drawing the geometric model of a tubular single cell, and setting the boundary conditions, physical field modules, and parameter settings. The simulation results, including gas concentration distribution, current density, temperature field, and performance curves, are analyzed accordingly. Furthermore, the paper investigates the impact of several factors on the cell performance, such as operating temperature, electrode layer porosity, and different tubular structures (anode-supported or cathode-supported), using the control variable method to explore the degree of influence on the numerical simulation.

2. Methodology

2.1. SOFC Mathematical Model Description

2.1.1. Momentum Transmission Model

Regarding the momentum transfer model of SOFCs, as the model belongs to the millimeter-scale structure and the initial flow velocities of fuel gas and air are relatively small, it is assumed that the flow is laminar and incompressible, without considering the density changes caused by variations in the inlet and outlet pipe diameters and gas flow processes [41]. The momentum conservation equation of the N-S motion equation for the incompressible flow is expressed as Equation (1):

$$\frac{\partial(\rho v_K)}{\partial t} + \nabla \cdot (\rho v_K v) = -\nabla P + \nabla \cdot (\mu \nabla v) + F \tag{1}$$

where ρ is the average density of each component (related to environmental temperature and pressure), v is the gas velocity, subscript k represents one of the x, y, and z directions, P is pressure, μ is viscosity coefficient, and F is volume force.

2.1.2. Material Transfer Model

Regarding the mass transfer model of SOFCs, several factors including concentration gradient, pressure gradient, temperature gradient, viscous resistance, and fluid inertial velocity affect mass transfer within the structure. To ensure the conservation of the total amount of each component, even when their concentrations change continuously, the Maxwell–Stefan mass transfer equation must be followed, which is given below:

$$\nabla \cdot \left(-\rho \cdot \omega_i \sum D_{ij} \cdot \nabla \chi_j + (\chi_j - \omega_j) \frac{\nabla P}{P} \cdot u - D_i^T \cdot \frac{\nabla T}{T} \right) + \rho \cdot u \cdot \nabla \omega_j = S_i \tag{2}$$

This equation includes the mass fraction ω of a component, the molar fraction χ of a component, the molar molecular weight M of a component, the diffusion coefficient D, and the thermal diffusion constant S_i, which is related to the heat source associated with electrochemical reactions and is never zero. This equation implies that in SOFCs, the total amount of reactants and products remains constant regardless of the reaction between fuel gas and air.

2.1.3. Electrochemical Model

Regarding the electrochemical model of the SOFC, it is well known that its working principle involves the flow of electrons generated by the decomposition of ammonia through the external circuit of the cell, and their return to the cathode through the load. Oxygen, after receiving electrons, reacts with hydrogen ions to produce oxygen ions. Currently, the derivation of the SOFC electromotive force is mainly based on the Nernst equation. The equation, which reflects the quantitative relationship between the electromotive force and the electrolyte concentration, is the core theory that explains the conversion between electrical and chemical energy. In the equation, E represents the theoretical Nernst voltage of the cell, η_{act} represents activation polarization loss, η_{conc} represents concentration polarization loss, and η_{ohm} represents ohmic polarization loss, which is also the reason why the actual output voltage of the cell is always less than the ideal open circuit voltage.

$$\begin{array}{c} V_{cell} = E - \eta_{act} - \eta_{conc} - \eta_{ohm} \\ E = \frac{-\Delta G}{nF} = \frac{RT}{nF} \ln\left[\frac{(p_{H_2O})^3 (p_{N_2})}{(p_{NH_3})^2 (p_{O_2})^{3/2}} \right] \end{array} \quad (3)$$

Activating polarization refers to the voltage loss caused by the activation energy that the fuel cell has to overcome in electrochemical reactions. Electrochemical reactions must be dominated by overpotential rather than equilibrium potential so that the reaction can proceed in the correct direction. This polarization can be expressed by the fundamental theory of electrode kinetics, the Butler–Volmer equation. In this equation, i and i_0 are the current density and gas exchange current density, respectively, and R is related to the gas component concentration and temperature. r is the reference exchange current density, P_i is the total pressure of reactants, P_j is the total pressure of products, P_{ref} can be defined as the pressure reference value, and E_a is the reaction activation energy.

$$\begin{array}{c} \eta_{act} = \frac{2RT}{n_e F} \sin h^{-1}\left(\frac{i}{2i_0}\right) \\ i_0 = \gamma\left(\frac{P_i}{P_{ref}}\right)\left(\frac{P_j}{P_{ref}}\right) \exp\left(-\frac{E_a}{RT}\right) \end{array} \quad (4)$$

2.1.4. Heat Transfer Model

Regarding the heat transfer model of SOFC, theoretically, fuel cells directly convert chemical energy into electrical energy, but inevitably, a part of it is converted into thermal energy. The three main heat transfer modes considered are heat conduction between solid porous media, heat convection between solid surfaces and gas, and heat exchange when gas flows through porous media. Due to the complexity of actual working conditions, the simplified steady-state heat transfer equation for the working of a uniform SOFC is as follows:

$$\begin{array}{c} Q = \rho C_p \frac{\partial T}{\partial \tau} + \nabla \cdot (-k_{eff} \nabla T) \\ k_{eff} = \phi_g k_f + (1 - \phi_g) k_s \end{array} \quad (5)$$

where C_p is the equivalent specific heat capacity of the reactant gas at constant pressure, k_{eff} is the thermal conductivity of the porous media, φ_g is the porosity of the porous media, k_f is the standard fluid heat transfer coefficient, and k_s is the standard solid thermal conductivity.

2.2. *SOFC Numerical Simulation*

Given that the structure of tubular SOFCs is generally divided into AS-SOFC and CS-SOFC designs, the modeling methods for both types are generally similar. However, the order of the layers in the structure can have a certain impact on the performance of the cell. In AS-SOFCs, the anode is typically the thickest layer and serves as the mechanical support for the cell. An SOFC consists of an anode, a cathode, and an electrolyte. The main difference between AS-SOFC and CS-SOFC is the thickness of the support layer. In AS-SOFCs, the thickest layer is the anode, which serves as the mechanical support and the active layer for fuel oxidation. In contrast, in CS-SOFCs, the thickest layer is the cathode, which serves as both the mechanical support and the active layer for oxygen reduction. Both types of SOFCs are three-layer structures with the electrolyte sandwiched between the anode and cathode. Despite these structural differences, both types of tubular SOFCs utilize similar electrochemical and transport phenomena for their operation, and therefore share similar modeling approaches. These approaches typically involve the use of coupled mass, momentum, and energy conservation equations to describe the transport and reaction processes occurring within the cell, as well as the use of various electrochemical models to describe the cell's performance. Additionally, modeling efforts for both types of tubular SOFCs may also incorporate considerations such as thermal and mechanical stresses, which can affect the cell's durability and performance over time.

2.2.1. Geometry Procedure

Based on practical considerations, the geometric model of a single tube-type AS-SOFC cell without an external connector is shown in Figure 1. The dimensions of the geometric structure are presented in Table 1. To facilitate subsequent calculations, the overall design was modeled in millimeters. The thickness of the anode diffusion layer is slightly larger than that of the other diffusion layers to ensure the stability of the supporting structure. After setting the boundary layer, the mesh was created along the surfaces covering the entire region of the tubular AS-SOFC, including each layer structure. It is recommended to refine the mesh as much as possible at the inlet and outlet of the gas channel and at the reaction interface to ensure accurate calculation results with the finite element method. Additionally, a mesh sensitivity analysis was conducted to ensure the reliability of the CFD results. The mesh was incrementally refined by dividing the cell into smaller elements to observe the effect of mesh density on the results. The analysis indicated that the results were sensitive to mesh density, and a sufficiently fine mesh was required to accurately capture the complex flow and heat transfer phenomena in the system. Furthermore, a mesh convergence study was conducted to ensure that the results were consistent with the number of elements in the mesh. In conclusion, the mesh sensitivity analysis confirmed the reliability of the CFD results, which could be used to optimize the design and operation of the tubular AS-SOFC.

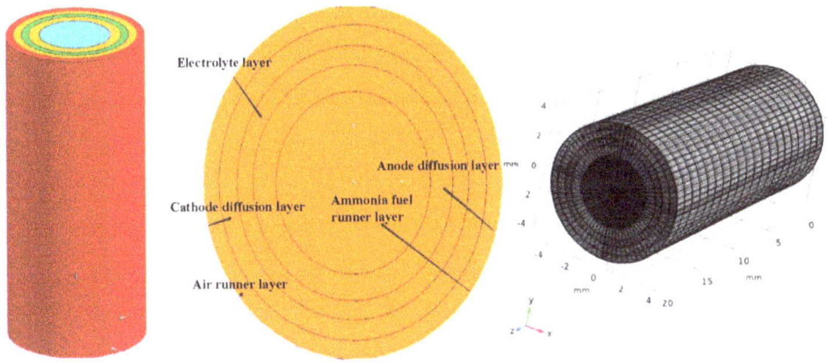

Figure 1. Geometric structure diagram of SOFC monomer.

Table 1. SOFC geometry dimensions in UAVs [30,42].

Items	Units	Value
Inner diameter of ammonia fuel runner layer	mm	4.62
Anode diffusion layer thickness	mm	0.78
Electrolyte layer thickness	mm	0.45
Cathode diffusion layer thickness	mm	0.32
Air layer thickness	mm	0.51
Pipe range	mm	20.00

2.2.2. Boundary Assumption Conditions

Considering the current situation and disregarding the production of SOFC cell stacks, no external connecting device was necessary. Figure 1 illustrates the geometric model of the tubular AS-SOFC single cell, and Table 1 shows the geometric dimensions of the structure. The design is entirely in millimeter scale for convenience in subsequent calculations, and the anode diffusion layer is slightly thicker than the other diffusion layers to maintain the stability of the support structure. The gases introduced into the tube were assumed to be ideal and incompressible, and the density of ammonia does not significantly decrease, regardless of how it reacts with oxygen. The shape of each layer of the gas channel and electrode is considered ideal and do not cause any structural deformation due to excessive usage. The gas always flows in laminar flow, even at the inlet or outlet, assuming standard atmospheric pressure. The heat transfer module only takes into account the thermal conduction of the porous medium and the convective heat transfer of the gas and disregards the influence of thermal radiation. All electrochemical reactions should occur at the reasonable boundaries, i.e., at the junction of the electrode layer and the electrolyte layer. Additionally, given the small size of this model, it can be assumed that the initial temperature distribution of each subsystem is constant, the partial pressures of the fluid in each layer are uniform, and the fluid viscosity remains constant.

With regard to electrical conductivity conditions, the assumption is that the interfacial potential and current density are uniform, resulting in partial derivatives of the boundary variables being considered zero, and the boundaries being primarily treated as insulating. For flow conditions, as previously stated, all fluid flows are assumed to be laminar, and the boundaries are continuous without any slip phenomena. In terms of mass transfer conditions, initial concentration boundaries were set for the fuel and air flow channels, while all other contact boundaries were continuous without considering any slip phenomena, in order to investigate the effect of mass transfer on the electrochemical reaction. For heat transfer conditions, except for the initial temperature set at the fuel inlet, contact boundaries involved in the reaction were set as continuous heat flow boundaries, while boundaries not related to the reaction were set as adiabatic boundaries, for temperature field analysis. Based on the actual situation and the aforementioned constraints, Table 2 shows the operating parameters required for simulation.

Table 2. Main working parameters [30,37,42,43].

Items	Units	Value
Initial polarization voltage	V	0.05
Anode equilibrium voltage	V	0
Cathode balance voltage	V	0.8
Anode exchange current density	A/m^2	1
Cathode exchange current density	A/m^2	0.1
Anode layer conductivity	S/m	10
Cathode layer conductivity	S/m	10
Conductivity of electrolyte layer	S/m	5
Anodic permeability	m^2	1×10^{-10}
Cathode permeability	m^2	1×10^{-10}
Porosity	—	0.35
Cathode surface area ratio	m^{-1}	2.8×10^4
Anode flow path viscosity	Pa·s	1.19×10^{-5}
Anode surface area ratio	m^{-1}	2.2×10^4
Viscosity of cathode flow path	Pa·s	2.46×10^{-5}
Air inlet velocity	m/s	1
Intake inlet temperature	K	1073.15
Operating pressure	atm	1
Reference diffusion coefficient	m^2/s	3.2×10^{-8}

2.3. *Verification*

The polarization and power density curves are key indicators of fuel cell performance, as shown in Figure 2. The actual voltage output of a fuel cell is affected by activation, concentration, and ohmic polarization. A smoother curve indicates less prominent polarization effects and better overall operational stability of the fuel cell, which can output more power at the corresponding operating voltage. The ammonia-fueled tubular AS-SOFC model's electrochemical performance was compared to the reference literature, and similar results were obtained for the polarization curve, with consistent overall slopes. Although there were differences in the peak values and amplitudes of the power density curve, the overall trend was still similar, validating the model established in this paper.

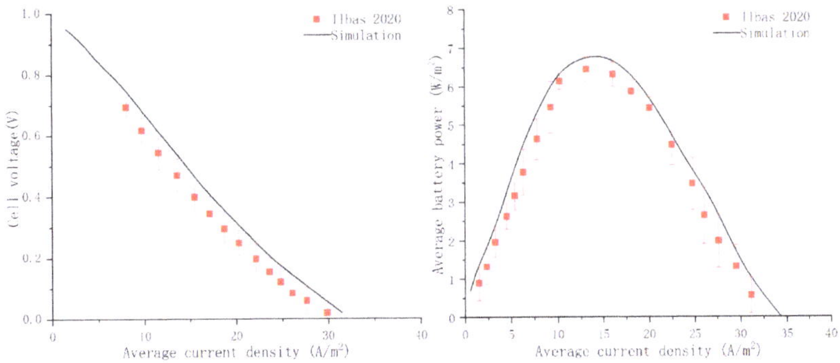

Figure 2. Comparison between experimental data [30] and simulation results.

Most of the material parameters in this model were obtained from experimental research on the electrochemical performance of an ammonia fuel tubular AS-SOFC model reported in reference [30]. However, differences in the performance curves may be attributed to several factors, such as the use of platinum and silver as anode electrodes in the reference literature, while ceramic flexible electrode materials such as supporting electrodes were used in the numerical simulations. Additionally, this paper stipulated more boundary assumptions for convenience, resulting in a relatively weak ohmic impedance of the electrode, making the polarization phenomenon very weak, and leading to more ideal output performance of the working cell. Nonetheless, this is a limitation of the simulation in this paper, and more realistic simulation results require considering various external environment and internal structural problems and establishing a completer and more accurate SOFC mathematical model.

3. Results and Discussion

The simulation results can be mainly focused on the concentration distribution of the reaction gases (NH_3, O_2, H_2), the distribution of electrode current density, the temperature distribution inside the single cell, and the polarization and power curves that reflect the performance of the cell. Based on these simulation results, further research can be conducted to optimize the performance of tubular SOFCs. For example, more reasonable cell structures or changes in operating conditions can be designed to address issues such as non-uniform electrode current density or unstable temperature fields, with the aim of improving the efficiency and stability of the cell. In addition, the impact of different materials and structures on cell performance can be studied to explore the application prospects of new materials and structures. Furthermore, by analyzing the polarization curve and power curve of the cell, the performance and stability of the cell can be evaluated, and its optimal operating range can be determined. These results are of great significance for further improving the practical application value of tubular SOFCs, and can provide strong support and reference for research and application in UAVs.

3.1. Analysis of Gas Concentration Distribution

According to Figure 3, the molar concentration distribution of the reaction gases shows a decreasing trend both radially (from outer to inner) and along the flow direction. The color stratification of oxygen along the radial direction is more distinct, indicating that the consumption rate of oxygen is faster than that along the flow direction. This is mainly because the oxygen in the gas channel layer can be continuously replenished, but for oxygen to participate in the reaction at the electrolyte interface, it must diffuse through the porous resistance layer. Generally, the reaction consumption rate is greater than the diffusion rate, which leads to the gradually decreasing trend along the radial direction. In practice, insufficient oxygen participation in the reaction may result in the inability of the cell to operate.

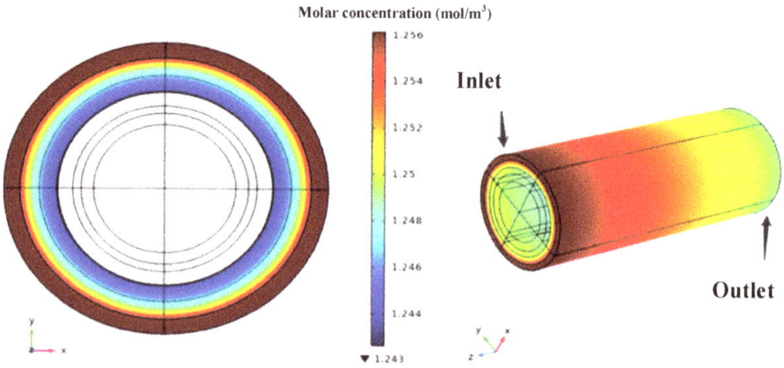

Figure 3. Oxygen molar concentration distribution.

Furthermore, it can be visually observed that the decrease in oxygen concentration is initially rapid and then slows down. The main reason for this is the change in reaction rate. According to chemical reaction kinetics theory, the reaction rate is directly proportional to the concentration of reactants. As oxygen is gradually consumed, the concentration of participating oxygen decreases, and the reaction rate slows down accordingly. The distribution of oxygen concentration is of great reference value for determining the operating performance of the cell. As mentioned earlier, in this model, oxygen is ionized at the cathode and enters the electrolyte and anode layer to participate in electrochemical reactions. Oxygen ions dominate in ion conductivity. Therefore, ensuring a sufficient oxygen supply is beneficial for improving reaction rate and ion conductivity.

According to Figure 4, both hydrogen and ammonia show a decreasing trend in concentration whether they are distributed radially (from inside to outside) or in the flow direction. The degree of change is similar for both species, primarily because the anode of the SOFC undergoes a gradual dehydrogenation reaction when ammonia is introduced to participate in the electrochemical reaction, allowing hydrogen ions to combine with oxygen ions to generate water. This dehydrogenation reaction is not equivalent to the external reaction of decomposing hydrogen due to the high-temperature environment. Additionally, as mentioned earlier in the setup, the reactant gas contains ammonia, hydrogen, and nitrogen, so the numerical value of hydrogen concentration should be much lower than that of ammonia concentration. However, in terms of numerical values, the degree of decrease for both species is much greater than that of oxygen. This is mainly because in the SOFC, the reaction rate at the anode is faster than that at the cathode, and from a microscopic perspective, the size of oxygen molecules is larger, so their diffusion rate in the porous medium is slower. Similarly, the concentration decrease rates of ammonia and hydrogen also show a fast-to-slow trend. According to the theory of chemical reaction kinetics, their reaction rates gradually decrease as the concentration of ammonia and hydrogen decreases, which is applicable to both the cathode and anode sides.

Considering that this model belongs to an anode-supported structure, where the anode layer thickness is greater than that of the electrolyte and cathode layers, although it needs to overcome the resistance of the porous medium, it does not have a significant impact on the diffusion of hydrogen. Compared with hydrogen, the thicker anode diffusion layer can make the reactant gas obtain more uniform distribution in the anode layer, thereby improving the reaction rate at the anode. On the other hand, in the cathode layer, the thinner, porous cathode layer means that oxygen loses more opportunities to participate in the reaction due to diffusion limitations caused by its larger molecular size. This results in most of the oxygen being discharged without diffusing sufficiently towards the interior.

This indirectly proves that if the amount of oxygen available for reaction is insufficient, its reaction rate with the anode does not increase, and ultimately the performance of the cell declines.

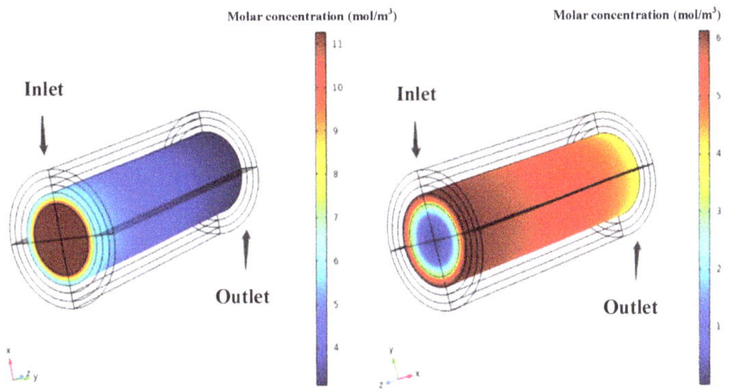

Figure 4. Ammonia and hydrogen fuels molar concentration distribution.

3.2. Current Density Analysis

The electrolyte current density distribution of this model at a working voltage of 0.8 V is shown in Figure 5. Generally, the current density distribution of an SOFC is related to the concentration distribution of hydrogen and oxygen. The uneven distribution of current density is a result of gas consumption, and the majority of the current is generated near the cathode inlet. Comparing the gas concentration distribution analyzed in the previous section, it can be seen that the decrease in current density follows a similar trend to the concentration distribution of oxygen, with a rapid decline followed by a slower decline. At the same time, another important consideration when simulating current density results is the reaction active area at the interface between the anode and the electrolyte layer.

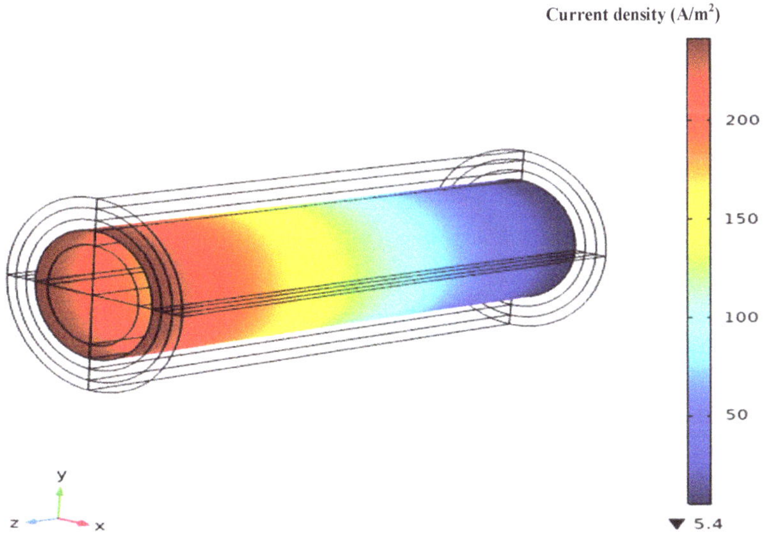

Figure 5. Electrolyte current density distribution.

Due to the presence of porous media, there are usually several active points where ion conduction occurs. If these active points are not uniformly distributed, it may also lead to uneven distribution of current density. However, because this model is at the millimeter scale and only calculates the average value of local current density, the numerical discrepancy of the oxygen concentration distribution is not significant enough, and this conclusion may not be sufficiently convincing. As is well known, current density reflects the output performance and operational stability of an SOFC. For an AS-SOFC using ammonia fuel, there are two main ways to improve performance: (1) increase the inlet oxygen flow and raise the inlet temperature of the anode to obtain more hydrogen decomposition, promoting electrochemical reaction rates by increasing reactant concentrations; or (2) increase the active surface area of the reaction (reaction sites between the electrode and the electrolyte layer), promoting electronic and ionic conduction activity.

3.3. Temperature Field Analysis

From Figure 6, it can be seen that both ammonia and air exhibit a temperature increase trend from the inlet side to the outlet side, as heat accumulates gradually during the electrochemical reactions. The temperature distribution from the fuel inlet side to the air inlet side shows an initial increase followed by a decrease, and the temperature in the reaction layer is higher than that in the flow channel layer. This is mainly due to the fact that in SOFCs, not only heat transfer occurs as fuel gas or air diffuses towards the reaction layer, but also Joule heat generated by the electrochemical reactions diffuses towards the periphery of the cell, where the heat generated by the electrochemical reactions dominates. In the heat transfer module of this AS-SOFC model, the reaction interface between the anode layer and the electrolyte layer has been set as the heat source surface, and all boundaries where reactions occur have heat flux continuity. There are no reaction interfaces where heat exchange cannot occur, and the anode layer itself is relatively thick and has good thermal conductivity. Heat can be transferred through the relatively ideal porous medium layer by conduction and convective heat transfer with the gas to the entire cell.

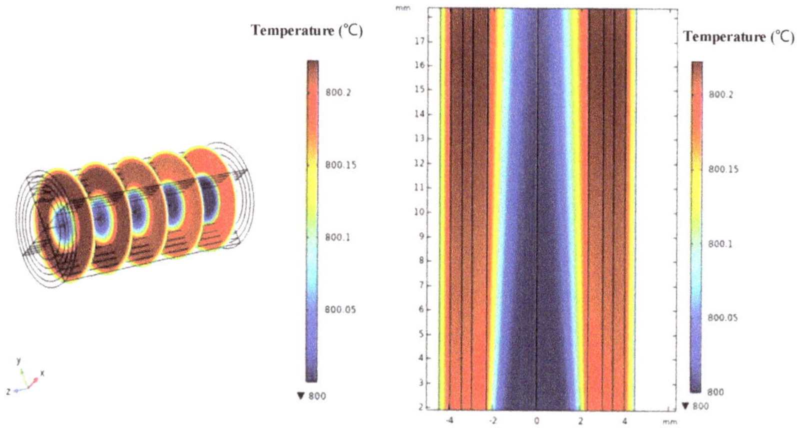

Figure 6. Temperature field distribution.

Research has shown that the temperature of the air in the cathode flow channel is generally increased by the heat exchange within the SOFC and is generally not directly affected by external interventions. Therefore, adjusting the airflow can also achieve the purpose of adjusting the SOFC temperature. Based on the current technological conditions, it is difficult to measure the internal temperature distribution of the cell in actual testing since the reaction region inside the tube is enclosed. Therefore, it is very meaningful to analyze the temperature distribution through numerical simulation. In this simulation analysis, the computation in the temperature field simulation was simplified, the given heat transfer coefficient and resistance values of the materials were small, and the temperature changes were not significant numerically. However, the overall trend of the temperature changes was similar to that observed in actual experiments, which indirectly demonstrates that numerical simulation can accurately analyze the temperature distribution inside the tubular cell and can replace the actual measurement method.

3.4. Performance Design

External environmental factors can have a significant impact on fuel cell performance, and one of the main factors is the temperature of the incoming gas. Figure 7 shows the polarization curve results of an AS-SOFC using ammonia fuel at operating temperatures of 800 °C, 900 °C, and 1000 °C, respectively. Although the numerical changes are not very significant, a comparison of the results indicates that, as the temperature increases, there can be a larger output current and thus higher power output for the same operating voltage. This increase in performance and stability is not limited to a single voltage range but is reflected in the overall output of the cell. This is mainly due to the enhanced conductivity of the electrolyte material at higher temperatures, the decrease in ohmic resistance, and the weakening of polarization effects. In addition, at high temperatures, more hydrogen can be released from the ammonia to produce larger amounts of hydrogen gas through decomposition. According to chemical reaction kinetics theory, the more intense diffusion motion of molecules and ions at high temperatures can promote the overall electrochemical reaction. Although increasing the temperature can enhance the performance of the fuel cell, the operating temperature of the SOFC is generally set below 1100 °C to prevent the possible occurrence of electrochemical side reactions that can disrupt the overall balance. Even solid-state electrolyte layers can undergo structural changes and deteriorate the performance of the cell at high temperatures, and current leakage can occur easily, which dramatically shortens the operating life of the SOFC. Therefore, in the process of exploring the optimal temperature conditions, the impact of temperature on the internal reactions and material characteristics of the cell must be taken into account.

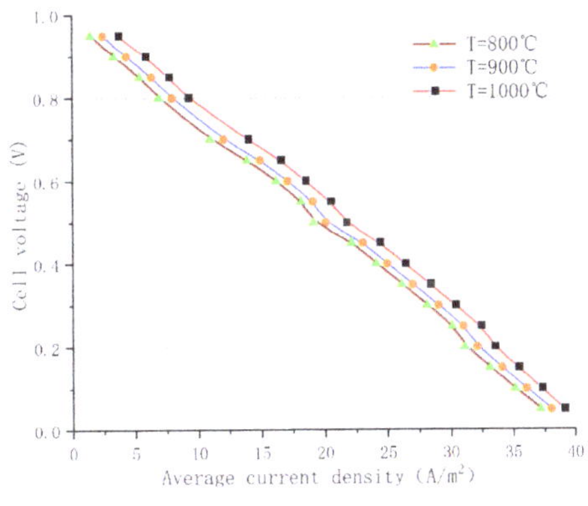

Figure 7. Polarization curves at different operating temperatures.

Among the various internal material properties of the cell, the porosity of the porous electrode is one of the main factors affecting fuel cell performance. Figure 8 shows the polarization curves of AS-SOFCs using ammonia fuel with a porosity of 0.35, 0.5, and 0.65, respectively. Comparative analysis of the results reveals that as porosity increases, fuel cell performance gradually improves. This is mainly because porosity affects the rate of gas diffusion. With larger porosity, gas diffusion resistance decreases, allowing more gas to quickly reach the reaction layer, thus accelerating the electrochemical reaction rate and weakening concentration polarization. However, the overall change in the numerical values is not significant, except for a relatively significant variation range in the working voltage of 0.6–0.8 V. This may be due to the strong electronic and ionic conductivity activity in this stage, resulting in a relatively weak polarization effect. Under the same porosity, it is easier to involve more gas in the electrochemical reaction. Therefore, it can be inferred that to further improve the fuel cell performance of the model by changing the porosity, it is necessary to keep the working voltage within this range. However, material properties, including porosity, should not be blindly pursued for enlargement and strengthening. Although increasing porosity optimizes fuel cell performance, excessive porosity can lead to unstable internal structure of the porous medium layer. Excessive porosity density can cause an inability to withstand gas flow pressure or to meet electrode reaction requirements. Similarly, material property parameters such as the conductivity of each layer and the viscosity coefficient of the flow channel also need to explore suitable values to optimize SOFC performance while meeting the stability requirements of the reaction.

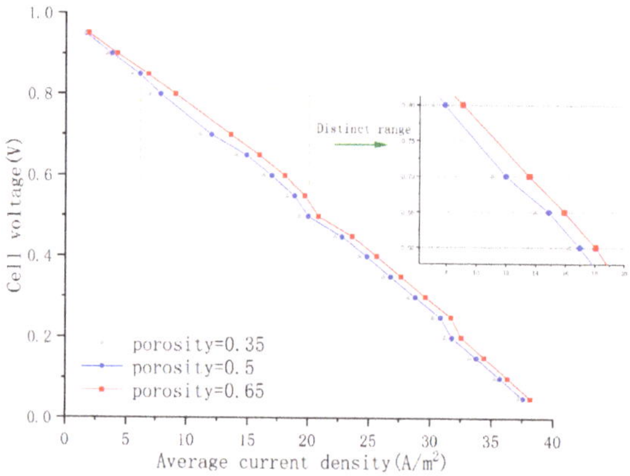

Figure 8. Polarization curves under different porosity.

In the current mainstream SOFC designs, there are two main types of cells: AS-SOFCs and CS-SOFCs. The main difference between these two types lies in the thicker support electrode layer inside the cell. Figure 9 shows the polarization curves of the two types of cells under the same operating conditions. After a comparative analysis, it is clear that CS-SOFC performs better overall than AS-SOFC in terms of working performance, with a significantly larger numerical change than the influence of temperature and porosity. This is mainly because the primary factor affecting the current density of the electrolyte layer is the oxygen concentration distribution in the cathode layer. For CS-SOFC, the thicker cathode support layer ensures sufficient oxygen supply and high cathode gas channel sealing, which has a significant positive effect on achieving uniform current density distribution and improving the performance of output current for tubular SOFCs.

Although CS-SOFC slightly outperforms AS-SOFC in numerical simulation results, when considering working stability in practical production applications, it is not sufficient to solely consider the working performance of the fuel cell. Both AS-SOFC and CS-SOFC are three-layer (sandwich) structures, with the electrolyte inserted between the anode and cathode. The only difference between the two is the thickness of the support layer, with the anode being the thickest layer in AS-SOFC and the cathode being the thickest layer in CS-SOFC. However, achieving the internal cathode support layer of CS-SOFC is challenging due to its high sintering temperature requirement. The mechanical strength of the cathode is a concern due to its high porosity, and the thermal expansion coefficient is normally higher than that of the anode and electrolyte. Therefore, there is still considerable controversy over which type of tubular SOFC will dominate the future industrial market. Overall, further research and development are needed to determine the most suitable SOFC type for practical production applications.

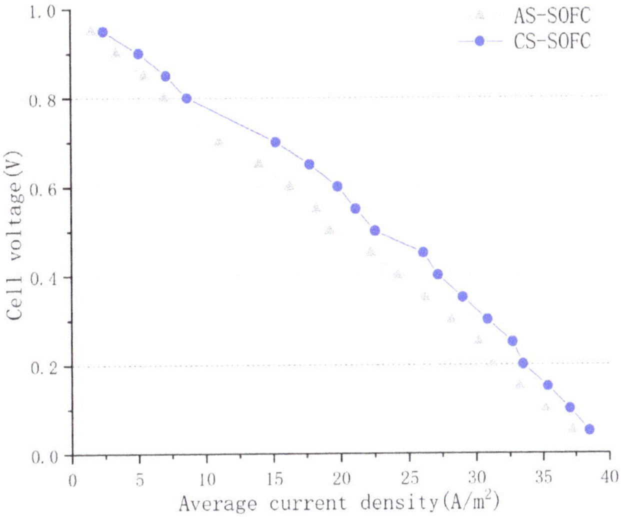

Figure 9. Polarization curves of anodic and cathodic SOFCs.

4. Conclusions

This study aimed to explore the viability of utilizing an ammonia fuel tubular SOFC for shipborne UAVs. To achieve this goal, a comprehensive three-dimensional model of a tubular anode-supported SOFC single cell was developed, which incorporated the necessary mathematical models for its structure and operation principles. To enhance the precision of the model, multiphysics fields were taken into account, and numerical simulations of the SOFC's working conditions under ammonia fuel were conducted. Through the simulations, several performance indicators were calculated and analyzed to determine the effects of different operating conditions on the SOFC's performance. To optimize the SOFC's performance, the paper employed the method of controlling variables to adjust the working parameters and overall structure of the cell. Through this optimization analysis, the factors that significantly impact cell performance were identified, which can guide future improvements in SOFC design and operation.

The study's results indicate that increasing the temperature of the SOFC enhances battery performance due to the facilitation of ammonia fuel dehydrogenation and improved rates of electrochemical reactions, electron, and ion conductivity. However, operating temperatures above 1100 °C can cause electrode microstructural deformation and electrochemical side reactions, leading to a decline in battery performance. In addition, increasing porosity can enhance battery performance due to reduced gas flow resistance, higher concentration of reaction gas, and reduced concentration polarization. A significant performance improvement was observed in the range of 0.6–0.8 V due to strong electron and ion conductivity and weak polarization effects. Nevertheless, excessive porosity can cause instability in the electrode layer structure.

Moreover, CS-SOFC showed improved performance compared to AS-SOFC mainly because of its thicker cathode layer, which provides sufficient oxygen supply and better sealing, resulting in more uniform current density distribution. However, AS-SOFC has greater practicality in practical production than CS-SOFC, and both have significant potential for development.

The findings of this study offer valuable insights into the potential application of ammonia fuel for shipborne UAVs and provide a foundation for future research and development in the field of SOFCs.

Author Contributions: Conceptualization, Z.W. and F.H.; methodology, F.Z.; software, F.Z.; validation, Y.M., R.X.; formal analysis, Y.M., R.X.; investigation, Y.M., R.X.; resources, F.Z.; data curation, F.Z.; writing—original draft preparation, Z.W.; writing—review and editing, F.H.; visualization, F.Z.; supervision, F.H.; project administration, F.H.; funding acquisition, F.H. All authors have read and agreed to the published version of the manuscript.

Funding: This work was supported by the Fundamental Research Funds for the Central Universities (3132023510), the Open Fund of National Center for International Research of Subsea Engineering Technology and Equipment (No. 3132023358), the CNOOC Marine Environment and Ecological Protection Public Welfare Foundation Project (CF-MEEC/TR/2023-9), the Dalian High Level Talent Innovation Support Program (2021RQ132), the Natural Science Foundation of Liaoning Province (2022-MS-154), the China Postdoctoral Science Foundation (2020M680928), and the 111 Project (B18009).

Data Availability Statement: Not applicable.

Conflicts of Interest: The authors declare no conflict of interest.

Nomenclature

C_p	Equivalent specific heat capacity
D	Diffusion coefficient
E_a	Reaction activation energy
E	Theoretical Nernst voltage (V)
F	Volume force
h	Heat transfer coefficient ($W{\cdot}m^{-2}{\cdot}K^{-1}$)
L	Heat transfer length (mm)
k	Thermal conductivity ($W{\cdot}m^{-1}{\cdot}K^{-1}$)
v	Gas velocity ($m{\cdot}s^{-1}$)
M	Molar molecular weight
ΔP	Pressure drop (kPa)
Pr	Prandtl number
q	Heat (W)
Q	Heat flux (W)
Re	Reynolds number
S_i	Thermal diffusion constant
T	Temperature (K)
U	Overall heat transfer coefficient ($W{\cdot}m^{-2}{\cdot}K^{-1}$)
v	Specific volume ($m^3{\cdot}kg^{-1}$)
P_w	Pump power (kW)
μ	Dynamic viscosity (Pa·s)
u	Velocity ($m{\cdot}s^{-1}$)
Greek symbols	
μ	Viscosity coefficient
ρ	Density ($kg{\cdot}m^{-3}$)
η_{act}	Activation polarization loss
η_{conc}	Concentration polarization loss
η_{ohm}	Ohmic polarization loss
φ_g	Porosity of the porous media
Subscripts	
0	Standard or parameter at equilibrium conditions
ref	Reference value
k	Coordinate variable
x, y, z	Directions
i, o	Inlet or outlet
Abbreviations	
AS-SOFC	Anode-supported solid oxide fuel cell
CS-SOFC	Cathode-supported solid oxide fuel cell
ICEs	Internal combustion engines
SOFC	Solid oxide fuel cells
UAVs	Unmanned aerial vehicles

References

1. Onifade, S.T.; Alola, A.A. Energy transition and environmental quality prospects in leading emerging economies: The role of environmental-related technological innovation. *Sustain. Dev.* **2022**, *30*, 1766–1778. [CrossRef]
2. Deng, Z.; Bing, F.; Guo, Z.; Wu, L. Rope-Hook Recovery Controller Designed for a Flying-Wing UAV. *Aerospace* **2021**, *8*, 384. [CrossRef]
3. Cipolla, V.; Dine, A.; Viti, A.; Binante, V. MDAO and Aeroelastic Analyses of Small Solar-Powered UAVs with Box-Wing and Tandem-Wing Architectures. *Aerospace* **2023**, *10*, 105. [CrossRef]
4. Zappa, C.J.; Brown, S.M.; Laxague, N.J.; Dhakal, T.; Harris, R.A.; Farber, A.M.; Subramaniam, A. Using ship-deployed high-endurance unmanned aerial vehicles for the study of ocean surface and atmospheric boundary layer processes. *Front. Mar. Sci.* **2020**, *6*, 777. [CrossRef]
5. Olabi, A.G.; Abdelkareem, M.A.; Wilberforce, T.; Alkhalidi, A.; Salameh, T.; Abo-Khalil, A.G.; Hassan, M.M.; Sayed, E.T. Battery electric vehicles: Progress, power electronic converters, strength (S), weakness (W), opportunity (O), and threats (T). *Int. J.* **2022**, *16*, 100212. [CrossRef]
6. Micallef, A.; Guerrero, J.M.; Vasquez, J.C. New Horizons for Microgrids: From Rural Electrification to Space Applications. *Energies* **2023**, *16*, 1966. [CrossRef]
7. Liu, F.; Ding, D.; Duan, C. Protonic Ceramic Electrochemical Cells for Synthesizing Sustainable Chemicals and Fuels. *Adv. Sci.* **2023**, *10*, 2206478. [CrossRef] [PubMed]
8. Liu, F.; Diercks, D.; Hussain, A.M.; Dale, N.; Furuya, Y.; Miura, Y.; Fukuyama, Y.; Duan, C. Nanocomposite Catalyst for High-Performance and Durable Intermediate-Temperature Methane-Fueled Metal-Supported Solid Oxide Fuel Cells. *ACS Appl. Mater. Interfaces* **2022**, *14*, 53840–53849. [CrossRef]
9. Tawalbeh, M.; Murtaza, S.Z.; Al-Othman, A.; Alami, A.H.; Singh, K.; Olabi, A.G. Ammonia: A versatile candidate for the use in energy storage systems. *Renew. Energy* **2022**, *194*, 955–977.
10. Rony, Z.I.; Mofijur, M.; Hasan, M.M.; Rasul, M.G.; Jahirul, M.I.; Ahmed, S.F.; Kalam, M.A.; Badruddin, I.A.; Khan, T.Y.; Show, P.L. Alternative fuels to reduce greenhouse gas emissions from marine transport and promote UN sustainable development goals. *Fuel* **2023**, *338*, 127220. [CrossRef]
11. Lei, T.; Min, Z.; Gao, Q.; Song, L.; Zhang, X.; Zhang, X. The Architecture Optimization and Energy Management Technology of Aircraft Power Systems: A Review and Future Trends. *Energies* **2022**, *15*, 4109. [CrossRef]
12. Çelik, D.; Meral, M.E.; Waseem, M. Investigation and analysis of effective approaches, opportunities, bottlenecks and future potential capabilities for digitalization of energy systems and sustainable development goals. *Electr. Power Syst. Res.* **2022**, *211*, 108251. [CrossRef]
13. Feng, Y.; Qu, J.; Zhu, Y.; Wu, B.; Wu, Y.; Xiao, Z.; Liu, J. Progress and prospect of the novel integrated SOFC-ICE hybrid power system: System design, mass and heat integration, system optimization and techno-economic analysis. *Energy Convers. Manag. X* **2023**, *18*, 100350. [CrossRef]
14. Ma, Y.; Wang, Z.; Yu, S.; Han, F.; Ji, Y.; Cai, W. Design and Simulation of Ammonia-Fueled SOFC-ICE Hybrid Power System for Ship Application. In Proceedings of the 2022 7th International Conference on Power and Renewable Energy (ICPRE), Shanghai, China, 23–26 September 2022; IEEE: New York, NY, USA, 2022; pp. 772–777.
15. Wu, S.; Miao, B.; Chan, S.H. Feasibility assessment of a container ship applying ammonia cracker-integrated solid oxide fuel cell technology. *Int. J. Hydrog. Energy* **2022**, *47*, 27166–27176. [CrossRef]
16. Xu, Q.; Guo, Z.; Xia, L.; He, Q.; Li, Z.; Bello, I.T.; Zheng, K.; Ni, M. A comprehensive review of solid oxide fuel cells operating on various promising alternative fuels. *Energy Convers. Manag.* **2022**, *253*, 115175.
17. Qureshi, F.; Yusuf, M.; Kamyab, H.; Vo, D.-V.N.; Chelliapan, S.; Joo, S.-W.; Vasseghian, Y. Latest eco-friendly avenues on hydrogen production towards a circular bioeconomy: Currents challenges, innovative insights, and future perspectives. *Renew. Sustain. Energy Rev.* **2022**, *168*, 112916. [CrossRef]
18. Qureshi, F.; Yusuf, M.; Kamyab, H.; Zaidi, S.; Khalil, M.J.; Khan, M.A.; Alam, M.A.; Masood, F.; Bazli, L.; Chelliapan, S.; et al. Current trends in hydrogen production, storage and applications in India: A review. *Sustain. Energy Technol. Assess.* **2022**, *53*, 102677. [CrossRef]
19. Sonker, M.; Tiwary, S.K.; Shreyash, N.; Bajpai, S.; Ray, M.; Kar, S.K.; Balathanigaimani, M. Ammonia as an alternative fuel for vehicular applications: Paving the way for adsorbed ammonia and direct ammonia fuel cells. *J. Clean. Prod.* **2022**, *376*, 133960. [CrossRef]
20. Ashrafi, M.; Lister, J.; Gillen, D. Toward a harmonization of sustainability criteria for alternative marine fuels. *Marit. Transp. Res.* **2022**, *3*, 100052. [CrossRef]
21. Hollmann, J.; Fuchs, M.; Spieker, C.; Gardemann, U.; Steffen, M.; Luo, X.; Kabelac, S. System simulation and analysis of an LNG-fueled SOFC system using additively manufactured high temperature heat exchangers. *Energies* **2022**, *15*, 941. [CrossRef]
22. Wu, Z.; Zhu, P.; Yao, J.; Abu Mansor, M.R.; Chen, J.; Huang, S.-M. Design of novel flow channels of tubular biomass-syngas fueled SOFC towards uniform temperature distribution. *Sustain. Energy Technol. Assess.* **2023**, *55*, 102947. [CrossRef]
23. Raza, T.; Yang, J.; Wang, R.; Xia, C.; Raza, R.; Zhu, B.; Yun, S. Recent advance in physical description and material development for single component SOFC: A mini-review. *Chem. Eng. J.* **2022**, *444*, 136533. [CrossRef]
24. Yang, Z.; Yu, X.; Dedman, S.; Rosso, M.; Zhu, J.; Yang, J.; Xia, Y.; Tian, Y.; Zhang, G.; Wang, J. UAV remote sensing applications in marine monitoring: Knowledge visualization and review. *Sci. Total Environ.* **2022**, *838*, 155939. [CrossRef]

25. Tran, T.L.C.; Huang, Z.C.; Tseng, K.H.; Chou, P.H. Detection of Bottle Marine Debris Using Unmanned Aerial Vehicles and Machine Learning Techniques. *Drones* **2022**, *6*, 401. [CrossRef]
26. Wang, X.; Zhu, J.; Han, M. Industrial Development Status and Prospects of the Marine Fuel Cell: A Review. *J. Mar. Sci. Eng.* **2023**, *11*, 238. [CrossRef]
27. Benet, Á.; Villalba-Herreros, A.; D'amore-Domenech, R.; Leo, T.J. Knowledge gaps in fuel cell-based maritime hybrid power plants and alternative fuels. *J. Power Sources* **2022**, *548*, 232066. [CrossRef]
28. Marciello, V.; Di Stasio, M.; Ruocco, M.; Trifari, V.; Nicolosi, F.; Meindl, M.; Lemoine, B.; Caliandro, P. Design Exploration for Sustainable Regional Hybrid-Electric Aircraft: A Study Based on Technology Forecasts. *Aerospace* **2023**, *10*, 165. [CrossRef]
29. Choi, Y.; Lee, J. Estimation of Liquid Hydrogen Fuels in Aviation. *Aerospace* **2022**, *9*, 564. [CrossRef]
30. Ilbas, M.; Kumuk, B.; Alemu, M.A.; Arslan, B. Numerical investigation of a direct ammonia tubular solid oxide fuel cell in comparison with hydrogen. *Int. J. Hydrog. Energy* **2020**, *45*, 35108–35117.
31. Barelli, L.; Bidini, G.; Cinti, G. Operation of a Solid Oxide Fuel Cell Based Power System with Ammonia as a Fuel: Experimental Test and System Design. *Energies* **2020**, *13*, 6173. [CrossRef]
32. Khodabandeh, E.; Akbari, O.A.; Akbari, S.; Taghizadeh, A.; Pour, M.S.; Ersson, M.; Jönsson, P.G. The effects of oil/MWCNT nanofluids and geometries on the solid oxide fuel cell cooling systems: A CFD study. *J. Therm. Anal. Calorim.* **2021**, *144*, 245–256. [CrossRef]
33. Rathore, S.S.; Biswas, S.; Fini, D.; Kulkarni, A.P.; Giddey, S. Direct ammonia solid-oxide fuel cells: A review of progress and prospects. *Int. J. Hydrog. Energy* **2021**, *46*, 35365–35384. [CrossRef]
34. Farhad, S.; Hamdullahpur, F. Conceptual design of a novel ammonia-fuelled portable solid oxide fuel cell system. *J. Power Sources* **2010**, *195*, 3084–3090. [CrossRef]
35. Chaudhary, T.N.; Mehmood, M.; Saleem, U.; Abbasi, M.S.; Chen, B. Modeling of thermal impacts in a single direct methane steam reforming solid oxide fuel cell. *J. Power Sources* **2020**, *472*, 228605. [CrossRef]
36. Buttler, A.; Koltun, R.; Wolf, R.; Spliethoff, H. A detailed techno-economic analysis of heat integration in high temperature electrolysis for efficient hydrogen production. *Int. J. Hydrog. Energy* **2015**, *40*, 38–50. [CrossRef]
37. Kishimoto, M.; Kume, T.; Iwai, H.; Yoshida, H. Numerical analysis of ammonia-fueled planar solid oxide fuel cells. *ECS Trans.* **2017**, *78*, 2845. [CrossRef]
38. Kee, R.J.; Zhu, H.; Sukeshini, A.M.; Jackson, G.S. Solid oxide fuel cells: Operating principles, current challenges, and the role of syngas. *Combust. Sci. Technol.* **2008**, *180*, 1207–1244. [CrossRef]
39. Tan, W.Y. Numerical simulation on the performance of a solid oxide fuel cell with direct ammonia internal decomposition. *Energy Sources Part A Recovery Util. Environ. Eff.* **2014**, *36*, 2410–2419. [CrossRef]
40. Hajimolana, S.A.; Hussain, M.A.; Daud, W.W.; Chakrabarti, M.H. Dynamic modelling and sensitivity analysis of a tubular SOFC fuelled with NH3 as a possible replacement for H2. *Chem. Eng. Res. Des.* **2012**, *90*, 1871–1882. [CrossRef]
41. COMSOL Multiphysics, Version 6. COMSOL Inc.: Burlington, MA, USA. Available online: https://www.comsol.com (accessed on 1 March 2023).
42. Lai, Y.; Wang, Z.; Cui, D.; Han, F.; Ji, Y.; Cai, W. Thermal impact performance study for the thermal management of ammonia-fueled single tubular solid oxide fuel cell. *Int. J. Hydrog. Energy* **2023**, *48*, 2351–2367. [CrossRef]
43. Wang, Y.; Gu, Y.; Zhang, H.; Yang, J.; Wang, J.; Guan, W.; Chen, J.; Chi, B.; Jia, L.; Muroyama, H.; et al. Efficient and durable ammonia power generation by symmetric flat-tube solid oxide fuel cells. *Appl. Energy* **2020**, *270*, 115185. [CrossRef]

MDPI
St. Alban-Anlage 66
4052 Basel
Switzerland
www.mdpi.com

Aerospace Editorial Office
E-mail: aerospace@mdpi.com
www.mdpi.com/journal/aerospace

www.ingramcontent.com/pod-product-compliance
Lightning Source LLC
Chambersburg PA
CBHW061445210926
51850CB00008B/1615
* 9 7 8 3 0 3 6 5 9 0 2 7 1 *